T0178014

Aircraft Engines and Gas Turbines

Aircraft Engines and Gas Turbines

second edition

Jack L. Kerrebrock

The MIT Press
Cambridge, Massachusetts
London, England

Set by Asco Trade Typesetting Ltd. from disks provided by the author.

Library of Congress Cataloging-in-Publication Data

Kerrebrock, Jack L.
    Aircraft engines and gas turbines / Jack L. Kerrebrock.—2nd ed.
      p.   cm.
    Includes bibliographical references and index.
    ISBN 978-0-262-11162-1 (hc : alk. paper), 978-0-262-53403-1 (pb)
    1. Aircraft gas-turbines.   I. Title.
  TL709.K46   1992
  629.134′35—dc20   91-41047
                                           CIP

Contents

Since the first edition of this text was completed, in 1977, the changes in aircraft engines have been mainly evolutionary. At that time the after-burning turbofan engine was firmly entrenched as the engine of choice for high-performance military aircraft, as was the high-bypass turbofan for commercial transports and military logistics aircraft. These engine types continue to dominate their application sectors, with continual refinements greatly improving their performance. The principal qualitative changes in emphasis in aircraft propulsion have been a revival of interest in hyper-sonic air-breathing propulsion and a new interest in ultra-high-bypass engines, such as the high-speed turboprop. Most of the content of the first edition remains valid. The technology has continued to mature, and to a certain extent my understanding of it has also matured and improved; thus, changes in details will be found throughout this edition.

With the benefit of many helpful comments from both students and professionals, a considerable number of errors have been corrected. Though it would be foolish to assert that all errors have been eliminated, it is probably true that this edition will suffer less from errors than the first edition did. One frequent comment has been that more expansive treat-ment of certain difficult points would be helpful, that the presentation was too terse. With hindsight, I agree; thus, more extensive discussions will be found at many places in the text, although I remain firmly of the view that it is better to state a logical proposition once correctly than many times imprecisely.

Extensive changes have been made in the chapter on compressors, to better reflect the state of the art of compressor design and to provide a guide to the vigorous ongoing research in this area. A section on the stability of compression systems has been added, as has a discussion of numerical techniques for computing the flow in compressors. In view of the current interest in and the potential importance of hypersonic air-breathing engines, the discussion of this topic has been greatly expanded and updated.

The general structure of the book has not been changed. It focuses on the aircraft engine as a system, rather than on the disciplines important to aircraft engines. Thus, after the brief introduction in chapter 1, chapter 2 deals with the engine as a whole from the viewpoint of thermodynamics, or cycle analysis; chapter 3 repeats this treatment in more quantitative form. Chapters 4–6 discuss the behavior of the individual components that make up the engine, showing how the thermodynamic characteristics are realized through the fluid-dynamic behavior of the components. Chapter 7 discusses some aspects of structures peculiar to engines, and chapter 8 shows how the behavior of a complete engine can be deduced from the behavior of the components. Noise continues to be an important consideration for commercial aircraft engines; chapter 9 is essentially unchanged from the first edition, except that the rules for certification and the discussion of the takeoff noise of supersonic transports have been updated.

Chapter 10 has been expanded to reflect the current interest in supersonic combustion ramjets. Some readers may feel it is now out of balance with the remainder of the text, considering the tentative nature of this technology and its applications. The scramjet is, however, a useful vehicle for discussion of the physical aspects of high-speed flow with heat addition, and of heat transfer and film cooling.

Preface to the First Edition (1977)

This book is intended to provide an introduction to the engineering of aircraft propulsion systems with the emphasis on the engine, rather than on the disciplines involved in engine design. Because of the remarkable advances that have occurred since the large-scale introduction of gas turbine power plants into military aircraft in the 1950s and into commercial aircraft in the 1960s, a clear understanding of the characteristics of these devices is needed at the undergraduate or early graduate level. Such an understanding is essential both for entrance into professional work in the industry and for graduate study in the field. The understanding of a sophisticated engineering system that involves the sciences of fluid mechanics, solid mechanics, chemistry, automatic control, and even psychology (because of the problem of aircraft noise) has intrinsic value apart from its practical applications. At present, the fundamental information required for such an understanding is widely dispersed in the technical literature and subliterature. The aim of this book is to draw the information together in a unified form, so that the student can appreciate why aircraft propulsion systems have evolved to their present form and can thus be better prepared to contribute to their further evolution.

Automotive and stationary applications of gas turbines are growing rapidly. The engines used in these applications share much of their technology with aircraft gas turbines; indeed, they have benefited greatly from the aircraft-engine developments of the last two decades. While this book is concerned primarily with aircraft engines, the discussions of component technology are equally applicable to these other applications. The discussion of cycles in chapters 2 and 3 gives some attention to automotive and stationary engines.

The approach taken in this book is to treat the propulsion system at increasing levels of sophistication, beginning in chapter 1 with a phenomenological discussion of the processes by which energy is converted from heat to mechanical energy to thrust.

In chapter 2, several types of engines are discussed in the framework of ideal cycle analysis, where the components of an actual engine are represented parametrically without quantitative reference to the engine structure. Here the dependence of the engine's performance on the compressor pressure ratio and the turbine inlet temperature is established, as well as the trends of thrust and specific impulse with flight Mach number. The arguments are repeated more quantitatively in chapter 3 for a narrower spectrum of engines to convey the influence of nonidealities in the engine cycles.

Chapters 4–7 examine the mechanical characteristics required of each major engine component to achieve the parametric behavior assumed in the cycle analysis. At this step the enormous literature of the field must be abstracted and interpreted to clarify the important physical limitations and trends without submerging the reader in vast analyses and data correlations. Naturally, the presentation is strongly influenced by my own viewpoint. If it errs in detail or by omission at some points, I hope that the overview will help the serious student to correct these inadequacies.

Chapter 8 synthesizes from the component characteristics evolved in chapters 4–6, a complete gas generator and a complete propulsion system. An attempt is made to treat in a reasonably uniform way the problems of engine control, inlet-to-engine and engine-to-nozzle matching, and inlet distortion, which so strongly dictate the ultimate performance of the system.

The mechanisms by which aircraft engines produce noise are discussed in chapter 9. At its present state of development, this subject is both highly mathematical and highly empirical. Although the mathematics in this chapter is somewhat more advanced than that in other chapters, it should be understandable to a well-prepared college junior or senior. In any case, some care has been taken to make the physical arguments independent of the mathematics.

Since flight at very high Mach numbers leads to complex chemical behavior of the air as it passes through the engine, the possibilities for air-breathing propulsion at Mach numbers above 6 are discussed separately in chapter 10, where the thermochemistry of high-temperature combustion products is included.

Finally, chapter 11 deals with some of the simpler techniques of propulsion systems analysis, the tool used by the preliminary designer to determine which engine should be committed to the lengthy and costly process of development.

To understand this text, a student needs good first courses in gas dynamics, thermodynamics, and solid mechanics, along with the appropriate mathematics. These subjects will not be reviewed here, but some of the results of compressible flow are collected at the beginning of chapter 4.

Though this book developed from a one-semester undergraduate course in aircraft engines at MIT, it contains more information than can reasonably be taught in one semester. A good one-semester undergraduate course in aircraft engines might consist of chapters 1 and 2 and the following sections and subsections of the remaining chapters:

3.1–3.5, 3.7
4.1, 4.2.2.1–4.2.2.3, 4.3, 4.4.3, 4.4.4
5.1, 5.2.2, 5.2.3, 5.2.5, 5.3–5.6
6.1–6.4
8.1, 8.2, 8.4

The text in its entirety is suitable for first-year graduate students with no prior exposure to aircraft engines.

Acknowledgements

My contacts within the aircraft propulsion community—industrial, government, and academic—have broadened since the writing of the first edition; the number of persons to whom I am indebted has increased well beyond my ability to accurately record them all here. But some certainly must be noted. Professors Alan H. Epstein and Edward F. Greitzer of MIT, Professor Nicholas Cumpsty of Cambridge University, and Professors Frank Marble and Edward Zukoski of Caltech have contributed greatly to my understanding of the fluid mechanics of aircraft propulsion. I have been privileged to work with many excellent doctoral students at MIT; each contributed more to my understanding than I to theirs. Those who influenced the contents of this book include Wai K. Cheng, Edward F. Crawley, Mohammed K. Durali, James Fabunmi, Alan H. Epstein, Jeffrey B. Gertz, John C. Kreatsoulas, Wing F. Ng, Arun Sehra, William T. Thompkins, and Vreg Yousefian.

The undergraduate students who have used the first edition in classes at MIT have lent a sense of perspective which has been very helpful in improving the treatment of several topics which are not so easy to understand, and have identified many errors in the first edition; I can only hope that more have not been introduced in the revision. My son Peter has enriched this students' view of the text with helpful comments from a viewpoint which would not otherwise have been available, as well as adding a touch of humor to the revision.

Very special thanks are again due my wife Vickie, who has helped immeasurably in preparing the new edition in (nearly) entirely computerized form, and has carefully edited the entire text.

Finally, I am indebted to the California Institute of Technology for support as Fairchild Distinguished Scholar during 1990, when much of the revision was done.

Aircraft Engines and Gas Turbines

1 Introduction to Concepts

The purpose of this chapter is to describe in simple physical terms the fundamental characteristics of gas turbines and related flight vehicle propulsion systems—the characteristics that control and limit their design and their application. Some of these characteristics are thermodynamic, some fluid-dynamic, some mechanical. It is important to realize that they all play important roles in the engineering choices that enter into the design of propulsion systems, whether the application is on the ground, in an aircraft, or in a launch vehicle.

All aircraft engines and gas turbines are heat engines, in which thermal energy derived from the combustion of fuel with air (or with an oxidizer carried on the vehicle) is converted to useful work in one way or another.

When the useful output of a gas turbine is in the form of shaft power used to drive a wheeled vehicle, a machine, or an electric generator, its efficiency will usually be characterized as *thermal efficiency*, defined as the fraction of the thermal-energy input converted to mechanical work. This concept should be familiar to those acquainted with thermodynamics.

In aircraft propulsion, of course, the useful work of the engine is work done in propelling the aircraft. It is appropriate then to define a second efficiency, the *propulsive efficiency*, as the ratio of the propulsive power to the total mechanical power produced by the engine. Analogous efficiencies of utilization can be defined for other applications of gas turbines, but the propulsive efficiency is particularly important to this discussion because it plays a dominant role in determining the configurations of aircraft engines. The different types of engines—turbojets, turbofans, and turboprops— result from choosing configurations that yield an appropriate compromise between high overall efficiency (which is the product of thermal efficiency and propulsive efficiency) and other factors (such as the ratio of thrust to weight) in the various flight regimes.

For aircraft engines, weight and size are also important—as is cost, which can be a deciding factor in competition between the providers of engines for any particular application. And with the continuing growth of airline traffic and of environmental pressure on airports, takeoff noise has become a major problem for commercial aircraft operators; thus, noise produced per unit of thrust has become an important criterion in engine design. Emission of smoke and gaseous pollutants is controlled in the certification process as well, and has become a matter of more concern since it has been suspected that pollutants may affect the atmospheric ozone layer.

For automotive applications, cost limits engines to much simpler and less efficient designs than those that have evolved for aircraft. For stationary applications, reliability, efficiency, and cost are controlling factors; size and weight are much less important. These ground-based applications of gas turbines will be dealt with only in an incidental way in this volume, which focuses on aircraft propulsion.

1.1 Thermal Efficiency

The conversion of thermal energy to mechanical energy is subject to the laws of thermodynamics. These laws determine an upper limit on the thermal efficiency that depends only on the temperatures at which heat is added to and rejected from the working fluid of the engine. Most gas turbines use the atmosphere as a heat sink, so for the usual situation the minimum available heat-rejection temperature is the atmospheric temperature, denoted here by T_0. In space, heat rejection by radiation may provide a lower sink temperature, although it poses additional problems. The maximum available heat-addition temperature is in principle limited only by the characteristics of the combustion process (or nuclear reactor). In practice it is usually limited by the temperature capabilities of materials. If this maximum heat-addition temperature is denoted by T_m, the *maximum possible* thermal efficiency is that attained by a Carnot cycle operating between these temperature extremes, and is expressed as

$$\eta_c = 1 - \frac{T_0}{T_m}. \tag{1.1}$$

In the stratosphere (between 11 and 30 km altitude) T_0 is approximately 217°K. Current aircraft gas turbines have peak temperatures near 1500°K, so that η_c is approximately 0.85. Automotive and stationary gas turbines generally have peak temperatures below 1300°K, for reasons of cost and durability, and they reject heat at about 300°K, so for them η_c is about 0.77. Actual engines have lower thermal efficiencies because the Brayton cycle on which they operate is less efficient than a Carnot cycle for the same maximum temperature, and because of losses in the components due to viscous effects. For comparison, the maximum possible efficiency for steam power plants, set by the temperature limit of the materials in the superheater, is near 0.66.

1.2 Propulsive Efficiency

Unlike thermal efficiency, the propulsive efficiency, representing conversion between two forms of mechanical energy, is limited only by the laws of mechanics and can in principle approach unity. It is defined as

$$\eta_p = \frac{\text{Thrust power delivered to vehicle}}{\text{Net mechanical power delivered to engine mass flow}}.$$

The numerator is equal to the thrust multiplied by the flight velocity; the denominator is the difference between the product of mass flow rate and kinetic energy per unit mass in the exhaust, and the product of mass flow rate and kinetic energy per unit mass in the airflow into the engine. The difference between the denominator and the total power input in the fuel is power rejected in the form of heat. For an aircraft engine or a land-based gas turbine, it normally appears as heat in the exhaust gases.

By the conservation of momentum, the net force acting on the engine due to the flow through it is equal to the time rate of change of the momentum of that flow. This is equal to the difference between the momentum fluxes out of the exhaust and into the inlet, each of which is the product of the mass flow rate and the velocity. Both momentum fluxes are described far enough from the engine that the pressure is equal to the ambient value, so the pressure forces are negligible. We shall see later that it is convenient to modify this condition for some calculations. For the sake of simplicity, we will sometimes also assume in this chapter that the mass flows into and out of the engine are equal. In fact the nozzle flow rate will exceed the inlet flow rate by the amount of the fuel flow, which is around 2–4 percent of the airflow for most aircraft engines. But this assumption will be removed in chapter 3.

In this approximation, if the inlet mass flow per unit time is $\dot{m}$, the flight velocity is u_0, and the exhaust velocity is u_e, the thrust is

$$F = \dot{m}(u_e - u_0) \tag{1.2}$$

and the propulsive efficiency is

$$\eta_p = \frac{\dot{m}(u_e - u_0)u_0}{\dot{m}(u_e^2/2 - u_0^2/2)} = \frac{2u_0}{u_e + u_0}. \tag{1.3}$$

The propulsive efficiency decreases as the ratio of exhaust velocity to flight velocity increases. From equation 1.2, we can see that for a given mass flow

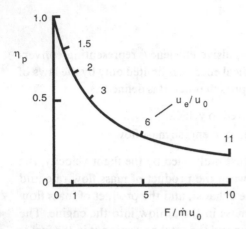

Figure 1.1
Propulsive efficiency as a function of thrust per unit of inlet air momentum, with ratio of exhaust velocity to flight velocity indicated.

and a given flight velocity the thrust increases with u_e/u_0. Thus, a definite tradeoff must be made between propulsive efficiency and thrust per unit mass flow. This relationship, plotted in figure 1.1, applies generally to all aircraft engines.

Increased mass flow in general implies increased engine size and weight, and it may also increase drag, so that there is a compromise to be struck between low overall engine size and weight and high propulsive efficiency. The choice depends on the application, with relatively small values of propulsive efficiency appropriate for military fighter engines and comparatively high values for commercial transport engines. These compromises will be discussed in subsequent descriptions of the various engine types, more quantitatively as we proceed through the successive chapters.

1.3   Specific Impulse and Range

The discussion of engine types in terms of thermal and propulsive efficiencies links cycle analysis to thermodynamics and mechanics, thus providing an intuitive grasp of the characteristics of propulsion systems in terms of familiar principles; however, for the purposes of performance analysis and design optimization, the efficiency of a propulsion system is usually characterized in terms of the *specific impulse*, defined as the number of units of thrust produced per unit of fuel (or fuel plus oxidizer) weight flow rate. This quantity enters directly into calculations of the fractional weight change of an aircraft during flight. It is generally denoted by I.

Suppose an aircraft is in steady, straight, level flight. The thrust, F, must then equal the drag, D. The aerodynamic performance of the airframe is characterized for these purposes by its ratio of lift to drag, L/D. Since the lift must equal the weight W of the aircraft, $F = W/(L/D)$. The weight of the aircraft decreases as fuel is consumed; the rate of change is $dW/dt = -F/I$, by the definition of I. Thus

$$\frac{dW}{dt} = -\frac{W}{I(L/D)},$$

and if I and L/D are constant in time the flight duration t is given by

$$t = I(L/D) \ln \frac{W_g}{W_g - W_f} \qquad (1.4)$$

where W_g is the initial (gross) weight and W_f is the weight of the fuel consumed. It is usual to present this result in terms of range, which is simply the product of the flight duration and the flight velocity u_0, so that

$$\text{Range} = u_0 I(L/D) \ln \frac{W_g}{W_g - W_f}. \qquad (1.5)$$

Historically, much effort has gone toward increasing the range of aircraft. For long-range transport aircraft, for bombers, and for logistic aircraft, the fuel weight is a substantial fraction of the gross weight, and the fraction $W_g/(W_g - W_f)$ is considerably larger than unity. In this case, whereas structural weight and engine weight affect the range logarithmically, I, u_0, and L/D affect it directly; thus, a premium is put on the latter factors. On the other hand, when $W_g/(W_g - W_f)$ is nearer unity, as for military fighters, helicopters, and other short-range aircraft, engine weight becomes as important as specific impulse, since it contributes to W_g; in fact, it contributes more than directly, because additional engine weight dictates some additional airframe weight to support it.

The specific impulse can be further related to the preceding discussion of efficiencies by noting that the overall propulsion system efficiency is simply

$$\eta = \frac{Fu_0}{(-dW/dt)h} = \frac{Fu_0}{(F/I)h} = \frac{u_0 I}{h}$$

where h is the energy content of the fuel (to be discussed in a later section of this chapter). Thus, the factor $u_0 I$ in equation 1.5 is simply ηh, the product of the energy content of the fuel (in units such as ft-lb per lb or m-kg per kg) and the efficiency with which it is used. The value of h for liquid hydro-carbon fuels is about 4800 km. For hydrogen it is 14,300 km, and for methane it is 5600 km.

When (as for a launch vehicle) most of the useful work of the propulsion system goes into increasing the kinetic energy of the vehicle rather than overcoming atmospheric drag, the role of specific impulse is readily seen by equating the thrust to the mass of the flight vehicle times its acceleration as follows:

$$m\frac{dV}{dt} = F = -g\frac{dm}{dt}I.$$

Integrating this for constant I yields

$$V(t) - V(0) = -gI \ln\frac{m(t)}{m(0)},$$

which is the equivalent of equation 1.5 for acceleration as opposed to sustained flight at constant speed.

1.4 Ramjets

Ramjets are conceptually the simplest of aircraft engines. Figure 1.2 is a schematic cross-sectional diagram of such an engine configured to fly at

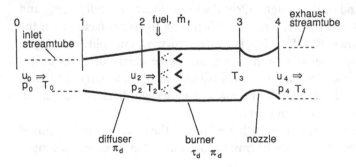

Figure 1.2
Schematic diagram of ramjet engine for supersonic flight.

moderate supersonic speeds. As we shall see, a different configuration is needed for hypersonic flight. Since we will focus for the present on the behavior of the airflow passing through the engine (indicated by the dashed inlet and exhaust streamtubes), the schematic shows only the internal functions of the engine (that is, those that influence the airflow through the engine). The engine consists of an inlet (or diffuser), a combustor (or burner), and a nozzle. The inlet decreases the flow velocity relative to the engine from the flight velocity u_0 to some smaller value, u_2. As we shall see later, the flow is subsonic from station 1 to station 2, although it is supersonic at station 0, and for subsonic flow a reduction in velocity implies an increase in flow area. The difference in kinetic energies of the air per unit mass $(u_0^2/2 - u_2^2/2)$ is converted to an increase in thermal energy, so that $T_2 > T_0$; at the same time, the pressure increases from p_0 to a higher value, p_2. Fuel is then mixed with the air, and the mixture is burned in the combustor. If the velocity u_2 is small relative to the local sonic velocity (the Mach number $M_2 \ll 1$), the combustion occurs at nearly constant pressure; the net result is that the thermal energy of the fluid increases and its density decreases. In the nozzle the flow is expanded, ideally to the original pressure, with a consequent drop in temperature from T_3 to T_4 and an increase in kinetic energy $u_4^2/2 - u_3^2/2$. Since T_3 is larger than T_2, the difference in thermal energies between stations 3 and 4 is larger than that between stations 2 and 0; therefore, the change in kinetic energy in the nozzle is larger than that in the inlet, and u_4 is larger than u_0. The change in momentum $u_4 - u_0$ per unit mass flow provides the thrust, as given by equation 1.2.

The conversion of thermal energy to mechanical energy is represented ideally by a Brayton cycle, as shown in figure 1.3. This cycle may be thought of as a superposition of a number of Carnot cycles, indicated by the small rectangles, each with a temperature ratio $T_2/T_0 = T_3/T_4$. Accordingly, the maximum possible efficiency of the cycle is

$$\eta_B = 1 - \frac{T_0}{T_2}. \tag{1.6}$$

The maximum efficiency can approach the limiting Carnot efficiency η_C only if T_2 approaches T_3—that is, if all of the temperature rise occurs in the inlet rather than in the combustor. The thermal efficiency of the ideal ramjet therefore is controlled by the inlet compression process, which governs the temperature ratio T_2/T_0. In the ideal case where $u_2 \ll u_0$ this ratio approaches the stagnation to static temperature ratio of the inlet flow,

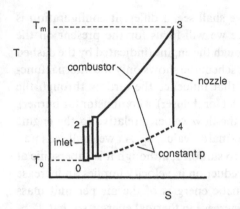

Figure 1.3
Temperature-entropy diagram of Brayton cycle for ramjet, with the elementary Carnot cycles of which it is composed.

$$\frac{T_2}{T_0} = 1 + \frac{\gamma - 1}{2}M_0^2,$$

so that

$$\eta_B = \frac{[(\gamma - 1)/2]M_0^2}{1 + [(\gamma - 1)/2]M_0^2} \tag{1.7}$$

where M_0 is the flight Mach number and $\gamma = c_p/c_v$ is the ratio of specific heats. Thus, for $M_0 < 1$, T_2/T_0 approaches unity and the thermal efficiency of the ramjet becomes small. It can be a highly efficient engine for $M_0 > 3$.

In the ideal ramjet, u_4/u_0 (and hence η_p) is determined by the combustor temperature ratio T_3/T_2. For a given u_0, increasing T_3 will increase thrust but reduce propulsive efficiency.

At very high flight Mach numbers, say above 6, the temperature rise that results from slowing the inlet airflow to subsonic speed is so large that adding fuel to the flow does not result in a significant temperature rise, because the normal products of combustion are highly dissociated. The conventional ramjet is therefore not effective in this flight regime. To meet the need for air-breathing propulsion at hypersonic speeds (Mach numbers above about 6), the supersonic combustion ramjet (scramjet) was conceived. In this type of ramjet, the hypersonic inlet flow is diffused only to

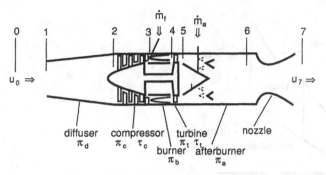

Figure 1.4
Schematic diagram of turbojet engine.

supersonic speed prior to fuel addition. The combustion takes place in the supersonic flow (hence the name), and the supersonic flow from the combustor is then expanded through the nozzle. The advantage of this scheme is that the temperature of the air flow and that of the combustion products need not rise above the range in which combustion of the fuel (usually hydrogen) can be substantially completed. The special characteristics of high speed propulsion systems are discussed in chapter 10.

1.5 Turbojets

The poor performance of the ramjet at low Mach numbers is improved by adding a compressor and its associated drive turbine to create a turbojet (figure 1.4). The compressor raises the air pressure and temperature prior to combustion and thus improves the cycle efficiency. The ideal Brayton cycle for the turbojet is shown by the solid lines in figure 1.5. The thermal efficiency is now given by

$$\eta_B = 1 - \frac{T_0}{T_3}. \tag{1.8}$$

If, for example, the compressor pressure ratio is 12, corresponding to an ideal compressor temperature ratio of 2.03, the ideal thermal efficiency is about 0.5. To drive the compressor, the turbine must have a temperature drop roughly equal to the temperature rise in the compressor. Because $T_4 > T_3$, the equality of the turbine and compressor powers results in

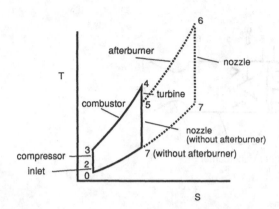

Figure 1.5
Temperature-entropy diagram for Brayton cycle of turbojet, with afterburning modification shown dashed.

$T_4/T_5 < T_3/T_2$, so that the turbine pressure ratio is smaller than the compressor pressure ratio; it follows that $p_5 > p_2$. Thus, the combination of compressor, combustor, and turbine, called a *gas generator*, produces a rise in the pressure as well as in the temperature of the airflow.

At present, limitations due to the materials used in the turbine restrict the turbine inlet temperature T_4 to values well below those corresponding to a stoichiometric mixture of fuel and air in the combustor, so that the turbine exhaust gas contains considerable residual oxygen. Additional thrust can be obtained by adding fuel in an afterburner. The cycle for this modification is shown dashed in figure 1.5. Because this fuel is added at lower pressure than the fuel in the primary combustor, it is used less efficiently. The penalty at subsonic speeds is so large that afterburning is used only for short bursts of extra thrust. At Mach numbers of 2.5 or more, the afterburning turbojet becomes highly efficient because the pressure rise associated with diffusion in the inlet raises the nozzle pressure ratio to a high value, as in the ramjet.

The propulsive efficiency of a turbojet is determined in the same way as that of a ramjet: by the combustor temperature ratio. For an ideal engine η_p can be made to approach unity by letting T_4 approach T_3; however, the thrust for a given engine size (or mass flow) becomes small, as indicated by figure 1.1, so in practice this option is not applied.

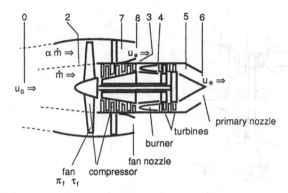

Figure 1.6
Schematic diagram of turbofan engine for subsonic flight.

1.6 Turbofans

A better way to improve the propulsive efficiency of the basic turbojet is offered by the turbofan, sketched in figure 1.6. Here, a second turbine is added downstream of the compressor-drive turbine, and the power from it is used to drive a fan that pumps air through a secondary nozzle. By this means a portion of the energy of the primary jet is removed, its velocity is reduced, and the energy is transferred to the fan airstream. Thus, the effective value of $F/\dot{m}u_0$ is reduced and η_p is increased as in figure 1.1; the penalty in engine weight is less than would be caused by decreasing T_4. Turbofans power most modern subsonic transports, having replaced turbojets in this application. Modified to include afterburning in the duct airflow or in the mixed primary and fan flows, they also power most high-performance military aircraft.

1.7 Turboprops and Other Shaft Engines: Regeneration

In low-speed flight vehicles, or when very high propulsive efficiency is desired, turboshaft engines are used. In such engines most of the useful work is extracted from the exhaust gas by a turbine, which often rotates on a separate shaft from the gas generator, as sketched in figure 1.7. The engine may drive a propeller, a helicopter rotor, the wheels of a truck, or any other machine. If it drives a propeller, the combination is termed a *turboprop*. Turboprops are used for small aircraft and for some logistic and special-

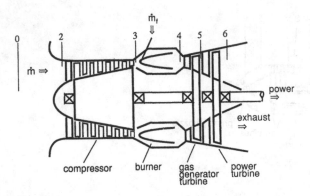

Figure 1.7
Schematic diagram of shaft turbine.

purpose aircraft, such as those used for patrol, for which long flight endurance is a major attribute. The maximum flight speed of turboprop-powered aircraft is limited by the occurrence of sonic flow at the propeller tips. The supersonic flow leads to shock waves which lower the efficiency of the propeller and also produce a great deal of noise. Recently a substantial effort has led to the development of high-speed turboprops, with thin, swept-back blades, which are capable of efficient operation at Mach numbers approaching 0.8; however, these engines have not yet been adopted for use in commercial transports.

In land or marine applications, where weight and size are not primary considerations and it is desirable to avoid the complexity and expense of a high-pressure-ratio compression system, *regeneration* may be used to increase the thermal efficiency beyond that attainable with a simple Brayton cycle. A regenerator is a heat exchanger that transfers heat from the exhaust gas to the compressor discharge air. As sketched in figure 1.8, this transfer can be accomplished with a rotating heat-storage matrix, or a fixed-surface heat exchanger can be used.

In a rotating heat exchanger, the matrix absorbs heat from the hot turbine exhaust gases and transfers it to the cooler air leaving the compressor. Ideally it will raise the compressor airflow to the turbine-exit gas temperature, but in order for this ideal condition to be achieved the mass and the surface area of the heat-transfer matrix would have to be very large. The regenerative Brayton cycle is illustrated in figure 1.9 for this ideal situation

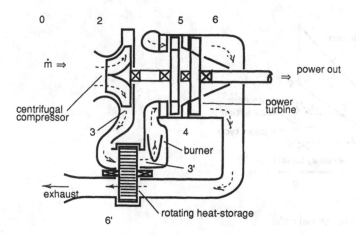

Figure 1.8
Shaft turbine with centrifugal compressor and rotating matrix regenerator.

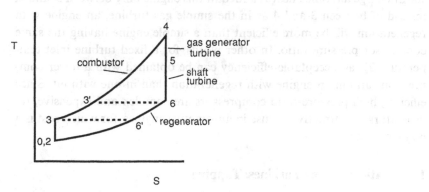

Figure 1.9
Temperature-entropy diagram for regenerative Brayton cycle.

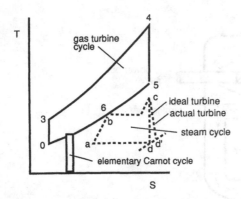

Figure 1.10
A gas turbine-steam combined cycle.

where the temperature of the exhaust gas is reduced to that of the compressor discharge air. Since heat is added in this engine only between 3' and 4, instead of between 3 and 4 as in the simple gas turbine, an engine with regeneration will be more efficient than a simple engine having the same compressor pressure ratio. In other words, for a fixed turbine inlet temperature T_4, an acceptable efficiency can be obtained with a lower compression ratio in an engine with regeneration than in one without. Since efficient, high-pressure-ratio compressors are complex and expensive, regenerators are attractive for use in automotive engines, for which cost is a major consideration.

1.8   Stationary Gas Turbines: Topping

The turbine inlet temperature of modern gas turbines is considerably higher than the peak steam temperature in steam power plants. Depending upon the compression ratio of the gas turbine, the turbine exhaust temperature may be high enough to permit efficient generation of steam using the waste heat from the gas turbine. Such an arrangement is referred to as a gas turbine-steam combined-cycle power plant. The cycle is illustrated in figure 1.10.

A combined-cycle power plant is capable of very high efficiency when the turbine inlet temperature of the gas turbine is high. One advantage over the regenerative gas turbine is that a steam boiler is easier to manufacture and maintain than a regenerator. A disadvantage relative to conventional steam plants is that gas turbines require clean fuel, such as natural gas or low-sulfur fuel oil.

1.9   Energy Exchange, Mach Number, and Reynolds Number

Four types of energy exchange have been implicitly involved in the above descriptions of engines: (1) the exchange within a flowing fluid of kinetic energy for thermal energy or vice versa, (2) the transfer of energy to or from a fluid by forces acting on moving blades, (3) the conversion of chemical energy to thermal energy, and (4) the transfer of thermal energy from solid bodies to flowing fluids.

The exchange from kinetic energy to thermal energy occurs when the momentum of a fluid is changed by pressure forces. The increasing pressure compresses the gas, and the compression work appears as an increase in internal (thermal) energy according to the first law of thermodynamics. The Mach number is defined as the ratio of the flow velocity to the velocity of sound in the fluid: $M = u/a$. When squared, it may be viewed as a measure of the ratio of kinetic energy to thermal energy of the fluid. Thus,

$$\frac{\gamma - 1}{2} M^2 = \frac{u^2/2}{c_p T}. \tag{1.9}$$

It follows that if process 1 is to be important, changes in $[(\gamma - 1)/2] M^2$ that are significant relative to unity must occur. The ramjet depends entirely on this process of energy exchange for its inlet compression, and this is the reason it must operate at Mach numbers above unity.

Process 2 appears in the turbojet, the turbofan, the turboprop, and all other devices using fluid-dynamic machinery. The airflow over a blade in a compressor, for example, exerts a force on the blade. If the blade moves in a direction opposite to the force, then the blade does work on the air, increasing its mechanical energy. Process 1 may take place at the same time, so the overall change in fluid energy appears partly as kinetic energy and partly as thermal energy. The force exerted on a body per unit area by a fluid is proportional to $\rho u^2/2$, where ρ is the fluid density and u is the velocity (which may be taken to be of the same order as the velocity of the body). The power delivered to the fluid by the body, per unit area, is then of the order of $\rho u^3/2$. Thermal energy of the fluid is convected by the body at

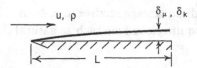

Figure 1.11
The penetration of viscous and thermal effects into a flowing fluid.

the rate $\rho u c_p T$ per unit area. Thus, the ratio of energy addition by the body to convected thermal energy per unit time and area is

$$\frac{\rho u^3/2}{\rho u c_p T} = \frac{u^2}{2c_p T} = \frac{\gamma - 1}{2} M^2, \tag{1.10}$$

and it can be seen that the Mach number plays the same key role in process 2 as in process 1. For the moving blades of the compressor or turbine to effectively exchange energy with the air, they should move at a Mach number on the order of unity.

Process 3 is so familiar that it requires no elaboration, but process 4 requires some discussion. In gas turbines we are concerned primarily with convective heat transfer—that is, heat transfer that occurs between a solid surface and a fluid because of the motion of the fluid over the surface. When the Reynolds number is large, the thermal effects of the surface on the fluid, like the viscous effects, are confined to a region near the surface that is thin relative to the characteristic length of the surface. That is, if we consider the flow over a flat plate of length L, as sketched in figure 1.11, with fluid density ρ, velocity u, and viscosity μ, then for laminar flow the viscous effects penetrate a distance δ_μ of order

$$\frac{\delta_\mu}{L} \approx \sqrt{\frac{\mu}{\rho u L}} = \sqrt{\frac{1}{\mathrm{Re}}}.$$

If the fluid has a Prandtl number $c_p \mu / k$ near unity (the value for air is about 0.7), where k is the thermal conductivity, or if the flow is turbulent, the thermal effect of the plate penetrates a distance $\delta_k \approx \delta_\mu$. In most of the components of a gas turbine, we wish to minimize viscous effects; hence we desire large Re and thin boundary layers. But in a regenerator the thermal effect must penetrate the entire flow, so either Re must be small or the ratio of spacing between heat-transfer surfaces to their flow length must be small, of order 1/Re. In either case the result tends to be a bulky and heavy

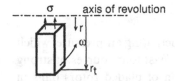

Figure 1.12
A prismatic bar rotating about an axis through its end, illustrating the origin of centrifugal stress in rotating parts of an engine.

device compared to the compressor and turbine. For this reason, regenerators are not used in aircraft engines.

1.10 Stresses

Since the speed of sound in air is about 340 m/sec at normal conditions, the blading of compressors and turbines should have velocities of 340 m/sec or more. This requirement has forced designers of gas turbine engines to cope with materials, vibration, and stress problems of a very high order. By contrast, the piston speed of a typical "high-speed" gasoline engine is only about 20 m/sec.

Some appreciation for the problem can be had by considering a prismatic bar rotating about an axis through its end, as in figure 1.12, with angular velocity ω. At any radius, the stress in the bar due to centrifugal force will be

$$\sigma = \int_r^{r_T} \rho\omega^2 r \, dr = (\rho\omega^2/2)(r_T^2 - r^2).$$

Near the axis of rotation, the stress is

$$\frac{\sigma}{\rho} = \frac{(\omega r_T)^2}{2}. \tag{1.11}$$

For ωr_T of the order of 340 m/sec, the material of the bar must have a strength-to-density ratio σ/ρ of the order of 6×10^4 m^2/sec^2. For steel, with a density of 8000 kg/m^3, this implies a stress of $\sigma = 4.8 \times 10^8$ N/m^2—close enough to the strength limit of the material that great sophistication and care in design are required. The problem is compounded in the case of turbines by the exposure of the rapidly rotating turbine blades to hot exhaust gases.

1.11   Noise

Acoustical noise is radiated from regions of fluctuating air pressure, which may be produced in many ways. There are at least four sources of strong unsteady flows in aircraft engines: the motion of bladed rotors (fans in particular), the passage of moving compressor and turbine blades past neighboring stationary blades, combustion (which results in local expansions of the burning gases), and turbulent mixing of high-velocity gases (such as the mixing of propulsive jets with the ambient air). The last, which leads to jet noise, is a direct manifestation of propulsive inefficiency, since the noise represents energy radiated from the jet as it dissipates its excess kinetic energy by turbulent mixing with the air. As the bypass ratios of turbofans have been increased to improve η_p, their jet noise has decreased. Noise from turbomachinery has its origins in unsteadiness, due in some cases simply to the rotation of the compressor and in other cases to the passage of moving blades past nearby stationary objects or through their wakes. The "buzz saw" noise of high-bypass engines on takeoff is in the former category; the high-pitched whine more usually associated with turbojet engines is in the latter.

1.12   Thrust and Drag

Conventionally, the forces acting on an aircraft in its direction of motion are divided into two parts: thrust and drag. The thrust is defined as the part of the force resulting from changes in the momentum (or pressure) of the air that *flows through the engine*. The drag is the force resulting from changes in the momentum of the air that *flows over the exterior of the vehicle*. In some cases this distinction is ambiguous, but in general it is useful and indeed essential to avoid confusion.

The definitions of thrust and drag do not imply that the drag is independent of the engine's operation or that the thrust is not influenced by the flow over the exterior of the aircraft. Especially in supersonic aircraft, one must account for the interaction of the internal (engine) and external airflows in determining either thrust or drag, and when the engines are embedded in the wing roots or in the fuselage the thrust and drag accounting requires an understanding of the flow over the entire aircraft.

Consider the nacelle-mounted engine shown schematically in figure 1.13. Assume that a net *engine force, $F - D_e$*, which represents the sum of

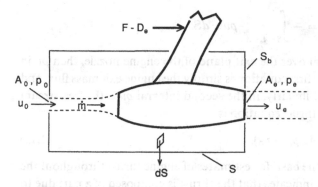

Figure 1.13
Schematic of nacelle-mounted engine, illustrating conventions for separating thrust (F) and drag (D_e) and method for computing thrust.

the thrust and the drag of the nacelle, is carried to the aircraft's structure by the strut, so that the thrust and the drag of the nacelle are defined independently of the aircraft. A control volume is constructed, closed at the front and sides by surfaces sufficiently far from the nacelle that the pressure and the flow velocity have their free-stream values, p_0 and u_0. The volume is closed at the rear by a plane through the exit nozzle of the engine. The force $F - D_e$ must equal the time rate of change of the momentum of the contents of the control volume. If the flow is steady, this balance may be expressed as

$$F - D_e - A_e(p_e - p_0) - \int_{S_b} (p - p_0)\, dS = \int_S \rho u \mathbf{u} \cdot d\mathbf{S},$$

where $\mathbf{u}$ is the (vector) flow velocity, u is its component in the flight direction, and dS is the differential element of the surface S. Because the aft plane of the control volume is close to the engine, the pressure on the aft plane may not be equal to the free-stream pressure. The pressure of the internal (engine) flow crossing this plane has been denoted p_e, where e stands for exit. In accordance with the definitions of thrust and drag, this balance is divided into two parts:

$$F - A_e(p_e - p_0) = \int_{A_e} \rho u \mathbf{u} \cdot d\mathbf{S} + \int_{A_0} \rho u \mathbf{u} \cdot d\mathbf{S}$$

and

$$D_e + \int_{S_b} (p - p_0)\, dS = - \int_{S - A_e - A_0} \rho u \mathbf{u} \cdot d\mathbf{S}.$$

If ρ and u are uniform over the exit plane of the engine nozzle, then ρu in the first integral of the first equation is simply the engine exit mass flux, and u is the exit velocity. Similarly, in the second integral ρu is the free-stream mass flux, so that the thrust equation is

$$F = \dot{m}_e u_e - \dot{m}_0 u_0 + A_e(p_e - p_0). \tag{1.12}$$

This result will form the basis for estimates of engine thrust throughout the following chapters. It indicates that the thrust is composed of a part due to the excess of momentum in the exhaust and a part due to the excess pressure. If the downstream closure of the control volume had been placed far downstream, where $p_e \to p_0$, the latter contribution would have been zero; but then the analysis of engine performance would have to include an analysis of the mixing of the exhaust jet with the external flow, so as to arrive at the velocity over the downstream plane. Placing the closure plane at the engine exit eliminates this problem. It introduces another problem, however: The difference between p on S_b and p_0 can be affected by the engine exhaust. In subsonic flight, potential-flow theory tells us that the pressure drag of the nacelle is zero, provided the external flow is parallel to the flight direction on S_b. The drag is then entirely due to viscous shear on the surface of the nacelle.

In supersonic flow the presence of shock waves in the external flow leads to an entropy rise, which appears in part as a pressure defect on the plane S_b and in part as a velocity defect there. Each will lead to an increase in drag. To the extent that deflections of the external airflow are caused by the engine airflow, the drag may be thought of as due to the engine rather than to the airframe. At times both airframe designers and engine designers have been loath to accept responsibility for this interface.

Spillage drag at the inlet and base drag at the exit are examples of such interaction problems. Figure 1.14 shows the flows that result in the excess drag. If the engine cannot accept all the flow in the streamtube with cross-sectional area equal to that of the engine, a shock forms that aids in turning the flow around the outside of the diffuser; but in the process it increases the entropy of the air, thus creating a drag in the external flow (termed *additive drag*). If the nozzle area of the engine must be reduced to a value smaller than the base area of the nacelle for the engine to operate properly,

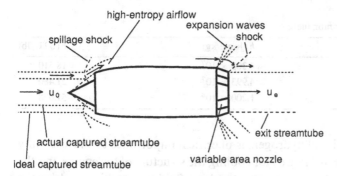

Figure 1.14
The flow over an engine at supersonic speed, showing origins of inlet spillage drag and base drag.

the external flow must turn through an expansion and shock system to fill the space not occupied by the jet and then realign to the axial direction. The shocks result in entropy and drag.

1.13 Fuels and Propellants

In the discussion thus far, the fuel has implicitly been treated as a source of energy input to the engine, the energy release occurring in the process of combustion of the fuel with air. For most aircraft engines this is a reasonable first approximation, because the normal ratio of fuel flow to air flow is quite small—well below the value of 1/15 corresponding to a stoichiometric (chemically correct) mixture, which leads to complete reaction of the oxygen to form carbon dioxide and water—and because the combustion heat release is so large as to dominate the thermal effects of the fuel on the engine.

But the fuel does have other roles in the engine, and those roles grow in importance as the aircraft's speed increases. At supersonic speeds, where the stagnation temperature of the engine airflow is sufficiently high to pose problems for organic materials such as the engine lubricants and for electronic systems, the fuel can serve as a coolant for the bearings and for the controls systems, before flowing to the combustor.

For very-high-speed propulsion systems, such as the supersonic combustion ramjet (scramjet), air-turborocket (ATR), and liquid-air collection (LACE) systems discussed in some detail in chapter 10, the heat capacity of

Table 1.1
Heating values of common fuels.

Fuel	h (joule/kg)	h (BTU/lb)
Kerosene	4.303×10^7	18,510
Methane	4.993×10^7	21,480
Hydrogen	1.200×10^8	51,608

the fuel, usually liquid hydrogen, is of critical importance to the feasibility of the engine. The fuel is used for cooling the structure of the engine and the airframe in all such concepts, as a working fluid in a turbine to drive the compressor in the air-turborocket, and as a coolant to liquefy the incoming airflow in the LACE system.

High-speed propulsion systems have also been proposed in which an on-board oxidizer (usually liquid oxygen) would be used to supplement the oxygen available from the atmosphere. In these cases the aircraft engine takes on some of the characteristics of a rocket engine; because the oxidizer flows through the nozzle with the fuel and air, it increases the momentum flux from the nozzle (i.e., the second term of equation 1.12). In the limit of a pure rocket, this term and the last pressure term express the thrust.

The complex process of the combustion of the fuel is discussed in some detail in chapter 4. Its effects on the gas flow can be represented at various levels of accuracy and realism, depending on the need. The combustion process results in a change in the chemical composition of the gas flow, as well as a change in its thermal energy; where accuracy is important, the change in specific heats, or more generally the change in the thermo-dynamic properties of the combustion products relative to those of air, must be accounted for. The techniques for doing this are explained in chapter 10, as the effects are most important for engines operating at high flight Mach numbers.

For lower-speed propulsion systems using conventional hydrocarbon fuels (such as gasoline or kerosene), it is usual to represent the effect of combustion by a *heating value*, h, defined as the energy added to the airflow per unit mass of fuel burned. In conventional usage, a distinction is made between the *upper heating value* of a fuel, which is the heat release available if the water formed in the process of combustion is condensed to liquid, and the *lower heating value*, which is the heat release available if the water remains as vapor in the exhaust. For aircraft engines the lower heating

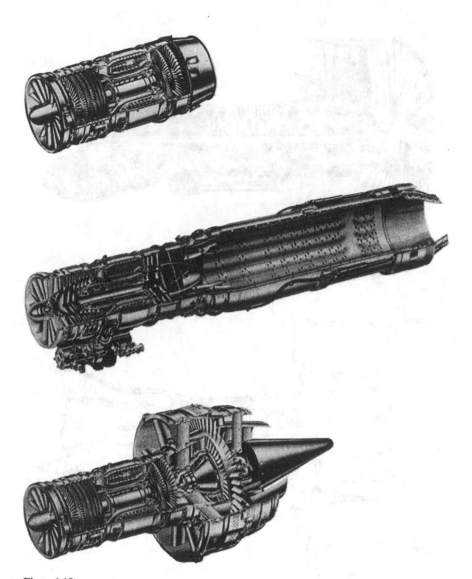

Figure 1.15
A small, early single-shaft gas generator (GE J85) and an afterburning turbojet and an aft-fan turbofan based on it.

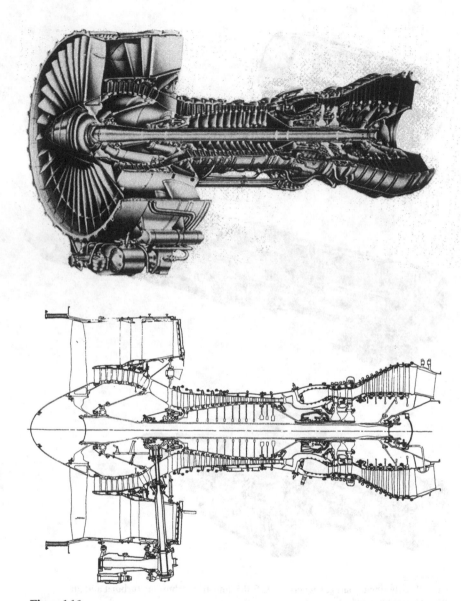

Figure 1.16
A modern high-bypass turbofan engine (Pratt & Whitney 2000 series) in cutaway and in
cross section.

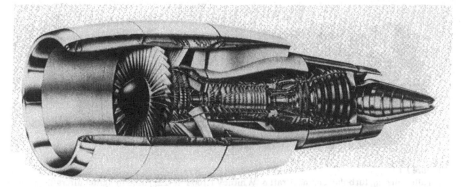

Figure 1.17
A modern large high-bypass turbofan (General Electric CF6-80C2) in a nacelle.

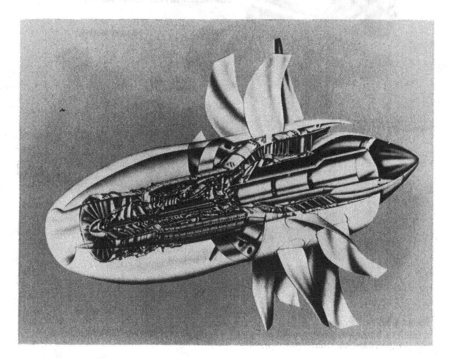

Figure 1.18
General Electric's Unducted Fan Engine, a turboprop with counterrotating swept
high-speed propellers.

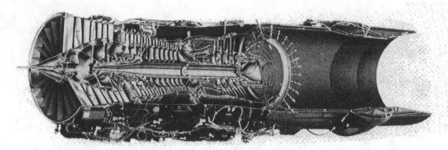

Figure 1.19
An afterburning turbofan engine (Pratt & Whitney F100 PW-222) used in fighter aircraft.

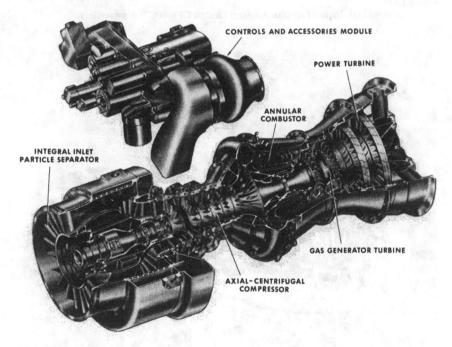

Figure 1.20
A small turboshaft engine for helicopter application (General Electric T 700) with dust separator and axial-centrifugal compressor.

value is the more useful one, since condensation of the water in the exhaust does not occur within the engine or in the nozzle. Some values of h are given in table 1.1.

1.14 Some Engines in Cutaway

The engine types mentioned here and their evolutionary trends will be discussed in much greater detail in the remainder of this book. It will be helpful to refer to the geometries of some actual engines. Fortunately, the engine manufacturers have developed elegant cutaway and cross-sectional drawings of many of their products. Only a few of these are presented here (figures 1.15–1.20), but the serious reader is urged to obtain more and to spend some time studying them and considering the functional reasons behind the shapes of the various components.

Problems

1.1 Assuming constant specific heats c_p and c_v, derive expressions for the thermal efficiency of the ideal ramjet and for the net work per unit of air flow. By equating the latter to the increase in jet kinetic energy, show that the thrust per unit of air mass flow is

$$\frac{F}{\dot{m}} = u_0 \sqrt{1 + \frac{2c_p T_0}{u_0{}^2}\left(\frac{T_2}{T_0} - 1\right)\left(\frac{T_3}{T_2} - 1\right)}.$$

1.2 Derive an expression for the specific impulse of an aircraft engine operating on a Carnot cycle as a function of peak-to-ambient temperature ratio, thrust per unit of mass flow, and flight velocity. Plot I as a function of M_0 with $F/\dot{m}a_0$ (where a_0 is the speed of sound) as a parameter for $T_m = 1500°K$, $T_0 = 217°K$, and $h = 4.30 \times 10^7$ joule/kg.

1.3 Write expressions for the engine pressure ratio p_5/p_2 and the engine temperature ratio T_5/T_2 of a turbojet in terms of the compressor temperature ratio T_3/T_2 and the turbine inlet temperature ratio T_4/T_2, where station 5 is at the exit of the turbine. Assume that the compression and the expansion are reversible and adiabatic. Compute p_5/p_2 and T_5/T_2 for $T_3/T_2 = 2$ and $T_4/T_2 = 5$.

1.4 Compute the maximum Mach number of the piston motion, based on the speed of sound in ambient air, for a piston engine with a stroke of 0.1 m and a rotation speed of 4000 rpm. Compare with equation 1.10.

1.5 Following the argument of equation 1.10, compute the stress in a thin ring rotating with angular speed ω about its axis of symmetry. If the ultimate strength is 1.36×10^9 Nm^{-2} and the density is 8×10^3 kg-m^{-3}, what is the limiting peripheral velocity? Compare the stress at the center of a bar with the same tip velocity.

1.6   The bypass ratio α as defined in figure 1.6 is a key design parameter for aircraft engines. Using the definitions of thermal and propulsive efficiencies, show that it can be expressed as

$$\alpha = \left(\frac{q}{c_p T_0}\right)\left(\frac{1}{(\gamma - 1)M_0{}^2}\right)(\eta)\left(\frac{\dot{m}_0 u_0}{F}\right) - 1$$

where q is the heat added per unit mass of airflow through the core engine and η is the overall efficiency. Assume that the core and fanjet velocities are equal and that all fan work is converted to fan-jet kinetic energy.

2 Ideal Cycle Analysis: Trends

Cycle analysis is the study of the thermodynamic behavior of air as it flows through an engine, without regard for the mechanical means used to effect its motion. Rather than deal with the inlet, the compressor and the turbine, themselves, we characterize them by the results they produce. Thus, for example, the compressor will be specified by a stagnation pressure ratio and efficiency. The behavior of a real engine is in fact determined by its geometry, in that for a given geometry the compressor will produce some well-defined pressure ratio—depending, of course, on operating parameters such as speed and airflow, the latter being controlled by some characteristic of the flow path, such as a downstream orifice. Thus, when we plot the results of cycle analysis in this chapter as curves of thrust and specific impulse versus, say, Mach number, we are not portraying the behavior of a real engine as a function of those variables. Rather, each point on such a curve represents a potential performance for some engine, which can be realized if an engine can be assembled from components that develop the airflows, pressure ratios, and efficiencies assumed in the calculation. There is no guarantee that this is possible. We sometimes say that the cycle analysis represents a "rubber engine" to emphasize the point that the computed performance characteristics are not those associated with an actual engine. The main purpose of cycle analysis is to determine which characteristics to choose for the various components of an engine to best satisfy a particular need. In chapters 3–7 we will consider the characteristics of actual components and what determines them; then, in chapter 8, we will examine the behavior of the components when assembled as an engine.

The value of cycle analysis depends directly on the realism with which the components are characterized. If a compressor is specified by pressure ratio and efficiency, for example, and the analysis purports to select the best pressure ratio for a particular mission, the choice may depend on the assumed variation of efficiency with pressure ratio. Unless a realistic variation of efficiency with pressure ratio is included in the analysis, its conclusions will be useful only in situations where the results are insensitive to efficiency. Fortunately this is often the case, because in practice the optimization of the cycle leads to parameter choices that minimize the effect of losses.

We shall develop the cycle analysis in stages; we begin with the simplest possible set of assumptions, where all components are taken to be ideal, and recognize that only conclusions insensitive to this assumption can be deduced. Our purpose is to portray the characteristics of the several types of aircraft engines and gas turbines as they depend on the major design parameters, so that they can be compared. Realistic assumptions as to component losses will be introduced in chapter 3, and the analysis will be repeated to demonstrate the methods for choosing loss-sensitive parameters.

Thus, we assume

• that compression and expansion processes in the inlet, the compressor, the turbine, and the nozzle are *isentropic*,
• that combustion occurs at constant static pressure,
• that the working fluid is a *thermally perfect gas*, with *constant specific heat*, and
• that the exhaust nozzles expand the engine exhausts completely to ambient pressure.

No attempt will be made to treat all possible engine cycles here; neither the author's nor the reader's patience would permit it. The hope is that the methods exemplified will allow the reader to carry out analyses according to need.

2.1   Stagnation Temperature and Pressure

In this book continual use will be made of the concepts of stagnation pressure and stagnation temperature. They are discussed extensively in texts on gas dynamics, but are of such importance to aircraft propulsion that some explanation is essential here. The stagnation temperature is defined as the temperature reached when a *steadily* flowing fluid is stagnated (brought to rest) *adiabatically*—that is, without transfer of heat to or from the fluid. The stagnation pressure is the pressure reached when this process is also *isentropic*—i.e., reversible. The stipulation that the flow be steady is important. In unsteady flow, energy can be transferred to or from the fluid without heat exchange. These definitions are applicable to all substances, whether ideal gases or not. In later discussions they will be generalized to account for chemical reactions and other nonideal aspects of the fluids. For the purposes of the present chapter, if T_t denotes the stagnation temperature, T the static (thermodynamic) temperature, and u the flow velocity, it follows from the conservation of energy that

$$c_p T_t = c_p T + u^2/2.$$

Introducing the Mach number, $M = u/\sqrt{\gamma R T}$ where R is the gas constant, we get

$$T_t = T\left(1 + \frac{\gamma - 1}{2}M^2\right). \tag{2.1}$$

As noted, the stagnation pressure, p_t, is defined as the pressure reached if the stream is brought to rest isentropically as well as adiabatically. Then, since for an isentropic process in an ideal gas $p_t/p = (T_t/T)^{\gamma/(\gamma-1)}$, we have

$$\frac{p_t}{p} = \left(1 + \frac{\gamma - 1}{2}M^2\right)^{\gamma/(\gamma-1)}. \tag{2.2}$$

Because ratios of stagnation temperatures and pressures will be used extensively, a special notation will be adopted for them. We denote a ratio of stagnation pressures across a component of the engine by π, with a subscript indicating the component: d for diffuser, c for compressor, b for burner, t for turbine, n for nozzle, f for fan. Similarly, τ will denote a ratio of stagnation temperatures.

Stagnation temperatures divided by ambient static temperature will be denoted by θ with a subscript. Thus,

$$\frac{T_{t0}}{T_0} = 1 + \frac{\gamma - 1}{2}M_0{}^2$$

will be θ_0, and T_{t4}/T_0 will be θ_t (for turbine inlet).

Stagnation pressures divided by ambient static pressure will be denoted δ, so that

$$\frac{p_{t0}}{p_0} = \left(1 + \frac{\gamma - 1}{2}M_0{}^2\right)^{\gamma/(\gamma-1)} \equiv \delta_0,$$

and we note that

$$\delta_0 = \theta_0{}^{\gamma/(\gamma-1)}.$$

As stated, cycle analysis deals with the thermodynamic behavior of the fluid as it flows through the engine, and hence with the thermodynamic state of the fluid, which is specified by the actual temperature and pressure (or by any other two properties, such as enthalpy and entropy). These are what we have termed the static properties, not the *stagnation* properties. The actual thermodynamic state of a flowing fluid and its local stagnation state are connected by well-defined adiabatic (and, in the case of pressure, isentropic) processes, however; thus, it is consistent to think of the compression, heating, and expansion processes as occurring between fluid states which are the stagnation states, even though the fluid in fact never reaches the stagnation state. Thus, for example, when we represent the compressor by stagnation pressure and temperature ratios, we are dealing with the pressures and temperatures which the fluid would attain if the velocity of air could be reduced to zero at the inlet and at the exit by means of adiabatic and isentropic processes. Since this is not possible in practice, the stagnation properties are not realizable. The extent to which this is the case will become clear in later discussions of the fluid dynamics of the components.

For the present argument, it is consistent and perhaps helpful to think of the stagnation pressures and temperatures as the actual thermodynamic quantities in the situation where the Mach number is small at all the points in the engine at which the thermodynamic quantities are evaluated.

2.2   The Ramjet

With the notation and the station numbers of figure 1.2, it follows from equation 1.2 that the thrust of the ideal ramjet is given by

$$\frac{F}{\dot{m}a_0} = M_0\left(\frac{u_4}{u_0} - 1\right)$$

and the specific impulse is

$$I = \frac{F}{g\dot{m}_f}$$

if $\dot{m}_f$ is the fuel mass flow rate and g is the acceleration of gravity. To compute u_4/u_0, we note first that the nozzle exit stagnation temperature, T_{t4}, is given by

$$T_{t4} = T_4\left(1 + \frac{\gamma - 1}{2}M_4{}^2\right) = T_0\theta_0\tau_b,$$

where the first equality follows from the definition of stagnation temperature and the second equality results from the chain of processes through the engine which effect the stagnation temperature. In the case of the ramjet, this chain consists only of the combustion process, since the inlet and nozzle processes conserve stagnation temperature.

Similarly, the exit stagnation pressure is

$$p_{t4} = p_4\left(1 + \frac{\gamma - 1}{2}M_4{}^2\right)^{\gamma/(\gamma-1)} = p_0\delta_0.$$

Here the stagnation pressure is constant throughout the engine, because the inlet and nozzle flows are assumed isentropic, and the combustion is assumed to occur at constant static pressure and at sufficiently low Mach number that the distinction between static and stagnation pressure in the combustor is nil.

Since the nozzle is ideally expanded, $p_4 = p_0$; thus, from the preceding relation, we have

$$\left(1 + \frac{\gamma - 1}{2}M_4{}^2\right)^{\gamma/(\gamma-1)} = \delta_0 = \left(1 + \frac{\gamma - 1}{2}M_0{}^2\right)^{\gamma/(\gamma-1)}$$

and $M_4 = M_0$. It follows that the static temperature ratio T_4/T_0 is equal to τ_b, so that

$$\frac{u_4}{u_0} = \frac{M_4}{M_0}\sqrt{\frac{T_4}{T_0}} = \sqrt{\tau_b}$$

and finally

$$\frac{F}{\dot{m}a_0} = M_0(\sqrt{\tau_b} - 1). \tag{2.3}$$

Define h as the heating value of the fuel (meaning the energy addition to the flow per unit of mass of fuel burned, as discussed in chapter 1); then, from an energy balance across the burner, we have

$$\dot{m}c_p T_0\theta_0 + \dot{m}_f h = (\dot{m} + \dot{m}_f)c_p T_0\theta_0\tau_b,$$

and if the fuel/air ratio $f = \dot{m}_f/\dot{m}$ is much less than unity we have

$$f = (c_p T_0\theta_0/h)(\tau_b - 1)$$

and the specific impulse is

$$I = \left(\frac{a_0 h}{gc_p T_0}\right)M_0\frac{\sqrt{\tau_b} - 1}{\theta_0(\tau_b - 1)}. \tag{2.4}$$

Equations 2.3 and 2.4 exhibit the dependence of the thrust and the specific impulse, F and I, on the Mach number and the burner temperature ratio, showing that for given τ_b both F and I increase linearly with M_0 for M_0 small, where θ_0 is near unity. For a given M_0, F increases monotonically with τ_b, while I decreases from a limiting value of $(a_0 h/gc_p T_0)(M_0/2\theta_0)$ attained for $\tau_b \to 1$.

A case of special interest is that where f has its stoichiometric value. For hydrocarbon fuels, T_{t3} is then of the order of 2500°K for low M_0. Increasing M_0 increases θ_0, but T_{t3} does not increase much because the combustion products (nominally carbon dioxide and water vapor) tend to dissociate. The result is that $\theta_b = T_{t3}/T_0$ is nearly independent of M_0, and we can find the dependence of F and I on M_0 by putting $\tau_b = \theta_b/\theta_0$, where θ_b is held constant in equation 2.3 and in the numerator of equation 2.4. But the $\tau_b - 1$ in the denominator of equation 2.4 represents the fuel/air ratio, so it must be replaced by

$$(\tau_b - 1)_{\text{stoich}} = f_{\text{stoich}} h/c_p T_0 \theta_0.$$

The result is that for the stoichiometric ramjet we have

$$\frac{F}{\dot{m}a_0} \approx \left(\sqrt{\frac{\theta_b}{\theta_0}} - 1 \right) M_0$$

and

$$I \approx \frac{a_0}{g f_{\text{stoich}}} \left(\sqrt{\frac{\theta_b}{\theta_0}} - 1 \right) M_0,$$

and I is simply a constant times $F/\dot{m}a_0$, which is shown in figure 2.1 for $\theta_b = 10$. We see that the thrust per unit mass flow and the specific impulse peak at about $M_0 = 2.5$, so that the best operating range for hydrocarbon-fueled ramjets is between $M_0 = 2$ and $M_0 = 4$. As was noted in chapter 1, the Mach-number range of the ramjet can be extended to much higher values if the combustion is carried on in supersonic flow, so that the temperature of the combustion products is maintained at a level where dissociation does not limit the heat release. This possibility is discussed at length in chapter 10; it cannot be treated usefully without consideration of stagnation pressure losses and other factors which are outside the domain of ideal cycle analysis.

In the stratosphere, a_0 is approximately 296 m/sec, so that $F/\dot{m}$, in units of kilograms of thrust per kilogram per second of airflow, is about 83 sec-

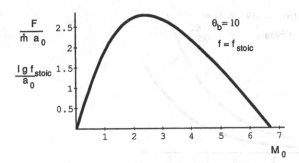

Figure 2.1
Thrust per unit mass flow and specific impulse for simple ramjet with stoichiometric combustion.

onds at $M_0 = 2.5$. For stoichiometric combustion of hydrocarbon fuels, such as kerosene, f is about 0.067, so the group a_0/gf_{stoich} is about 452, and the peak I is about 1310 seconds.

The significance of the variation of $F/\dot{m}a_0$ with Mach number, shown in figure 2.1, can be better understood if we realize that the mass flow which an engine of given size will ingest is a strong function of Mach number and altitude. The density of the atmosphere varies roughly exponentially with altitude; a useful approximation is

$$\frac{\rho_0}{\rho_0(h = 0)} = \exp(-h/9144),$$

where h is measured in meters. If an engine captured a streamtube (figure 1.2) of a constant area A_0 as M_0 varied, then the mass flow would be simply $\dot{m} = \rho_0 u_0 A_0$. But it is more nearly true that the Mach number of the flow inside the engine at some point is constant, the size of the inlet streamtube adjusting accordingly. In the ramjet the point that controls the flow might be the nozzle "throat" (i.e., the point of minimum flow area in the nozzle); in a turbojet, it might be the compressor inlet. In either case, the mass flow will depend on the Mach number at that internal point, the flow area, and the stagnation pressure and temperature. The ratio of ρu to $(\rho u)^*$, the value for $M = 1$, is a function only of M, the familiar A^*/A of channel flow theory. If the stagnation pressure and the stagnation temperature correspond to standard sea-level values, $(\rho u)^*_{h=0}$ is approximately 239.2 kg/m²sec. The magnitude of $(\rho u)^*$ is proportional to the stagnation pressure and to

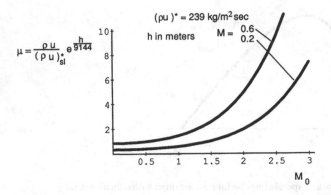

Figure 2.2
Engine mass flow per unit area divided by the value for choked flow at sea-level static conditions, as a function of flight Mach number M_0 and Mach number in engine M. The altitude h is in meters.

the negative square root of the stagnation temperature, which introduces a dependence on M_0. Thus,

$$\frac{\rho u}{(\rho u)^*_{h=0}} = M\left(\frac{(\gamma+1)/2}{1+[(\gamma-1)/2]M^2}\right)^{(\gamma+1)/2(\gamma-1)}\left(1+\frac{\gamma-1}{2}M_0{}^2\right)^{(\gamma+1)/2(\gamma-1)}$$

$$\times \exp(-h/9144)$$

$$\equiv \mu(M, M_0)\cdot\exp(-h/9144).$$

The variation of this mass flow capability with M_0 is shown for two values of M in figure 2.2 for M_0 up to 3. Above this value, nonideal behavior of the diffuser completely destroys the validity of the trend.

Multiplying the thrust curve of figure 2.1 by this mass flow factor, we see that the thrust of a ramjet actually should rise very steeply with Mach number up to values well above 3, for a given altitude. In practice, a ramjet-powered vehicle would normally fly a trajectory of increasing altitude as the Mach number increases, such that the thrust is more or less constant.

2.3   The Turbojet

We define π_c and τ_c as the stagnation pressure and temperature ratios across the compressor (that is, $\pi_c = p_{t3}/p_{t2}$ and $\tau_c = T_{t3}/T_{t2}$), and π_t and τ_t as the similar ratios across the turbine, and proceed to analyze the turbojet as we did the ramjet. From the station numbers of figure 1.4, the nozzle outlet stagnation temperature is

$$T_{t7} = T_7\left(1 + \frac{\gamma - 1}{2}M_7{}^2\right) = T_0\theta_0\tau_c\tau_b\tau_t\tau_a$$

where the nozzle has been assumed adiabatic, so $\tau_n = 1$. Similarly,

$$p_{t7} = p_7\left(1 + \frac{\gamma - 1}{2}M_7{}^2\right)^{\gamma/(\gamma-1)} = p_0\delta_0\pi_c\pi_t.$$

Since the nozzle is ideally expanded, $p_7 = p_0$; thus

$$1 + \frac{\gamma - 1}{2}M_7{}^2 = (\delta_0\pi_c\pi_t)^{(\gamma-1)/\gamma} \tag{2.5}$$

and

$$\frac{T_7}{T_0} = \frac{\theta_0\tau_c\tau_b\tau_t\tau_a}{(\delta_0\pi_c\pi_t)^{(\gamma-1)/\gamma}}.$$

Since the flows through the compressor and the turbine are isentropic,

$$\tau_c = \pi_c{}^{(\gamma-1)/\gamma}, \quad \tau_t = \pi_t{}^{(\gamma-1)/\gamma}, \quad \theta_0 = \delta_0{}^{(\gamma-1)/\gamma}$$

and we find $T_7/T_0 = \tau_b$. From equation 2.5,

$$M_7{}^2 = \frac{2}{\gamma - 1}(\theta_0\tau_c\tau_t - 1),$$

and from the definition of θ_0,

$$M_0{}^2 = \frac{2}{\gamma - 1}(\theta_0 - 1),$$

so that

$$\frac{u_7}{u_0} = \frac{M_7}{M_0}\sqrt{\frac{T_7}{T_0}} = \sqrt{\frac{\tau_b(\theta_0\tau_c\tau_t - 1)}{\theta_0 - 1}}.$$

The thrust of the turbojet is then given by $F = \dot{m}u_0(u_7/u_0 - 1)$ or

$$\frac{F}{\dot{m}a_0} = \sqrt{\frac{2\tau_b(\theta_0\tau_c\tau_t - 1)}{\gamma - 1}} - M_0. \tag{2.6}$$

This expression is not yet complete, as we must recognize that the power of the turbine equals that of the compressor. Since it is assumed here that the fuel mass flow is negligible relative to the air flow, and that the specific heat of the working fluid is constant, this condition can be written as

$$\dot{m}c_p(T_{t3} - T_{t2}) = \dot{m}c_p(T_{t4} - T_{t5}).$$

Because the absolute magnitude of T_{t4} is generally limited by the temperature and stress capabilities of the materials or by the cooling technology, it is useful to define a dimensionless temperature that represents this limitation. Let $\theta_t = T_{t4}/T_0$; then the compressor-turbine power balance can be solved for τ_t:

$$\tau_t = 1 - \frac{\theta_0}{\theta_t}(\tau_c - 1).$$

Substituting this expression in equation 2.6 and rearranging somewhat gives

$$\frac{F}{\dot{m}a_0} = \sqrt{\frac{2\theta_0}{\gamma - 1}\left(\frac{\theta_t}{\theta_0\tau_c} - 1\right)(\tau_c - 1) + \frac{\theta_t M_0{}^2}{\theta_0\tau_c}} - M_0, \qquad (2.7)$$

which gives the thrust per unit of mass flow as a function only of M_0, τ_c, and θ_t.

The advantage of this particular arrangement of equation 2.7 is that it exhibits the effects of the several parameters in an especially clear way. Thus, it is clear that if the ratio $\theta_t/\theta_0\tau_c$ approaches unity, implying that no heat is added in the combustor, the thrust approaches zero because the first term under the square root approaches zero, while the second approaches $M_0{}^2$. Similarly, if the compressor temperature ratio τ_c approaches unity this first term again goes to zero, while the second gives $\theta_t M_0{}^2/\theta_0 = M_0{}^2\tau_b$, in agreement with equation 2.3 for the ramjet.

A number of students have opined that the algebraic steps required to arrive at equation 2.7 are sufficiently obscure to make their discovery and the verification of the result somewhat burdensome. A clue which may be helpful is to both add to and subtract from the quantity under the square root the quantity

$$\frac{2\theta_0}{\gamma - 1}\left(\frac{\theta_t}{\theta_0\tau_c}\right).$$

To obtain an expression for the specific impulse, an energy balance across the combustion chamber is needed:

$$\dot{m}c_p(T_{t4} - T_{t3}) = \dot{m}_f h,$$

where, as for the ramjet, $\dot{m}_f$ is the fuel mass flow and h is its heating value. Thus

$$\frac{\dot{m}_f}{\dot{m}} = \frac{c_p T_0}{h}(\theta_t - \theta_0 \tau_c),$$

$$I = \frac{F}{g\dot{m}_f},$$

and

$$I = \frac{\dfrac{a_0 h}{g c_p T_0}\left(\sqrt{\dfrac{2\theta_0}{\gamma - 1}\left(\dfrac{\theta_t}{\theta_0 \tau_c} - 1\right)(\tau_c - 1) + \dfrac{\theta_t M_0{}^2}{\theta_0 \tau_c}} - M_0\right)}{\theta_t - \theta_0 \tau_c} \qquad (2.8)$$

so that I depends on the fuel and air properties through the first (dimensional) factor and otherwise only on the same parameters as does the thrust. The dimensional factor has the dimensions of time, and I is normally expressed in seconds.

A good deal can be learned from study of equations 2.7 and 2.8. First, putting $\theta_0 = 1$ for $M_0 = 0$, we find

$$\frac{F}{\dot{m}a_0} = \sqrt{\frac{2}{\gamma - 1}\left(\frac{\theta_t}{\tau_c} - 1\right)(\tau_c - 1)} \qquad (M_0 = 0) \qquad (2.9)$$

and

$$I = \frac{a_0 h}{g c_p T_0}\frac{\sqrt{\dfrac{2}{\gamma - 1}\left(\dfrac{\theta_t}{\tau_c} - 1\right)(\tau_c - 1)}}{\theta_t - \tau_c} \qquad (M_0 = 0). \qquad (2.10)$$

Both the thrust and the impulse are nonzero at $M_0 = 0$, in contrast to the ramjet. Note that for $\tau_c \to 1$ both F and I approach zero, as this is the limit of the ramjet. A little study of equations 2.7 and 2.8 will show that they revert to equations 2.3 and 2.4 when $\tau_c \to 1$.

From the first two factors involving τ_c in equation 2.7, it is clear that for given M_0 and θ_t there is a value of τ_c that maximizes $F/\dot{m}a_0$, since the first factor decreases with increasing τ_c and the second increases. The first represents the decrease in burner temperature rise, for fixed θ_t, as τ_c is increased. The second represents the improving thermal efficiency of the cycle with increasing τ_c. For most applications of gas turbine engines it is desirable to select cycle parameters near those that maximize the amount of work produced per unit of airflow. This obviously maximizes the power of the engine (or, in the case of the turbojet, the thrust) for a given airflow, and this is usually desirable for aircraft engines. It is less obvious, although true, that for real components with losses the compressor pressure ratio that yields the best efficiency is not so different from the ratio that produces the most power. To find the compressor temperature ratio that maximizes the thrust, we differentiate equation 2.7 with respect to τ_c, set the result to zero, and solve for τ_c to find

$$\tau_c = \frac{\sqrt{\theta_t}}{\theta_0}\left(\text{maximum } \frac{F}{\dot{m}a_0}\right). \tag{2.11}$$

The resultant value of $F/\dot{m}a_0$ is

$$\left(\frac{F}{\dot{m}a_0}\right)_{\text{max}} = M_0\left(\sqrt{1 + \frac{(\sqrt{\theta_t}-1)^2}{\theta_0 - 1}} - 1\right), \tag{2.12}$$

and the corresponding specific impulse is

$$I = \frac{a_0 h}{gc_p T_0}\frac{M_0\left(\sqrt{1 + \frac{(\sqrt{\theta_t}-1)^2}{\theta_0 - 1}} - 1\right)}{\theta_t - \sqrt{\theta_t}}. \tag{2.13}$$

There is no similar optimum value of τ_c for the specific impulse. A little study of equation 2.8 shows that I increases monotonically as τ_c is increased to the value of θ_t/θ_0 that just reduces the burner temperature rise to zero at this limit, a limiting process is necessary to determine the value of I when the burner temperature rise approaches zero. Putting $\tau_c = (\theta_t/\theta_0)(1 - \varepsilon)$, expanding equation 2.8 in ε, and letting ε tend to zero shows that the limiting value is

$$I = \left(\frac{a_0 h}{gc_p T_0}\right)\left(\frac{M_0(\theta_t - 1)}{2\theta_t(\theta_0 - 1)}\right) \quad (\tau_c \to \theta_t/\theta_0). \tag{2.14}$$

Recalling the discussion in chapter 1, note that this limit corresponds to a choice of cycle (see figure 1.5) in which point 3 approaches point 4, maxi-

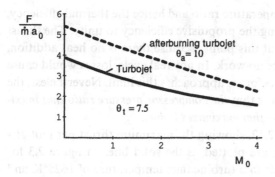

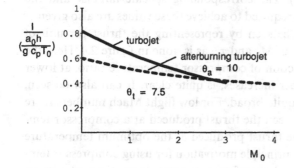

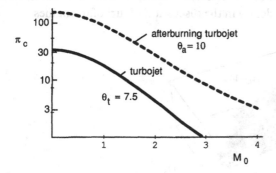

Figure 2.3
Solid lines indicate thrust per unit mass flow, specific impulse, and compressor pressure ratio
for maximum thrust for a simple turbojet with turbine inlet temperature 7.5 times ambient
temperature. Dashed lines show afterburning turbojet with afterburner temperature 10 times
ambient temperature.

mizing the compressor temperature ratio and hence the thermal efficiency, and at the same time driving the propulsive efficiency to unity. The thrust per unit of airflow is zero at this point, because there is no heat addition, and the engine can produce no work. In any real cycle, losses would cause the thrust to be small long before τ_c approaches this limit. Nevertheless, the argument does serve to show that *the compressor pressure ratio that maximizes I is larger than the one that maximizes $F/\dot{m}a_0$.*

The results of equation 2.12, showing the maximum thrust per unit airflow that can be attained, are plotted as the solid lines in figure 2.3 for $\theta_t = 7.5$, which corresponds to a turbine inlet temperature of 1625°K and $T_0 = 216$°K (stratosphere). The corresponding specific impulse and the compressor pressure ratio required to achieve these values are also given.

These points may also be seen by representing the thrust per unit of airflow as a function of both M_0 and τ_c, as is done in figure 2.4. Here the peaking of thrust as a function of compressor temperature ratio, at lower values as the Mach number increases, is quite clear. It can also be seen, however, that the peak is quite broad. For low flight Mach numbers there is very little difference between the thrust produced at a compressor temperature ratio of 2 and the thrust produced at the optimum temperature ratio, which is near 3. The principle motivation for using compressor temperature ratios above about 2, therefore, is to improve the thermal efficiency; we will see this more clearly in the discussion of turbofan engines.

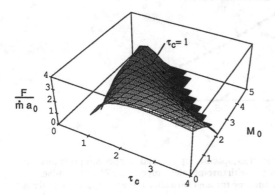

Figure 2.4
Thrust per unit of airflow for an ideal turbojet, as a function of both compressor temperature ratio, τ_c, and flight Mach number, M_0, for a turbine inlet temperature ratio of 7.5.

It is interesting to note from both figures 2.3 and 2.4 that the (optimum) compressor pressure ratio drops rapidly above $M_0 = 2$, and reaches 1.0 at $M_0 = 2.95$; at this Mach number the optimum turbojet has become a ramjet! Beyond $M_0 = 2.95$, the optimum τ_c is less than unity; thus, in this range the turbine should be before the combustor and the compressor after it. Normally in this range of M_0 one should remove the limit on T_{t4} along with the turbomachinery, so that the calculation may seem meaningless beyond the value of M_0 where $\tau_c = 1$. There is, however, a possible use of configurations where the turbine precedes the compressor: as a means for cooling the aircraft structure, the avionics, or the cockpit while minimizing the penalty associated with the cooling. This is discussed in section 2.9.

2.4   The Afterburning Turbojet

Suppose that we now add to the turbojet a second combustion chamber—an afterburner—downstream of the turbine, as indicated in figure 1.4. Temperatures higher than those in the combustor can be used here, because the surface area is small enough to permit cooling and there are no moving blades exposed to the hot gas. Let T_{t7}/T_0 be denoted θ_a. Retracing the cycle analysis, we find that $F/\dot{m}a_0$ is modified only by replacing τ_b with $\tau_b\tau_a$, where $\tau_a = T_{t7}/T_{t5}$. It is more useful, however, to have the expression explicit in terms of θ_a. Thus, writing $\tau_b = \theta_t/\theta_0\tau_c$ and $\tau_a = \theta_a/\theta_t\tau_t$ and substituting in equation 2.7, we find for the afterburning turbojet

$$\frac{F}{\dot{m}a_0} = \sqrt{\frac{2\theta_a}{\gamma - 1}\left(1 - \frac{\theta_t/\theta_0\tau_c}{\theta_t - \theta_0(\tau_c - 1)}\right)} - M_0. \tag{2.15}$$

From an energy balance across the entire engine, we see that

$$(\dot{m}_f + \dot{m}_a)h = \dot{m}c_p(T_{t7} - T_{t0});$$

thus,

$$\dot{m}_f + \dot{m}_a = \dot{m}(c_p T_0/h)(\theta_a - \theta_0),$$

and the specific impulse is simply

$$I = \frac{a_0 h}{g c_p T_0} \frac{\sqrt{\dfrac{2\theta_a}{\gamma - 1}\left(1 - \dfrac{\theta_t/\theta_0\tau_c}{\theta_t - \theta_0(\tau_c - 1)}\right)} - M_0}{\theta_a - \theta_0}. \tag{2.16}$$

Differentiating equation 2.15 with respect to τ_c, we find that the τ_c that maximizes $F/\dot{m}a_0$ is

$$\tau_c = \frac{1}{2}\left(1 + \frac{\theta_t}{\theta_0}\right),\tag{2.17}$$

and the resultant maximum $F/\dot{m}a_0$ is

$$\left(\frac{F}{\dot{m}a_0}\right)_{\max} = \sqrt{\frac{2\theta_a}{\gamma - 1}\left(1 - \frac{4\theta_t}{(\theta_0 + \theta_t)^2}\right)} - M_0.\tag{2.18}$$

The optimum τ_c given by equation 2.17 is larger than that for a nonafter-burning turbojet, given by equation 2.11. The optimum values of $\pi_c = \tau_c^{\gamma/(\gamma-1)}$ are shown in figure 2.3 (dashed lines) in comparison to those for the simple turbojet. For $M_0 = 2$, the optimum π_c for the afterburning turbojet is about 30; for the simple turbojet it is only about 4. Although the simple cycle analysis exaggerates this difference, it is true that for supersonic flight the optimum compressor pressure ratio is higher with afterburning than without. This is fortunate, for a given engine with a fairly high compression ratio can operate almost optimally with afterburning at flight Mach numbers near 2, where maximum thrust is required, and without afterburning subsonically, where the thrust demand is lower and good specific impulse is desired.

The thrust per unit of airflow of the afterburning turbojet is plotted as a function of both compressor temperature ratio and flight Mach number in figure 2.5. Here it can be seen that the thrust increase with compressor temperature ratio at low Mach numbers is rather slight above a temperature ratio of about 2. Since a temperature ratio below 2 yields the highest thrust at Mach numbers in the range 2–3, a value of about 2 is a good compromise for aircraft for which subsonic and supersonic afterburning thrust are both major performance criteria.

For an SST (supersonic transport), to cruise at $M_0 > 2$ is a critical engine requirement. Our calculation indicates that an afterburning turbojet with $\pi_c \approx 12$ ($\tau_c = 2.03$) should have as good specific impulse as an optimum simple turbojet (with $\pi_c \approx 14$), and about 50 percent higher thrust for a given airflow. The much better subsonic performance of the higher-pressure-ratio engine clearly makes it the better choice of these two possibilities. The General Electric GE-4 afterburning engine, which was once under development for the Boeing 2707 supersonic transport, had a pressure ratio of 12 at takeoff and somewhat less at cruise condition. The

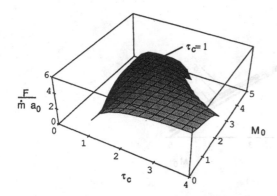

Figure 2.5
Thrust per unit of airflow for an ideal turbojet with afterburning as a function of both
compressor temperature, τ_c, and flight Mach number, M_0, for a turbine inlet temperature
ratio of 7.5.

Concorde also uses an afterburning turbojet with similar characteristics.
Takeoff-noise criteria have made such engines unsatisfactory for modern
civil supersonic aircraft, dictating the use of turbofan engines with some
degree of variability of the bypass ratio to allow acceptably low jet velocity
and hence noise at takeoff, together with efficient supersonic cruising.

The results given in figure 2.3 portray the behavior of families of engines,
each having the optimum compressor pressure ratio for the particular
Mach number. For an actual engine the compression ratio is set by the
configuration of the compressor and its rotational speed. More precisely,
as we shall see in the discussion of compressors in chapter 5, the compressor temperature ratio is set by the rotational Mach number—that is, by
the tangential velocity of the blades divided by the speed of sound in the
inlet air flow. To a good approximation we can say that

$$\tau_c = 1 + \frac{(\tau_c - 1)_{M_0=0}}{\theta_0}. \tag{2.19}$$

By substituting this expression for τ_c in equations 2.7 and 2.8 we get a more
realistic, though still approximate, representation of the variation of the
performance of an actual turbojet as a function of flight Mach number.
Figure 2.6 gives thrust and specific impulse in this way for a family of
turbojets with $\pi_c = 6$, 12, and 24 at $M_0 = 0$. Below $M_0 = 1.5$ the high-
pressure-ratio engine is superior in both thrust and specific impulse. Above

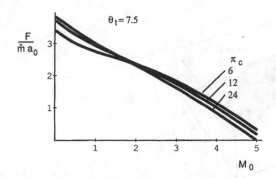

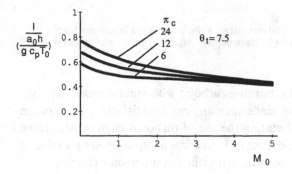

Figure 2.6
Thrust per unit mass flow and specific impulse for turbojet engines of various compressor
pressure ratios at $M_0 = 0$, with rotational speed fixed as M_0 varies.

$M_0 = 1.5$ there is a tradeoff; low π_c is better for thrust and high π_c is better
for specific impulse.

The thrust per unit airflow of the simple turbojet falls off badly at high
M_0 because the compressor outlet temperature rises to meet the turbine
inlet temperature, limiting the temperature rise of the combustor. An afterburner improves this situation greatly, as may be seen from figure 2.7,
where the variation of $F/\dot{m}a_0$ and I with M_0 is compared for an engine with
$\pi_c = 12$ at $M_0 = 0$ with and without an afterburner. The afterburning engine has about 30 percent more thrust at $M_0 = 0$ and maintains its thrust
well up to $M_0 = 4$, where the simple turbojet has dropped off severely. The
penalty in I shown for afterburning is realistic for $M_0 < 2$; however, for M_0
near the value where $\theta_0 \tau_c \to \theta_t$ the specific impulse of a real simple turbojet
would fall rapidly to zero, so that the afterburning engine is in fact superior
in specific impulse also for high M_0.

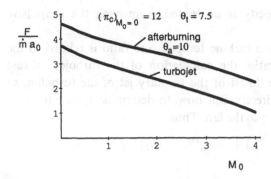

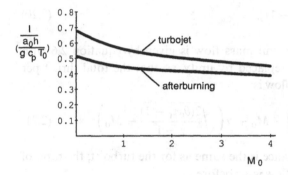

Figure 2.7
Comparison of thrust per unit mass flow and specific impulse of afterburning and simple turbojets, with a compressor pressure ratio of 12 at $M_0 = 0$ and constant rotational speed as in figure 2.6.

2.5   The Turbofan

In the turbofan engine shown schematically in figure 1.6, a part of the airflow through the fan is discharged directly through a nozzle to produce thrust; the remainder passes through the compressor, the combustor, and the turbine of the gas generator, then is exhausted through the primary nozzle. For convenience in the cycle analysis, we denote the overall compression ratio through the fan and the compressor by π_c, recognizing that this would in fact be the product of the fan and compressor pressure ratios. The pressure ratio of the fan alone will be π_f. A key parameter of the turbofan is the bypass ratio α, defined as the ratio of airflow through the bypass duct to that through the gas generator (compressor). As α is increased, more power is taken from the primary jet and put into the bypass jet, and the mean jet velocity is decreased, improving the propulsive efficiency.

Up to the point where the turbine temperature ratio is related to the compressor temperature ratio, the calculation of the turbojet's thrust applies equally well to the thrust of the primary jet of the turbofan, so that equation 2.6 applies directly. But now, to determine τ_t we must take account of the power flowing to the fan. Thus,

$$\dot{m}c_p(T_{t4} - T_{t5}) = \dot{m}c_p(T_{t3} - T_{t2}) + \alpha\dot{m}c_p(T_{t7} - T_{t2})$$

or

$$\tau_t = 1 - \frac{\theta_0}{\theta_t}[\tau_c - 1 + \alpha(\tau_f - 1)]. \tag{2.20}$$

The thrust of the duct per unit mass flow is given by equation 2.6 with $\tau_c\tau_t$ replaced by τ_f and τ_b replaced by unity, so that the total thrust per unit of gas generator mass flow is

$$\frac{F}{\dot{m}a_0} = \sqrt{\frac{2\theta_t}{\theta_0\tau_c}\left(\frac{\theta_0\tau_c\tau_t - 1}{\gamma - 1}\right)} - M_0 + \alpha\left(\sqrt{\frac{2(\theta_0\tau_f - 1)}{\gamma - 1}} - M_0\right). \tag{2.21}$$

The combustor energy balance is the same as for the turbojet; the ratio of fuel flow to compressor airflow is, as before,

$$\frac{\dot{m}_f}{\dot{m}} = \frac{c_p T_0}{h}(\theta_t - \theta_0\tau_c),$$

and the specific impulse is

$$I = \frac{(a_0 h/gc_p T_0)(F/\dot{m}a_0)}{\theta_t - \theta_0\tau_c}. \tag{2.22}$$

There seems to be little point in combining equations 2.20, 2.21, and 2.22, as no real simplification results. We are now faced with the fact that, in addition to the parameters M_0, τ_c and θ_t that characterized the turbojet, we have α and τ_f to consider.

It is a complex parametric problem to determine the best choice of θ_t, τ_c, α, and τ_f for a given application, but a little reflection will yield some simplifications and generalizations. We recall first that the basic reason for adding the fan to the turbojet is to improve the propulsive efficiency, that is, to make the jet velocity more nearly equal to the flight velocity. For a given total airflow and total energy in the duct jet and the primary jet, the propulsive efficiency will be highest if the two jets have the same velocity. (This follows from the fact that jet energy varies as the square of the velocity, whereas thrust varies directly as the velocity.) The number of parameters can therefore be reduced by considering only engines for which the two jet velocities are equal. In equation 2.21 this condition is equivalent to putting the two terms in square roots equal, subject to τ_t being given by equation 2.20. The result is the following expression for the fan temperature ratio:

$$\tau_f = \frac{1 + \theta_t + \theta_0(1 + \alpha - \tau_c) - (\theta_t/\theta_0\tau_c)}{\theta_0(1 + \alpha)}. \tag{2.23}$$

Corresponding to this choice of fan temperature ratio, the thrust per unit of *total airflow* is

$$\frac{F}{\dot{m}a_0(1 + \alpha)} = \sqrt{\frac{\theta_t - (\theta_t/\theta_0\tau_c) - \theta_0(\tau_c - 1) + \alpha(\theta_0 - 1)}{[(\gamma - 1)/2](1 + \alpha)}} - M_0. \tag{2.24}$$

One of the most important applications of turbofans is to transport aircraft that cruise at high subsonic Mach numbers. Here the fuel consumption in cruise is a major consideration, but engine weight is also important. To see how these criteria interact, we may plot the thrust per unit airflow and the specific impulse from equation 2.24 as functions of bypass ratio, for fixed θ_t, for the τ_c for maximum thrust, and for $M_0 = 0.8$. Such plots are given in figure 2.8 for $\theta_t = 7.5$, which gives $\pi_c = 22.3$; they show that the thrust per unit of airflow drops off rapidly as α increases from zero, but there is a very significant improvement in I at the same time. A good part of this improvement was achieved by increasing α to 5, the value selected for the first generation of large commercial transport engines such as the Pratt & Whitney JT9-D and the General Electric CF-6. Our plot shows that these engines should have fan pressure ratios near 3. As indicated, with current technology this would have necessitated two or more fan stages. The actual engines have only one stage to reduce fan noise and weight, so their core jet velocities are appreciably higher than their duct jet velocities. A military engine, the GE TF-39 for the C5A heavy logistics transport, was not subject to the same noise restrictions; it has a bypass ratio of 8, and "one and a half" fan stages.

In general, the optimum bypass ratio for any particular application is determined by a tradeoff involving fuel weight (reduced by increasing α),

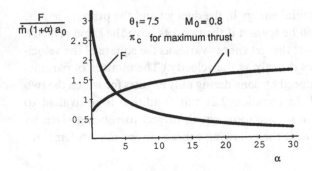

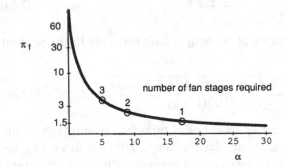

Figure 2.8
Thrust per unit of total mass flow and specific impulse for turbofan engines as functions of bypass ratio α at flight Mach number $M_0 = 0.8$, showing also the required fan pressure ratio.

engine weight (which increases with α for a given thrust), noise (which decreases with increasing bypass ratio), and installation drag (which increases with total engine airflow).

2.6   The Afterburning Turbofan

From section 2.5 it is clear that the turbofan is an excellent subsonic cruise engine. It also has much to offer aircraft that must cruise subsonically and also fly supersonically, provided an afterburner is added.

Ordinarily, the airflows of the gas generator and the fan are mixed before they enter the afterburner, as indicated in figure 2.9, and burning takes place in the entire airflow. This imposes on the fan pressure ratio the condition that $p_{t7} = p_{t5}$, or $\pi_f = \pi_c \pi_t$. For ideal components this implies that

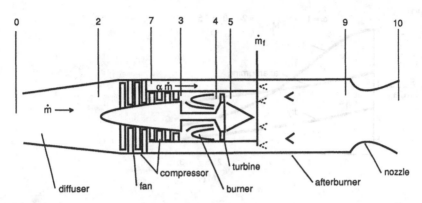

Figure 2.9
Schematic diagram of an afterburning turbofan engine.

$\tau_f = \tau_c \tau_t$. If burning raises the fan and gas generator exhaust streams to the same temperature at station 6, denoted by θ_a, the two streams will have the same exhaust velocity, and from equation 2.21 we may write the thrust simply as

$$\frac{F}{\dot{m}a_0(1 + \alpha)} = \sqrt{\frac{2\theta_a}{\theta_0\tau_f}\left(\frac{\theta_0\tau_f - 1}{\gamma - 1}\right)} - M_0. \tag{2.25}$$

The power balance between the turbine and the compressor and fan expressed by equation 2.20 must apply, but with the condition that $\tau_f = \tau_c \tau_t$. Eliminating τ_t from equation 2.20 with this expression gives the following relation for τ_f as a function of α:

$$\tau_f = \frac{\theta_t + \theta_0(1 + \alpha - \tau_c)}{\theta_t/\tau_c + \alpha\theta_0}. \tag{2.26}$$

Since the gas generator stream and the duct stream come to the same temperature at station 9, the combustion energy balance is very simple,

$$\dot{m}_f h = \dot{m}(1 + \alpha)c_p T_0(\theta_a - \theta_0),$$

and the specific impulse is

$$I = \frac{a_0 h}{g c_p T_0} \frac{\sqrt{\dfrac{2\theta_a}{\theta_0\tau_f}\left(\dfrac{\theta_0\tau_f - 1}{\gamma - 1}\right)} - M_0}{\theta_a - \theta_0}. \tag{2.27}$$

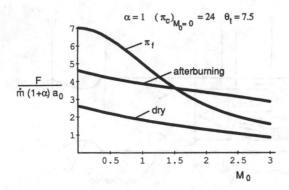

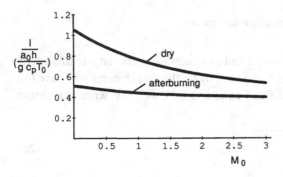

Figure 2.10
Comparison of thrust per unit of total mass flow and specific impulse for afterburning and nonafterburning turbofans, with fan temperature ratio required to match fan and turbine outlet pressures.

The performance of this engine, for $\alpha = 1$ and $(\pi_c)_{M_0=0} = 24$, is plotted in figure 2.10 along with the nonafterburning performance given by equation 2.21 and 2.22. (The latter is not quite a correct comparison, because the calculation does not include the mixing of the core and fan flows, which have different temperatures in the engine, represented by equations 2.21 and 2.22.) Several points should be noted. As the left graph in figure 2.10 shows, π_f varies considerably with M_0 for the matching conditions imposed, namely $\pi_f = \pi_c \pi_t$ and $\alpha = 1$. This would imply for a fixed engine that π_f must decrease with increasing M_0 relative to π_c, a requirement that can be met only with some effort. In practice, α would also vary somewhat with M_0, reducing the required variation of π_f. To gain some understanding of this, we may prescribe τ_f and its variation with M_0, and solve equation 2.26 for α, which is then allowed to vary. This changes the behavior

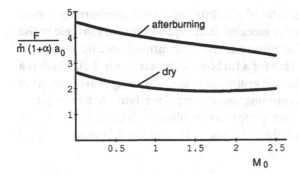

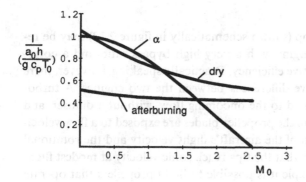

Figure 2.11
Thrust, specific impulse, and bypass ratio for afterburning turbofan with (nominally) constant fan temperature ratio.

somewhat, as is shown in figure 2.11, where α has been set to 1 at $M_0 = 0$ while τ_f varies as

$$\tau_f = 1 + \frac{(\tau_f - 1)_{M_0=0}}{\theta_0} \qquad (2.28)$$

in analogy to equation 2.19.

From equation 2.26,

$$\alpha = \frac{\theta_t(1 - \tau_f/\tau_c) - \theta_0(\tau_c - 1)}{\theta_0(\tau_f - 1)}. \qquad (2.29)$$

For this matching, α decreases to zero at $M_0 = 2.5$. The engine, being then a turbojet, offers better dry thrust than the turbofan with $\alpha = 1$.

The engine has a large ratio of afterburning to nonafterburning thrust. This can be advantageous for meeting dual requirements of subsonic cruise and supersonic dash. The subsonic, nonafterburning specific impulse is considerably better than that for a turbojet (compare figure 2.5), which is a further advantage for subsonic cruise. But these advantages are somewhat offset by a rather low afterburning specific impulse relative to a turbojet.

The engines for most high-performance military aircraft, including the F-111, the F-14 , the F-15, the F-16, the F-18, and the Advanced Tactical Fighter, are of this type.

2.7   The Turboprop

In a sense, the turboprop (shown schematically in figure 2.12) may be regarded as a turbofan engine with a very high bypass ratio and a correspondingly high propulsive efficiency. Practically speaking, however, there are very large qualitative differences between the two engines. A turboprop's propeller is exposed to the oncoming flow, without a diffuser, and this means that the tips of the propeller blades are exposed to a flow velocity that is the vector sum of the aircraft's flight velocity and the rotational tip velocity. The result is that the tips reach sonic velocity at modest flight speeds. Though in principle it is possible to build propellers that operate supersonically (in this sense), experience has shown that they are noisy and

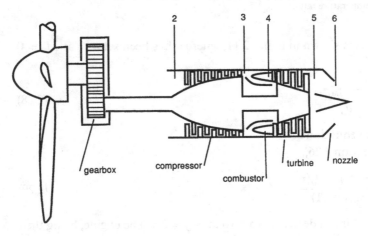

Figure 2.12
Schematic of turboprop engine.

inefficient. For this reason, most turboprop aircraft have been limited to flight Mach numbers not much above 0.6.

In the late 1970s and the 1980s, development began on high-speed turboprops, intended initially for flight Mach numbers as high as 0.8. In these designs the propellers have very thin, swept blades—as many as ten. Such propellers have shown good efficiencies to Mach numbers as high as 0.8.

The mechanical implementations proposed for high-speed turboprops are quite varied. In one, the high-speed propeller is driven through a gearbox, as is a more conventional propeller. In another, termed the *Unducted Fan* and shown in figure 1.18, the counterrotating propeller is driven by integral multi-stage turbines which expand the gas from the gas generator.

The flexibility that results from variation of the pitch of the propeller makes the turboprop unique among turbine engines. By varying the pitch, the efficiency of the propeller, defined as its propulsive power divided by the power supplied to it, can be made reasonably constant over a considerable speed range. (Of course this is not true at $M_0 = 0$, where this efficiency is necessarily zero.) This characteristic has led to the treatment of turboprops as devices for producing shaft power rather than thrust, and this practice will be followed here with due regard for any jet thrust that may be produced in addition to the propeller power. It should be borne in mind, however, that the turboprop can be treated as a turbofan with a very high bypass ratio; indeed, this has been the practice for high-speed turboprops.

The total propulsive power of the turboprop is the sum of the jet propulsive power Fu_0 plus the shaft power P delivered to the propeller times the propeller efficiency. For the ideal engine, the propeller efficiency is taken to be unity. It is convenient to define a total work coefficient, W, as

$$W = \frac{P}{\dot{m}c_p T_0} + \frac{u_0 F}{\dot{m}c_p T_0}, \tag{2.30}$$

that is, as the ratio of the propulsive power to the thermal energy in the airflow into the engine.

The jet thrust is given by equation 2.6, with the appropriate value of τ_t to account for the propeller work; thus,

$$W = \frac{P}{\dot{m}c_p T_0} + (\gamma - 1)\left[\sqrt{\frac{2\theta_t}{\theta_0 \tau_c}\left(\frac{\theta_0 \tau_c \tau_t - 1}{\gamma - 1}\right)} - M_0\right].$$

Balancing the compressor and propeller powers with the turbine power gives

$$P + \dot{m}c_p(T_{t3} - T_{t2}) = \dot{m}c_p(T_{t4} - T_{t5}),$$

or

$$\theta_t(1 - \tau_t) = \theta_0(\tau_c - 1) + \frac{P}{\dot{m}c_p T_0}.$$

Rather than substitute this result directly for τ_t, it is helpful to define a parameter,

$$\chi = \left(\frac{p_{t6}}{p_0}\right)^{(\gamma-1)/\gamma} = \delta_6^{(\gamma-1)/\gamma},$$

characterizing the extent to which the available energy has been taken from the exhaust flow. Clearly χ must be greater than unity if the engine is to operate, since p_{t6} must be greater than p_0 for the exhaust gas to flow from the nozzle. Forming ratios of pressures through the engine gives

$$\chi = (\delta_0 \pi_c \pi_t)^{(\gamma-1)/\gamma} = \theta_0 \tau_c \tau_t;$$

so

$$\tau_t = \frac{\chi}{\theta_0 \tau_c}$$

and

$$\frac{P}{\dot{m}c_p T_0} = \theta_t\left(1 - \frac{\chi}{\theta_0 \tau_c}\right) - \theta_0(\tau_c - 1).$$

Substituting in the expression for W gives

$$W = \theta_t\left(1 - \frac{\chi}{\theta_0 \tau_c}\right) - \theta_0(\tau_c - 1) + (\gamma - 1)\left(\sqrt{\frac{2\theta_t}{\theta_0 \tau_c}\frac{\chi - 1}{\gamma - 1}} - M_0\right). \quad (2.31)$$

The rate of fuel consumption per unit of propulsive work is conventionally used to characterize the efficiency of the engine. Denoting this "specific fuel consumption" by

$$s = \dot{m}_f/\dot{m}c_p T_0 W$$

and writing a heat balance for the combustor,

$$\dot{m}_f h = \dot{m}c_p T_0(\theta_t - \theta_0 \tau_c),$$

gives

$$s = \frac{\theta_t - \theta_0 \tau_c}{hW}.$$ (2.32)

If the energy content of the fuel, h, is expressed in rational units, say joules/kg, then s has units of kg/joule. It is more conventional to use thermal units for h (for example, BTU per lb), and to give s in lb of fuel per horsepower-hour. This leads to

$$s = \frac{2545(\theta_t - \theta_0 \tau_c)}{hW} \text{ lb fuel/hp hr,}$$

where the dimensions of h are BTU/lb.

The choice of χ determines the distribution of power between the jet and the propeller. What value of χ maximizes W, for given values of M_0, τ_c, and θ_t? Differentiating W with respect to χ and setting the result to zero will show that this value, denoted χ^*, is given by

$$\sqrt{\frac{2\theta_t}{\theta_0 \tau_c} \frac{\chi^* - 1}{\gamma - 1}} = M_0.$$ (2.33)

This is precisely the condition that makes u_6/u_0 unity, the jet thrust zero, and *the overall propulsive efficiency of the engine unity*. Because the propeller has been assumed to be perfectly efficient, it is best to shift all available energy from the jet to the propeller. If an efficiency $\eta_{\text{propeller}}$ is assigned to the propeller, the optimum χ^* is given by

$$\sqrt{\frac{2\theta_t}{\theta_0 \tau_c} \frac{\chi^* - 1}{\gamma - 1}} = \frac{M_0}{\eta_{\text{propeller}}}.$$

If we retain the simple result given above as equation 2.33, the maximum value of W and the corresponding s are given by

$$W^* = (\theta_t - \theta_0 \tau_c) \frac{\theta_0 \tau_c - 1}{\theta_0 \tau_c}$$ (2.34)

and

$$hs^* = \frac{\theta_0 \tau_c}{\theta_0 \tau_c - 1}.$$ (2.35)

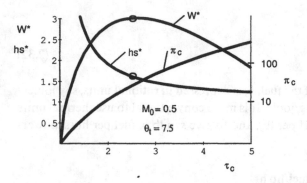

Figure 2.13
Propulsive work per unit of inlet enthalpy flux and specific fuel consumption for turboprop engine, as functions of compressor temperature ratio. (Multiply $hs*$ by $2545/18,500 = 0.14$ to find s in lb / hp hr.)

These results are plotted in figure 2.13 as functions of τ_c for $M_0 = 0.5$ and $\theta_t = 7.5$. A sharp peaking of the power is noted, with a continuous decrease of the specific fuel consumption. The inclusion of losses would of course cause $s*$ to turn up again at large τ_c.

Differentiating equation 2.34 with respect to τ_c shows that the value of τ_c that maximizes $W*$ is

$$\tau_c = \frac{\sqrt{\theta_t}}{\theta_0} \quad \text{(maximum } W*\text{)},$$

the same as for the simple turbojet. The peak $W*$ is

$$(W*)_{max} = (\sqrt{\theta_t} - 1)^2, \tag{2.36}$$

and we have

$$h(s*)_{max} = \frac{\sqrt{\theta_t}}{\sqrt{\theta_t} - 1}. \tag{2.37}$$

These points are indicated in figure 2.13 by the circles.

Evidently, for $M_0 = 0.5$ these results do not depend strongly on M_0, as only

$$\theta_0 = 1 + \left(\frac{\gamma - 1}{2}\right)M_0^2$$

enters. They depend mostly on θ_t and τ_c. Further, since the jet thrust has been set to zero, they apply as well to turboshaft engines, such as might power helicopters and trains.

2.8   Thrust Lapse

An important characteristic of aircraft engines is the variation of thrust with altitude and speed. The expressions for thrust developed here are for the dimensionless quantity $F/\dot{m}a_0(1 + \alpha)$, where $\dot{m}(1 + \alpha)$ is the total engine mass flow. As was pointed out in section 2.2, for a given engine this mass flow varies with flight Mach number, with atmospheric density (altitude), and with some controlling Mach number within the inlet or the engine (called simply M in figure 2.2). As was described in the discussion leading to equation 2.19, for turbine engines this limiting Mach number depends on the rotative speed and the speed of sound (i.e., temperature) of the inlet air. For constant rotative speed, as a first approximation,

$$M = \frac{M(0)}{\sqrt{\theta_0}}.$$

It then follows that

$$\frac{F(M_0, h)}{F(0,0)} = \frac{\dfrac{F}{\dot{m}a_0(1 + \alpha)}(M_0)}{\dfrac{F}{\dot{m}a_0(1 + \alpha)}(0)} \frac{\mu(M, M_0)e^{(-h/9144)}}{\mu(M,0)}. \tag{2.38}$$

This thrust ratio is plotted as a function of M_0 and α for fixed altitude in figure 2.14, from which it can be seen that the thrust of a high-bypass engine decreases with increasing flight Mach number whereas that of a turbojet increases. This may be understood in terms of the variation of propulsive efficiency with Mach number for the different bypass ratios. At one extreme, the turboprop delivers the power produced by its core engine to a propeller, which converts a more or less constant fraction of the power to propulsive work. For a fixed core power (and hence a fixed propulsive power), the thrust would vary inversely as the flight velocity. Actually the thrust lapse is a bit slower, because of the increase in engine mass flow due to rising ram pressure. At the other extreme, a turbojet has very low pro-

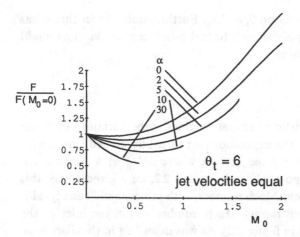

Figure 2.14
Variation of thrust with flight Mach number for various bypass ratios, showing decrease (lapse) of thrust for high bypass ratios.

pulsive efficiency at low speeds, improving with speed so that a larger fraction of the jet kinetic energy appears as thrust work at high Mach numbers. The increase in mass flow due to rising ram pressure augments this effect, and becomes the dominant trend at Mach numbers above 1.

One consequence of these trends is that, if the engines are sized by take-off requirements, a turbojet-powered aircraft can cruise at higher altitude than one with turboprop or high-bypass power, both experiencing the exponential decay in thrust with altitude indicated by equation 2.38. For example, compare a turbojet and a turbofan with bypass ratio of 5 at a flight Mach number of 0.8. The former has a thrust ratio of 1.20, the latter 0.75. If the turbofan has a cruise ceiling of 10 km, the turbojet can cruise at 14.3 km.

2.9   Cooling Cycles

At high flight speeds, the stagnation temperature of the airflow relative to the vehicle exceeds first the comfort level for the pilot, then the tolerance level of avionics and polymers, and eventually that of metals. In the strato-sphere, where $T_0 = 216°K$, the stagnation temperature reaches the standard temperature of $298.16°K$ at $M_0 = 1.38$.

One means of cooling the passengers, the avionics, and even the structure is to expand ram air through a turbine, lowering its stagnation temperature. After it is used for cooling, it must then be discharged overboard. An interesting question is what drag penalty must be incurred for such cooling.

The answer depends on the pressure drop experienced by the air in the cooling circuit and on the amount of heat added to it; however, it is interesting to examine the limiting case where, after expansion through a turbine, the air is reheated to the free stream stagnation temperature (perhaps by cooling the airframe), then passed through a compressor (which absorbs the power produced by the turbine), and finally discharged overboard through a nozzle. The cycle is then the analogue of a turbojet with the compressor and the turbine interchanged and with a compressor inlet temperature equal to the stagnation temperature.

We can find the thrust per unit of mass flow for this cycle from equation 2.7 by putting $\tau_c \to \tau_t$ (or $\pi_c \to \pi_t$), and $\theta_t \to \theta_0$, to find

$$\frac{F}{\dot{m}a_0} = \sqrt{\frac{2}{\gamma - 1}\left(\theta_0(2 - \tau_t) - \frac{1}{\tau_t}\right)} - M_0. \tag{2.39}$$

Differentiating this with respect to τ_t shows that the maximum thrust occurs for $\tau_t = 1/\sqrt{\theta_0}$, and is

$$\left(\frac{F}{\dot{m}a_0}\right)_{max} = \sqrt{\frac{4}{\gamma - 1}(\theta_0 - \sqrt{\theta_0})} - M_0. \tag{2.40}$$

This expression is shown in figure 2.15 for two fixed values of π_t and for the π_t that produces maximum thrust. Perhaps surprisingly, the cycle produces appreciable thrust above a flight Mach number of about 2.

One practical limit on use of this concept may be the temperatures in the compressor. The compressor discharge temperature is shown in the lower graph of figure 2.15 for the maximum-thrust condition. If we accept the same limit as for the turbine of a turbojet engine, the cycle may be operated to about a flight Mach number of 4.5.

As is shown at the bottom in figure 2.15, the turbine outlet temperature ratio that produces maximum thrust is somewhat above the ratio corresponding to standard temperature (1.38). If the turbine outlet temperature is set at 298°K, the thrust is lower than the maximum value, as shown in the upper graph.

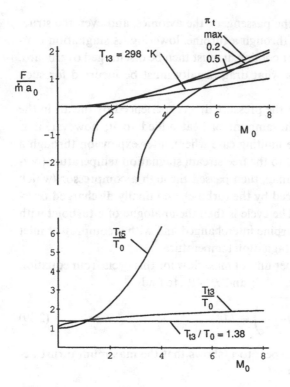

Figure 2.15
Thrust per unit of airflow for cooling cycles for the expansion turbine pressure ratio that
produces maximum thrust (max), and for two fixed pressure ratios.

2.10   The Regenerative Gas Turbine

An important modification of the turboprop (or turboshaft) engine is the
addition of a regenerator. This is a heat exchanger that withdraws heat
from the turbine exhaust gas and adds it to the compressor outlet air ahead
of the combustor, as indicated in figure 1.8. At low compressor pressure
ratios, regeneration yields large reductions in specific fuel consumption. It
is therefore very useful in automotive gas turbines, for example, where cost
militates against an efficient high-pressure-ratio compressor. However, its
weight has so far ruled it out for aircraft engines, where a high compression
ratio is preferred.

To estimate the performance characteristics of the ideal regenerative
engine, we note first that an ideal regenerator would raise the compressor

outlet airflow from T_3 to $T_{3'}$ of figure 1.9 by transferring heat from the turbine exhaust, thus lowering its temperature from T_6 to $T_{6'}$. If the specific heat and the mass flow of the turbine exhaust gases were equal to those of the compressor outlet air, $T_{6'}$ would then equal T_3. Actually, $T_{3'}$ must always be below T_6; the difference is determined by the regenerator's "effectiveness." For the ideal regenerator, we have

$$T_{3'} = T_6 = T_0\theta_t\tau_t.$$

The analysis of the turboprop applies to this engine with a modification to the combustor heat balance, which reads $\dot{m}_f h = \dot{m}c_p T_0(\theta_t - \theta_t\tau_t)$, so that the specific fuel consumption is

$$s = \frac{\theta_t(1 - \tau_t)}{hW} = \frac{\theta_t}{hW}\left(1 - \frac{\chi}{\theta_0\tau_c}\right) \tag{2.41}$$

where W is again given by equation 2.31.

Since regenerative gas turbines are used mostly in stationary applications or in low-speed vehicles, we can specialize equations 2.32 and 2.41 to $M_0 = 0$, whence

$$W = \theta_t\left(1 - \frac{\chi}{\tau_c}\right) - (\tau_c - 1)$$

and

$$s = \frac{\theta_t(1 - \chi/\tau_c)}{h[\theta_t(1 - \chi/\tau_c) - (\tau_c - 1)]}.$$

From equation 2.33 we see that $\chi^* \to 1$ as $\theta_0 \to 1$ for any value of $\theta_t/\theta_0\tau_c$. This is reasonable, since it corresponds to $p_{t6}/p_0 = 1$—that is, to the exhaust total pressure equaling the ambient pressure. With this condition we have finally

$$W^* = \frac{(\theta_t - \tau_c)(\tau_c - 1)}{\tau_c}, \tag{2.42}$$

which is identical to equation 2.31 for $\theta_0 = 1$, and

$$s^* = \frac{\theta_t}{h(\theta_t - \tau_c)}. \tag{2.43}$$

The thermal efficiency is $\eta^*_{\text{thermal}} \equiv P/\dot{m}_f h$; in the present case this becomes simply

$$\eta^*_{\text{thermal}} = \frac{1}{s^*h} = \frac{\theta_t - \tau_c}{\theta_t} = 1 - \frac{T_{t3}}{T_{t4}}. \tag{2.44}$$

Thus, as $\tau_c \to 1$ the efficiency of the regenerative engine approaches

$$\frac{\theta_t - 1}{\theta_t} = 1 - \frac{T_0}{T_{t4}},$$

which is the Carnot efficiency for the temperature limits T_{t4} and T_0. Of course, from equation 2.42 the work per unit mass of air goes to zero in this limit; figure 1.9 shows that this happens where $T_{3'} \to T_4$ and $T_{6'} \to T_0$. Thus, all heat is added at T_4 and rejected at T_0.

If we choose τ_c to maximize work for any given level of θ_t we find $\tau_c = \sqrt{\theta_t}$, and for this value

$$(W^*)_{\text{max}} = (\sqrt{\theta_t} - 1)^2 \quad (\text{maximum } W),$$

$$h(s^*)_{\text{max}} = \frac{\sqrt{\theta_t}}{\sqrt{\theta_t} - 1},$$

exactly as for the nonregenerated engine (see equations 2.36 and 2.37). In fact, for this condition of maximum work, $T_3 = T_6$ and there is no regeneration.

Thus we see that for the regenerated engine, as for the simple gas turbine, there is a tradeoff between power and efficiency; but for the regenerated engine, the efficiency is improved by lowering τ_c from the value for maximum power, whereas for the simple engine it was improved by increasing it from that for maximum power. In either case the efficiency approaches that of a Carnot cycle. This may be seen by comparing the cycles on T and S coordinates, as in figure 2.16, where the high-pressure-ratio cycle and the low-pressure-ratio regenerated cycle are compared against the maximum-work cycle, which is the same for regenerated and nonregenerated engines, all for a fixed maximum temperature. The average temperature of heat addition is higher for the high-pressure-ratio cycle than for the maximum-work cycle; it is higher for the regenerated cycle as well. It is also clear that the area of the cycle, which represents the work per unit mass of fluid, is smaller for both the high-pressure-ratio cycle and the regenerated cycle than it is for the maximum-work cycle.

The variations of W^* and hs^* for the simple and regenerative cycles as functions of compressor temperature ratio are plotted in figure 2.17 for a

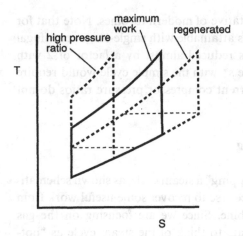

Figure 2.16
Comparison of high-pressure-ratio and regenerated cycles against maximum-work cycle, for fixed maximum temperature.

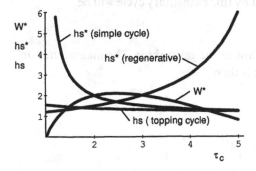

Figure 2.17
Comparison of specific work and specific fuel consumption of simple and regenerated shaft turbines.

value of $\theta_t = 6$, which is representative of modern turbines. Note that for low π_c (in the range of 5, such as is attainable with single-stage centrifugal compressors) fuel consumption is reduced almost by a factor of 2 with regeneration. To achieve the same s^* with the simple cycle would require $\pi_c \approx 60$, a high value indeed. Current compressor pressure ratios do not exceed 30.

2.11   Gas Turbines for Topping

When a gas turbine is used for "topping" a steam cycle, as shown schematically in figure 1.10, the idea is of course to recover some useful work from the heat rejected by the gas turbine. Since we are focusing on the gas turbine here, it is more appropriate to think of the steam cycle as "bottoming" the gas turbine. From this viewpoint the ideal bottoming cycle is that which achieves the Carnot efficiency for each point along the heat-rejection process, from 5 to 0. An elementary cycle of this sort is shown in figure 1.10. The power produced by this elementary cycle will be

$$dP = \dot{m}c_p(1 - T_0/T)\,dT,$$

where $\dot{m}$ is, as before, the mass flow in the gas turbine. The maximum total power from the bottoming cycle is then

$$P_{\text{bottom}} = \dot{m}c_p \int_{T_0}^{T_5} \left(1 - \frac{T_0}{T}\right) dT$$

$$= \dot{m}c_p T_0 \left[\frac{T_5}{T_0} - 1 - \ln\left(\frac{T_5}{T_0}\right)\right].$$

In the previous notation,

$$W_{\text{bottom}} = \theta_t \tau_t - 1 - \ln(\theta_t \tau_t)$$

$$= \frac{\theta_t}{\tau_c} - 1 - \ln\left(\frac{\theta_t}{\tau_c}\right).$$

The work of the gas turbine is

$$W_{\text{gas turbine}} = \theta_t(1 - 1/\tau_c) - (\tau_c - 1).$$

The work of the combined cycle is the sum of these, or

$$W = \theta_t - \tau_c - \ln(\theta_t/\tau_c), \tag{2.45}$$

while the thermal efficiency is

$$\eta = \frac{W}{\theta_t - \tau_c} = 1 - \frac{\ln(\theta_t/\tau_c)}{\theta_t - \tau_c}. \tag{2.46}$$

The efficiency is shown as $hs = 1/\eta$ for comparison with the simple and regenerative cycles (figure 2.17). The efficiency (or specific fuel consumption) does not vary as much with changes of τ_c as for either the simple or the regenerated cycle. Further, the fuel consumption is lower than for either of the basic cycles except for values of τ_c very near unity.

Of course, the gain in efficiency that can be realized by bottoming a gas turbine with a real cycle (see figure 1.10) is less than we have found, because even an ideal vapor cycle does not achieve full Carnot efficiency and so recovers only a part of the available energy in the gas turbine exhaust. But when cycle comparisons are made with realistic estimates of losses, as will be done in chapter 3, this cycle retains its attractiveness. Indeed, it is a major contender for application in modern fossil-fuel power plants.

2.12   The Importance of Turbine Inlet Temperature

In the preceding discussion of various engines and their performance characteristics, the turbine inlet temperature T_{t4}, or its dimensionless equivalent θ_t, was assigned a "typical value" while the other cycle parameters, such as τ_c and α, were varied to illustrate the effect of such variation on the performance and to show how optimum values of these parameters may be determined. θ_t was held constant because performance of the gas turbine improves continually with θ_t; thus, the maximum value which the turbine materials and the cooling system will allow is usually selected. In fact, much of the steady improvement in gas turbine performance has been due to improvements in oxidation-resistant alloys of nickel and cobalt and, more recently, to the development of air-cooled turbine blades. The factors that limit this latter important development will be discussed in some detail in chapter 6. Our purpose here is to examine the trends of performance with turbine inlet temperature.

Beginning with the simple turbojet, suppose we consider the variations of specific thrust $F/\dot{m}a_0$ and specific impulse with θ_t for a family of engines

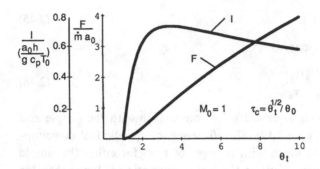

Figure 2.18
The effect of turbine inlet temperature ratio on thrust and specific impulse for the simple turbojet.

in which the compressor pressure ratio varies with θ_t to maximize $F/\dot{m}a_0$. This choice is arbitrary, but experience shows that real engines do not deviate greatly from it.

Figure 2.18, plotted from equations 2.7 and 2.8, shows that, for $M_0 = 1$, $F/\dot{m}a_0$ rises almost linearly with θ_t, while I first rises steeply then gradually decreases in the range of θ_t from 4 to 10. The variation of I may be understood as a result of two competing trends. As θ_t increases, the thermal efficiency of the engine increases continuously while the propulsive efficiency decreases because of the increase in jet velocity (which leads to the increase in $F/\dot{m}a_0$). Thus, over the interesting range of θ_t, the specific impulse of a simple turbojet would deteriorate with increased turbine inlet temperature, although the dramatic increase in thrust per unit of airflow might still justify such an increase.

With the turbofan we can have increased thrust and also improved specific impulse, since the jet velocity can be lowered by increasing the bypass ratio for a given θ_t. To see this we return to equation 2.24 and take $\tau_c = \sqrt{\theta_t}/\theta_0$, as for the turbojet. In equation 2.24 the square root represents the ratio of jet velocity to flight velocity, u_e/u_0. Suppose we hold u_e/u_0 constant as we vary θ_t. We find that the bypass ratio must then vary according to

$$1 + \alpha = \frac{(\sqrt{\theta_t} - 1)^2}{(\theta_0 - 1)\left[(u_e/u_0)^2 - 1\right]},\tag{2.47}$$

which is obtained by equating the square root in equation 2.24 to u_e/u_0. The thrust per unit of total airflow is simply

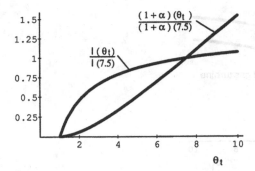

Figure 2.19
The effect of turbine inlet temperature ratio on specific impulse of turbofan engines of constant propulsive efficiency. Also plotted is the bypass ratio required to give constant propulsive efficiency (see equation 2.47).

$$\frac{F}{\dot{m}a_0(1 + \alpha)} = M_0\left(\frac{u_e}{u_0} - 1\right).$$ (2.48)

From equation 2.22, the specific impulse is

$$I = \left(\frac{a_0 h}{gc_p T_0}\right)\left(\frac{2}{(\gamma - 1)M_0}\right)\left(\frac{\sqrt{\theta_t} - 1}{\sqrt{\theta_t}}\right)\left(\frac{1}{u_e/u_0 + 1}\right).$$ (2.49)

Equation 2.49 shows clearly the effects of flight Mach number, turbine inlet temperature, and jet velocity on the specific impulse.

The effect of θ_t alone is displayed in figure 2.19 by plotting the ratios of I and $1 + \alpha$ to their values for a nominal θ_t of 7.5. I increases continually with increasing θ_t, reflecting the improving thermal efficiency of the cycle, for now η_p is constant since u_e/u_0 is held constant. The increased power of the core engine is absorbed by a larger fan mass flow, as indicated by the increase of $1 + \alpha$ with θ_t.

A reasonable upper limit for θ_t is about 10; for this value we might realize a 7 percent increase in I over the value for $\theta_t = 7.5$, which represents advanced technology as of this writing. For long-range aircraft this is a significant improvement.

There are similar trends for stationary and automotive gas turbines. Thus, from equations 2.44 and 2.46 we find the variations of η (relative to their values for $\theta_t = 6$) for regenerative gas turbines and gas turbine topping cycles (figure 2.20). Equally important, the power per unit of airflow

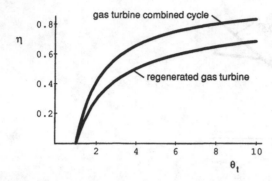

Figure 2.20
Effect of turbine inlet temperature ratio on efficiency of regenerated and ideally bottomed gas
turbines (gas turbine combined cycle).

(hence, roughly that for a given compressor size and cost) increases almost
linearly with θ_t.

Problems

2.1 Using the results of figure 2.1, compute the thrust (in newtons) of an ideal ramjet with
a combustor inlet cross-sectional area of $1\ m^2$ and a combustor inlet Mach number of
$M_2 = 0.2$, at an altitude of $h = 10$ km. Plot F versus M_0.

2.2 Very often air is "bled" from the compressor outlet of an aircraft engine for purposes
such as deicing and boundary-layer control. Rederive the expressions for $F/\dot{m}a_0$ and
$I/(a_0 h/gc_p T_0)$ for a simple turbojet (equations 2.7 and 2.8), assuming that a fraction ε of the
compressor airflow is bled off. Compute values of $F/\dot{m}a_0$ and I for $\varepsilon = 0.1$, $M_0 = 1$, $\theta_t = 7.5$,
and $\pi_c = 24$ and compare against figure 2.6.

2.3 Consider two jets of mass flow rates $\dot{m}_1$ and $\dot{m}_2$ with jet velocities u_1 and u_2. If the sum of
the jet kinetic powers is fixed at a value P, show that the sum of the jet thrusts is maximum for
$u_2 = u_1$.

2.4 Using the results of figure 2.6, compute the thrust (in newtons) of an ideal turbojet with a
compressor inlet area of $1\ m^2$, with $\pi_c = 12$, and with Mach number $M_2 = 0.5$ at the compressor inlet. Plot F versus M_0 for $h = 0$ and for $h = 10$ km.

2.5 A turboprop engine with $\theta_t = 6$ and $\tau_c = 2$ is to fly at $M_0 = 0.8$. According to the ideal
cycle analysis, what is the optimum core jet Mach number if the propeller efficiency is 1.0? If it
is 0.8?

2.6   For an ideal turbojet engine with fixed θ_t and π_c, does the nozzle-exit static temperature increase or decrease with increasing M_0? What does this imply about the thermal efficiency of the engine?

2.7   In the afterburning turbofan discussed in section 2.6, the fan and core flows are mixed before afterburning. An alternative is to "afterburn" only in the fan airflow and exhaust the fan and core flows through separate nozzles. Carry out an ideal cycle analysis for such an engine, choosing the fan pressure ratio to make the fan and core exit velocities equal.

2.8   Show for the ideal ramjet that, for fixed θ_b, $F/\dot{m}a_0$ is a maximum for $\theta_0 = \theta_b^{1/3}$.

3 Cycle Analysis with Losses

The most important deviations from the ideal behavior described in chapter 2 result from

imperfect diffusion of the free-stream flow from flight to engine-inlet conditions,
nonisentropic compression and expansion in the compressor and turbine,
incomplete combustion and stagnation pressure loss in the burners,
variation of the gas properties through the engine due to temperature and composition changes,
incomplete expansion (or overexpansion) to ambient pressure in the nozzle, and
extraction of compressor discharge air for turbine cooling or for use by the airframe.

Nozzle losses due to under- or overexpansion can be eliminated by design for proper expansion, but many engines use simple convergent nozzles for simplicity and weight savings. This particular loss (really a penalty due to fixed geometry, and qualitatively different from the other nonidealities listed) will be included in the cycle analysis even though it would be more logical to consider it when the other ramifications of engine geometry are considered.

The aim at this point is to characterize each of these mechanisms for deviation from ideality so that their effects can be included in a more realistic cycle analysis. More detailed discussion of the sources of the losses and the means of minimizing them will follow in subsequent chapters.

3.1 Variation in Gas Properties

As the air temperature rises in the compressor and as combustion changes the molecular composition of the gas as well as its temperature, its thermodynamic properties change. In the compressor the specific heat c_v rises with increasing temperature so that $\gamma = c_p/c_v$ decreases. A larger change occurs in the burner because of the large temperature rise and the formation of polyatomic gases (such as CO_2 and H_2O) which have low values of γ.

It is possible to use tabulated values of the thermodynamic properties of air and of combustion gases and thus take accurate account of these effects in the engine cycle analysis; this method must be followed in computing the performance of actual engines to the highest possible accuracy, and it will

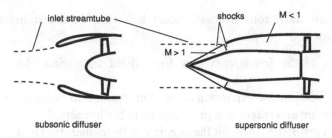

Figure 3.1
Schematic diagrams of diffusers.

be used in chapter 10. A different procedure will be followed here to retain as much as possible of the simplicity of the ideal cycle analysis. A mean c_{pc} and a mean γ_c will be defined for the compressor, and another pair of values, c_{pt} and γ_t, for the turbine. The first pair of values will be used for all processes occurring in the air ahead of the burner. The second set will be used for all processes in the combustion gases downstream of the burner. A mean specific heat, $\bar{c}_p$, will be defined for the range of temperatures in the burner.

3.2   Diffuser Pressure Recovery

As the engine airflow is brought from the free-stream conditions ahead of the aircraft to the conditions required at entrance to the engine, it may be smoothly decelerated as in the subsonic inlet at the left in figure 3.1, or it may be decelerated through shock waves, then further decelerated in a divergent passage as in the supersonic inlet shown at the right in the figure. In the subsonic inlets, viscous shear on the wall results in the growth of boundary layers that for this purpose may be thought of as regions in which the stagnation pressure of the fluid is low. Mixing this fluid with the inviscid core flow results in some reduction in the average stagnation pressure, below the value p_{t0} of the free stream. The ratio of this average stagnation pressure at the entrance to the engine (denoted p_{t2}) to the free-stream value will be termed the *diffuser pressure recovery* and denoted π_d. Thus,

$$\pi_d = \frac{p_{t2}}{p_{t0}} = \frac{p_{t2}}{p_0 \delta_0}.$$

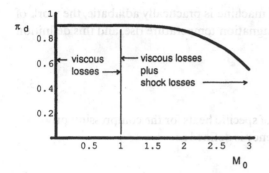

Figure 3.2
Typical diffuser pressure recovery as a function of flight Mach number.

When the flight velocity is supersonic, a further mechanism for loss of stagnation pressure exists in the compression through a series of shocks. Such losses vary markedly with M_0, and for $M_0 > 2$ they may constitute the principal source of diffuser pressure drop.

A typical variation of π_d with M_0 is shown in figure 3.2 for $0 < M_0 < 3$. For $M_0 > 1$, it includes the loss through a single oblique shock and a normal shock, as sketched in figure 3.1.

3.3 Compressor and Turbine Efficiencies

Losses in compressors and turbines originate primarily in regions of viscous shear on the blades and on the walls of the flow passages; these regions represent flows of lower stagnation pressure than the inviscid flow, as in the diffuser. The low-energy fluid becomes mixed into the base flow, and at the compressor (or turbine) outlet there is an average stagnation pressure and an average stagnation temperature.

Shock losses are also important in fan stages and in the first stage of modern transonic compressors.

For a given stagnation pressure ratio from inlet to outlet, the result of losses in a compressor is to require more energy input than for an ideal (isentropic) compressor. The efficiency is therefore defined as

$$\eta_c = \frac{\text{Ideal work of compression for a given } \pi_c}{\text{Actual work of compression for a given } \pi_c}.$$

Because the flow through the machine is practically adiabatic, the work of compression all appears as stagnation temperature rise, and this definition is equivalent to

$$\eta_c = \frac{\pi_c^{(\gamma_c - 1)/\gamma_c} - 1}{\tau_c - 1},\tag{3.1}$$

where now γ_c is a mean ratio of specific heats for the compression process.

Similarly, the turbine efficiency is defined as

$$\eta_t = \frac{\text{Actual work for given } \pi_t}{\text{Ideal work for given } \pi_t}.$$

For *uncooled* turbines, the flow is nearly adiabatic and this definition is equivalent to

$$\eta_t = \frac{1 - \tau_t}{1 - \pi_t^{(\gamma_t - 1)/\gamma_t}}.\tag{3.2}$$

Some modifications to this relation are in order for turbines in which cooler air is introduced to cool the blades; discussion of this rather complex subject will be deferred to chapter 6.

This definition of efficiency is *not* the one used in discussing steam turbines, for example, where the ratio of turbine output to inlet enthalpy in the steam above condenser conditions is taken as turbine efficiency. This is an energy extraction efficiency. η_c and η_t and as defined here are measures of the approach of the actual processes to adiabatic, isentropic processes.

Apropos of the need for consistency between assumptions, it must be noted that there are relationships between π_c and η_c and between π_t and η_t. As will be explained in chapter 5, what is nearly constant between compressors of different π_c is the efficiency for a small pressure change and a correspondingly small temperature change. This is termed the *polytropic efficiency*, and it will be denoted η_{pol}. Writing equation 3.1 for small changes in p_t and T_t gives

$$\eta_{pol} = \frac{(1 + \Delta p_t/p_t)^{(\gamma - 1)/\gamma} - 1}{1 + \Delta T_t/T_t - 1} \approx \frac{\gamma - 1}{\gamma} \frac{\Delta p_t/p_t}{\Delta T_t/T_t} = \frac{\gamma - 1}{\gamma} \frac{d \ln p_t}{d \ln T_t}.$$

Now, integrating and using the result to eliminate τ_c from equation 3.1 gives

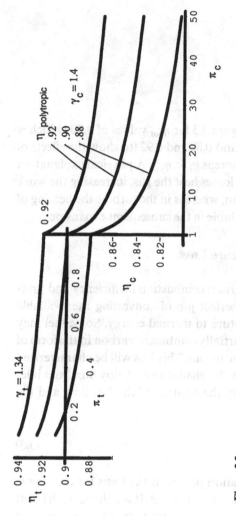

Figure 3.3
Compressor and turbine efficiencies as functions of pressure ratio for fixed polytropic efficiencies of 0.88, 0.90, and 0.92 and for typical values of ratio of specific heats.

$$\eta_c = \frac{\pi_c^{(\gamma_c-1)/\gamma_c} - 1}{\pi_c^{(\gamma_c-1)/\gamma_c\eta_{pol}} - 1};$$

similarly, from equation 3.2,

$$\eta_t = \frac{1 - \pi_t^{(\gamma_t-1)(\eta_{pol})/\gamma_t}}{1 - \pi_t^{(\gamma_t-1)/\gamma_t}}.$$

Values of η_c and η_t are shown in figure 3.3 for η_{pol} values of 0.90 (which is typical of modern turbomachines) and 0.88 and 0.92 (to show the effects of varying η_{pol}). Note that $\eta_t > \eta_{pol}$ whereas $\eta_c < \eta_{pol}$. A physical explanation of this is that in the compressor the losses heat the gas, increasing the work required for subsequent compression, whereas in the turbine the heating of the gas makes additional work available in the subsequent expansion.

3.4 Burner Efficiency and Pressure Loss

Two distinct losses occur in the burner: combustion inefficiency and pressure loss. The first implies an imperfect job of converting the available chemical energy of the fuel-air mixture to thermal energy. Some fuel may remain as soot particles, some as partially combusted carbon in the form of CO, some as other unburned hydrocarbons. This loss will be characterized by a burner efficiency η_b, defined as the change in enthalpy flux from inlet to exhaust of the burner divided by the product of the fuel flow and the energy content of the fuel:

$$\eta_b = \frac{\bar{c}_p[(\dot{m} + \dot{m}_f)T_{t4} - \dot{m}T_{t3}]}{\dot{m}_f h}. \tag{3.3}$$

One of the subtleties in the combustion process is the basis for defining h, conventionally termed the *heating value* of the fuel. It is defined as the heat released when the fuel in stoichiometric mixture with air is burned in a steady flow process (at constant atmospheric pressure), then cooled to the original ambient temperature. The heating values so determined will differ by about 10 percent depending upon whether the water vapor in the exhaust gas is condensed. If it is condensed, h includes its latent heat and is termed the *upper heating value*. If the water leaves as vapor, h is termed the *lower heating value*. This lower value is appropriate in gas turbines, since the water always leaves the burner as vapor. For a typical kerosene, h is about 4.30×10^7 joule/kg. The upper heating value is nearer 4.65×10^7 joule/kg.

The second loss is a drop in stagnation pressure, due partially to viscous effects and partially to the stagnation pressure loss that occurs whenever heat is added to a flowing gas. The latter is not very important in turbine engine burners but plays a dominant role in supersonic combustion ramjet burners; this subject is discussed in chapter 10. A very intense mixing process is essential to achieving the very high levels of heat release which are necessary in aircraft gas turbines, as will be explained in chapter 4. To produce this mixing, a pressure drop of the order of the dynamic pressure at burner entrance is necessary. The net effect of the two types of losses will be represented by a burner stagnation pressure ratio π_b.

3.5 Imperfect Expansion Loss

Perfect expansion of the exhaust nozzle implies that the flow is expanded isentropically to the final (ambient) pressure p_0 within the nozzle. If the flow is underexpanded, as may occur if the nozzle pressure ratio is larger than that which produces sonic velocity at the exit ($p_e > p_0$ in equation 1.12), then further expansion occurs downstream of the nozzle, but since the flow has no surfaces to expand against, this expansion does not produce an increase in thrust over that represented by the momentum and pressure at the nozzle exit, and the thrust is reduced from that for $p_e = p_0$.

Physically, the value of p_e/p_0 is controlled by the nozzle geometry and the ratio of stagnation pressure to ambient pressure. Since no other component of the engine is described geometrically in the cycle analysis, it makes no sense to describe the nozzle in detail. But two limiting cases can be discussed without reference to the nozzle geometry. These are the ideal nozzle, which corresponds to $p_e = p_0$ and gives an upper limit on thrust, and the simple convergent nozzle, which gives a lower limit for thrust and is also ideally expanded for nozzle pressure ratios less than that which produces sonic flow at the exit. The exhaust nozzles of turbojets, and often those of turbofans, are choked (they have pressure ratios larger than that required to produce sonic velocity at the point of minimum area) under most flight conditions, and the condition at the exit plane of a convergent nozzle is then simply that $M_e = 1$, so this condition can be specified independent of the nozzle geometry. If the nozzle is not choked, then (as noted) the two limits coincide but $M_e < 1$.

3.6   Heat Exchanger Effectiveness and Pressure Loss

In the regenerative gas turbine of figure 1.8, the compressor discharge air is heated before entering the combustor by heat exchange with the turbine exhaust gas. This exchange would be carried out by either passing the air and exhaust gas alternately through a packed bed or honeycomb, as in figure 1.8, or by counterflow in a more conventional heat exchanger, where the hot low-pressure gases would be separated from the cool high-pressure air by metal tubes or plates. Either case produces two kinds of deviation from ideal regeneration. First, the air and the gas suffer stagnation pressure losses as they flow through the regenerator. The fractional pressure drop on the air side may be different from that on the gas side of the regenerator, but (as we shall see) only the product of the two pressure ratios will appear in the cycle analysis. This product we denote π_r, so that in the notation of figures 1.8 and 1.9

$$\pi_r = \left(\frac{p_{t6'}}{p_{t6}}\right)\left(\frac{p_{t3'}}{p_{t3}}\right). \tag{3.4}$$

The regenerator would be thermally perfect if it raised the temperature of the compressor outlet air to the temperature of the turbine exhaust (if $T_{t3'} = T_{t6}$ in figure 1.9). Actually, there will always be some temperature drop across the heat exchanger plates or surfaces, so that $T_{t3'} < T_{t6}$. The effectiveness ε of the regenerator is defined as the ratio of the actual rise in compressor outlet air temperature to the ideal or the maximum possible. Thus,

$$\varepsilon = \frac{T_{t3'} - T_{t3}}{T_{t6} - T_{t3}}. \tag{3.5}$$

π_r and ε are related because the larger the flow Mach number through the regenerator, the larger the pressure drop for a given effectiveness. From Reynolds' analogy between heat transfer and shear in turbulent flows, which is explained in subsection 4.1.4, we find

$$\pi_r = 1 - \alpha M_r^2 \left(\frac{\varepsilon}{1 - \varepsilon}\right), \tag{3.6}$$

where α is a constant, best evaluated empirically but of order unity, and M_r is the Mach number of the flow through the regenerator. Clearly, the larger the flow passages for a given mass flow, the smaller M_r and the lower the pressure drop. (See problem 4.1.)

3.7 Turbojet with Losses

The simplest engine cycle that includes most of these effects is the turbojet cycle. Using the station numbers of figure 1.4, we find that the thrust is

$$F = \dot{m}_7 u_7 - \dot{m}u_0 + A_7(p_7 - p_0).$$

Let the overall fuel/air ratio (including primary combustor and afterburner) be f, and let ε be the fraction of compressor airflow bled from the engine at the compressor exit (for example, for use as an auxiliary power source); then $\dot{m}_7/\dot{m} = 1 + f - \varepsilon$. The exit area can then be related to other parameters by conservation of mass,

$$A_7 = \left(\frac{\dot{m}(1 + f - \varepsilon)}{\rho_0 u_0}\right)\left(\frac{R_t}{R_c}\right)\left(\frac{p_0}{p_7}\right)\left(\frac{T_7}{T_0}\right)\left(\frac{u_0}{u_7}\right),$$

and the thrust may be written as

$$\frac{F}{\dot{m}u_0} = (1 + f - \varepsilon)\frac{u_7}{u_0} - 1 + \frac{1 + f - \varepsilon}{\gamma_c M_0^2}\frac{R_t}{R_c}\frac{T_7}{T_0}\frac{u_0}{u_7}\left(1 - \frac{p_0}{p_7}\right). \tag{3.7}$$

The task now is to determine the ratios T_7/T_0, p_7/p_0, and u_7/u_0. Tracing the variations in stagnation temperature and pressure through the engine yields

$$T_{t7} = T_7\left[1 + \left(\frac{\gamma_t - 1}{2}\right)M_7^2\right] = T_0\theta_0\tau_c\tau_b\tau_t$$

$$= T_0\theta_a \quad \text{(afterburning)} \tag{3.8}$$

and

$$p_{t7} = p_7\left[1 + \left(\frac{\gamma_t - 1}{2}\right)M_7^2\right]^{\gamma_t/(\gamma_t - 1)} = p_0\delta_0\pi_d\pi_c\pi_b\pi_t\pi_a. \tag{3.9}$$

The first of these gives

$$\frac{T_7}{T_0} = \frac{\theta_t\tau_t}{1 + [(\gamma_t - 1)/2]M_7^2} \tag{3.10a}$$

and

$$\frac{T_7}{T_0} = \frac{\theta_a}{1 + [(\gamma_t - 1)/2]M_7^2} \quad \text{(afterburning)};$$ (3.10b)

the second gives

$$1 + \left(\frac{\gamma_t - 1}{2}\right)M_7^2 = \left(\frac{p_0}{p_7}\delta_0\pi_d\pi_b\pi_c\pi_t\right)^{(\gamma_t-1)/\gamma_t}.$$ (3.11)

The compressor-turbine power balance is

$$\dot{m}c_{pc}(T_{t3} - T_{t2}) = \dot{m}(1 + f_b - \varepsilon)c_{pt}(T_{t4} - T_{t5}),$$

where f_b is the fuel/air ratio of the primary burner. This can be written

$$\tau_t = 1 - \left(\frac{c_{pc}/c_{pt}}{1 + f_b - \varepsilon}\right)\left(\frac{\tau_c - 1}{\theta_t/\theta_0}\right);$$ (3.12)

finally, then,

$$\frac{u_7}{u_0} = \frac{M_7}{M_0}\sqrt{\left(\frac{\gamma_t R_t}{\gamma_c R_c}\right)\left(\frac{T_7}{T_0}\right)}.$$ (3.13)

From this point, it is convenient to recognize two limiting cases:

3.7.1 Convergent, Choked Nozzle

In this case, $M_7 = 1$, and u_7/u_0 is given directly by equation 3.13, while T_7/T_0 is determined by equation 3.10. The exit-to-entrance pressure ratio is controlled by the pressure balance through the engine. Thus, from equation 3.9,

$$\frac{p_7}{p_0} = \left(\frac{2}{\gamma_t + 1}\right)^{\gamma_t/(\gamma_t-1)}\delta_0\pi_d\pi_c\pi_b\pi_t\pi_a.$$

From the definitions of η_c and η_t,

$$\pi_c = [1 + \eta_c(\tau_c - 1)]^{\gamma_c/(\gamma_c-1)}$$

and

$$\pi_t = [1 - (1 - \tau_t)/\eta_t]^{\gamma_t/(\gamma_t-1)}.$$

Using equation 3.12 to eliminate τ_t, we get

$$\frac{p_7}{p_0} = \left(\frac{2}{\gamma_t + 1}\right)^{\gamma_t/(\gamma_t-1)} \delta_0 \pi_d \pi_c \pi_b \pi_a \left(1 - \frac{c_{pc}/c_{pt}}{1 + f_b - \varepsilon} \frac{\theta_0(\tau_c - 1)}{\eta_t \theta_t}\right)^{\gamma_t/(\gamma_t-1)}. \qquad (3.14)$$

There is little to be gained from substitution of these results into equation 3.7, so they will be left in this open form. Note that $F/\dot{m}a_0$ is a function only of M_0, τ_c, θ_t, and θ_a.

3.7.2 Ideally Expanded Nozzle

For this limit, $p_7/p_0 = 1$, and the whole last term of equation 3.7 is zero, but now M_7 must be determined from the pressure balance (equation 3.9). Thus,

$$\left(\frac{\gamma_t - 1}{2}\right) M_7{}^2 = (\delta_0 \pi_d \pi_c \pi_b \pi_a)^{(\gamma_t-1)/\gamma_t} \left[1 - \frac{c_{pc}/c_{pt}}{1 + f_b - \varepsilon}\left(\frac{\theta_0(\tau_c - 1)}{\eta_t \theta_t}\right)\right] - 1.$$

With equation 3.13, this determines $F/\dot{m}u_0$.

In either of the above two cases the specific impulse is given by

$$I = \frac{F}{g\dot{m}_f} = \frac{u_0(F/\dot{m}u_0)}{gf}, \qquad (3.15)$$

where $f = f_b + f_a$ is the total fuel/air ratio, f_b is the fuel/air ratio for the primary burner, and f_a is the fuel/air ratio for the afterburner. Referring to the definition of burner efficiency (equation 3.3), we get

$$f_b = \frac{\bar{c}_p T_0(1 - \varepsilon)}{\eta_b h}[(1 + f_b)\theta_t - \theta_0 \tau_c] \quad \text{(nonafterburning)}$$

and $\qquad\qquad\qquad\qquad\qquad\qquad\qquad\qquad\qquad\qquad\qquad (3.16)$

$$f = f_b + f_a = \frac{\bar{c}_p T_0(1 - \varepsilon)}{\eta_b h}[(1 + f)\theta_a - \theta_0] \quad \text{(afterburning)}$$

if the efficiencies of the primary burner and the afterburner are equal.

3.7.3 Effects of η_c, η_t, π_b, and π_d—Analytical Approach

Reasonably accurate numerical estimates of thrust and specific impulse can be obtained from the above results by straightforward computation. Before we proceed to this, however, it is useful to ask what insight can be obtained from the explicit formulas by analytical manipulation. Consider first the effects of pressure losses, compressor efficiency, and turbine efficiency for ideal nozzle expansion. To eliminate some complexity, we put

$c_{pc} = c_{pt}$, $\gamma_c = \gamma_t$, and $f_b \approx 0$. Equations 3.7 and 3.13 may then be combined to give

$$\frac{F}{\dot{m}u_0} + 1 = \left(\left(\frac{\theta_t - \theta_0(\tau_c - 1)}{\theta_0 - 1}\right)\right.$$

$$\left.\times \left\{1 - \left[\theta_0 \pi_1^{(\gamma-1)/\gamma}[1 + \eta_c(\tau_c - 1)]\left(1 - \frac{\theta_0(\tau_c - 1)}{\eta_t \theta_t}\right)\right]^{-1}\right\}\right)^{1/2}$$

and

$$I = \frac{(u_0 h \eta_b / \bar{c}_p T_0)(F/\dot{m}u_0)}{(1 + f_b)\theta_t - \theta_0 \tau_c},$$

where

$$\pi_1 = \pi_d \pi_b \pi_a.$$

Note first that if τ_c is held constant while η_c or η_t or π_t is varied then the variation of I is a constant times the variation of $F/\dot{m}u_0$. Letting $(F/\dot{m}u_0) + 1 = Z$, we find by differentiation that

$$\frac{\partial Z}{\partial \eta_c} = \frac{Q(\tau_c - 1)}{1 + \eta_c(\tau_c - 1)},$$

$$\frac{\partial Z}{\partial \eta_t} = \frac{Q\theta_0(\tau_c - 1)}{\eta_t[\eta_t \theta_t - \theta_0(\tau_c - 1)]},$$

and

$$\frac{\partial Z}{\partial \pi_1^{(\gamma-1)/\gamma}} = \frac{Q}{\pi_1^{(\gamma-1)/\gamma}},$$

where

$$Q = \frac{1}{2Z} \frac{\theta_t - \theta_0(\tau_c - 1)}{\theta_0(\theta_0 - 1)\pi_1^{(\gamma-1)/\gamma}\left(1 - \frac{\theta_0(\tau_c - 1)}{\theta_t \eta_t}\right)[1 + \eta_c(\tau_c - 1)]}.$$

Some interesting points are immediately apparent. First, the relative sensitivity of Z to η_c and η_t is

$$\frac{\partial Z}{\partial \eta_c} \bigg/ \frac{\partial Z}{\partial \eta_t} = \eta_t \left[\frac{\eta_t \theta_t}{\theta_0} - (\tau_c - 1)\right] \bigg/ [1 + \eta_c(\tau_c - 1)].$$

For typical values, such as $M_0 = 1$, $\theta_t = 7.5$, $\tau_c = 2.5$, and $\eta_c = \eta_t = 0.85$, this gives a value of about 1.54, showing that *the compressor efficiency is more important than turbine efficiency in maximizing $F/\dot{m}u_0$ (and I)*, for the turbojet. This conclusion does not carry over to the high-bypass turbofan, where the turbine work is much larger than it is in the turbojet, as we shall see from numerical results for the turbofan.

Forming a similar ratio of the sensitivities to $\pi_1^{(\gamma-1)/\gamma}$ and η_c gives

$$\pi_1^{(\gamma-1)/\gamma} \frac{\partial Z}{\partial \pi_1^{(\gamma-1)/\gamma}} \bigg/ \frac{\partial Z}{\partial \eta_c} = \frac{\tau_c}{1 + \eta_c(\tau_c - 1)};$$

for the above typical values this is 1.1, so that a percent improvement in $\pi_1^{(\gamma-1)/\gamma} = (\pi_d \pi_b \pi_a)^{(\gamma-1)/\gamma}$ is about as important as a percent improvement in η_c.

There are many possible uses of this type of analysis. Although the calculations tend to become complex algebraically, the added insight given by explicit expressions for the desired quantities often makes this approach more rewarding than direct numerical calculations.

3.7.4   Typical Results for Turbojets

Some typical numerical results for the turbojet with losses are presented in figure 3.4. The thrust per unit of airflow is shown in the top figure for a range of flight Mach numbers, for engines with convergent and ideally expanded nozzles, and with afterburning. The values of the loss parameters given on the figure are typical. For comparison, the ideal cycle result is also given. At low Mach numbers, the effects of both incomplete expansion and other losses are relatively small, but at Mach numbers near 3 the nonafter-burning engine with convergent nozzle has only about half the thrust of that with ideal expansion.

The same trends are reflected in the specific impulse, shown in the lower figure, but the effects of losses on the specific impulse are larger than on thrust.

Nozzle performance is critical at the higher values of M_0; the lowest curve in the top figure describes an engine with simple convergent nozzle. The loss in thrust (and in I) is about 60 percent at $M_0 = 3$. Note, however, that the loss due to imperfect expansion is small for $M_0 < 1$. Engines for subsonic aircraft characteristically have convergent nozzles for this reason.

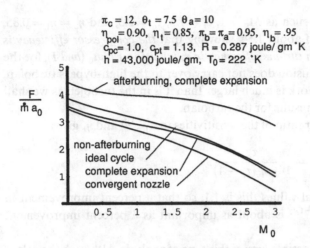

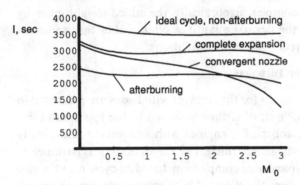

Figure 3.4
Thrust and specific impulse of turbojet with losses as a function of M_0, showing effects of nonideal nozzle expansion, and differences from ideal analysis.

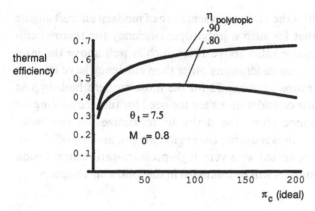

Figure 3.5
Thermal efficiency of a turbojet core engine as a function of ideal compression ratio, for two values of compressor polytropic efficiency. Other parameters are as in figure 3.4.

3.7.5 Thermal Efficiency of the Core

As was noted in section 1.3, the specific impulse reflects both the thermal efficiency and the propulsive efficiency. It is the most relevant measure of efficiency for a jet propulsion device. However, it is also interesting to examine the dependence of the thermal efficiency of the engine on the key cycle parameters. We noted in chapter 2 that for the ideal cycle the thermal efficiency increases continuously with increasing compression ratio, up to the limit where compressor outlet temperature equals turbine inlet temperature. It also increases with increasing turbine inlet temperature if the compression ratio is kept at or near the value for maximum power.

This second trend is still present for the real cycle with losses. However, the first is not. This is most clearly shown by calculating the thermal efficiency as a function of compression ratio. For the turbojet, we can represent the thermal efficiency as the difference in fluxes of kinetic energy between the exhaust and the inlet, divided by the energy input in the fuel flow. Thus,

$$\eta_{thermal} = \frac{u_0{}^2}{2hf}\left[(1+f)\left(\frac{u_6}{u_0}\right)^2 - 1\right].$$

This expression is plotted in figure 3.5 for a wide range of compression ratios, for two values of compressor and turbine polytropic efficiency, of which the value of 0.90 is the more representative of modern aircraft engine technology. We see that for such a polytropic efficiency the thermal efficiency peaks at a pressure ratio above 100, which is well above the level used in practice. In fact, considerations other than efficiency have thus far limited compression ratios to about 35 for the most modern high-bypass turbofans. Among these considerations are the need for turbine cooling air at reasonably low temperature, the desire to maximize the power-to-weight ratio of the core, thus reducing the engine weight, and the off-design operating problems connected with very-high-pressure-ratio compression systems. All of these matters will be discussed in the following chapters.

3.8 Turbofan with Losses

Because of the great commercial importance of the high-bypass turbofan, the analysis of its cycle with losses will be outlined, using the station numbers in figure 1.6. Up to equation 3.12, which expresses the compressor-turbine work balance, the expressions for the core flow are unchanged except that the core exit is station 6 rather than station 7. The turbine-compressor-fan power balance is modified by inclusion of the fan power,

$$\dot{m}c_{pc}(T_{t3} - T_{t2}) + \alpha\dot{m}c_{pc}(T_{t7} - T_{t2}) = \dot{m}(1 + f)c_{pt}(T_{t4} - T_{t5}),$$

so that

$$\tau_t = 1 - \left(\frac{\theta_0}{\theta_t}\right)\left(\frac{c_{pc}}{c_{pt}}\right)\left(\frac{1}{1+f}\right)[\tau_c - 1 + \alpha(\tau_f - 1)]. \tag{3.17}$$

By analogy to equation 3.7, the thrust of the fan flow is

$$\frac{F_8}{\dot{m}u_0} = \alpha\left[\left(\frac{u_8}{u_0}\right) - 1 + \left(\frac{1}{\gamma_c M_0^2}\right)\left(\frac{T_8}{T_0}\right)\left(\frac{u_0}{u_8}\right)\left(1 - \frac{p_0}{p_8}\right)\right], \tag{3.18}$$

where from equation 3.10

$$\frac{T_8}{T_0} = \frac{\theta_0\tau_f}{1 + [(\gamma_c - 1)/2]M_8^2} \tag{3.19}$$

and from equation 3.9

$$1 + \left(\frac{\gamma_c - 1}{2}\right)M_8^2 = \left(\frac{p_0}{p_8}\delta_0\pi_d\pi_f\right)^{(\gamma_c - 1)/\gamma_c}. \tag{3.20}$$

The fan jet velocity ratio is then

$$\frac{u_8}{u_0} = \frac{M_8}{M_0}\sqrt{\frac{T_8}{T_0}}. \tag{3.21}$$

The corresponding expressions for the core flow, making the appropriate changes in station numbers in equations 3.7, 3.10, and 3.13, are

$$\frac{F_6}{\dot{m}u_0} = (1+f)\frac{u_6}{u_0} - 1 + \frac{1+f}{\gamma_c M_0^2}\frac{R_t}{R_c}\frac{T_6}{T_0}\frac{u_0}{u_6}\left(1 - \frac{p_0}{p_6}\right), \tag{3.22}$$

$$\frac{T_6}{T_0} = \frac{\theta_t \tau_t}{1 + [(\gamma_t - 1)/2]M_6^2}, \tag{3.23}$$

$$1 + \frac{\gamma_t - 1}{2}M_6^2 = \left(\frac{p_0}{p_6}\delta_0 \pi_d \pi_c \pi_b \pi_t\right)^{(\gamma_t-1)/\gamma_t}, \tag{3.24}$$

and

$$\frac{u_6}{u_0} = \frac{M_6}{M_0}\sqrt{\frac{\gamma_t R_t}{\gamma_c R_c}\frac{T_6}{T_0}}. \tag{3.25}$$

The total thrust per unit of total airflow is

$$\frac{F}{\dot{m}a_0(1+\alpha)} = \frac{M_0}{1+\alpha}\left(\frac{F_6}{\dot{m}u_0} + \frac{F_8}{\dot{m}u_0}\right) \tag{3.26}$$

and the specific impulse is

$$I = \frac{F}{g\dot{m}_f} = \frac{a_0(1+\alpha)}{g}\left(\frac{F}{\dot{m}a_0(1+\alpha)}\right)\frac{1}{f}, \tag{3.27}$$

where

$$f = \frac{\bar{c}_p T_0}{\eta_b h}[(1+f)\theta_t - \theta_0\tau_c]. \tag{3.28}$$

As for the turbojet, it is convenient to recognize two limiting cases: the case where the (normally convergent) nozzles are ideally expanded and the case where they are choked. For most high-bypass engines, the ideal-expansion case will apply at takeoff, and the choked case at cruise (where the ram pressure rise in the inlet gives a pressure ratio on the order of 1.5, which multiplies the fan and engine pressure ratios).

3.8.1 Ideal Expansion

Here $p_0 = p_8 = p_6$, so the last terms in equations 3.18 and 3.22 are zero, while equations 3.20 and 3.24 give M_8 and M_6.

3.8.2 Choked Nozzle

Here $M_6 = M_8 = 1$, and p_8/p_0 and p_6/p_0 are found from equations 3.20 and 3.24.

It should be borne in mind in applying this last case that the scheme is valid only if p_{t6}/p_0 and p_{t8}/p_0 are larger than or equal to the values required to choke the nozzles, namely

$$\frac{p_{t6}}{p_0} = \left(\frac{\gamma_t + 1}{2}\right)^{\gamma_t/(\gamma_t - 1)}, \quad \frac{p_{t8}}{p_0} = \left(\frac{\gamma_c + 1}{2}\right)^{\gamma_c/(\gamma_c - 1)}.$$

3.8.3 Performance of Turbofans

Thrust per unit of total airflow and specific impulse are plotted in figure 3.6 as functions of bypass ratio for Mach number 0.8, a typical cruise number in the stratosphere. Results are shown for three fan pressure ratios. The nominal (or ideal) fan pressure ratio is shown, the actual pressure ratio being reduced somewhat by the inefficiency. That is, for each case the stagnation temperature ratio has been set at the value corresponding to the indicated ideal compression ratio. For both the fan and the compressor, the polytropic efficiency has been set at 0.90. Other parameters are as listed in the figure. The nozzles are assumed to be perfectly expanded for this example.

High-bypass commercial turbofans use single-stage fans to minimize noise generation. The fan pressure ratio has been limited to about 1.6. It can be seen from figure 3.6 that some gain in thrust and specific impulse could be had by using a fan pressure ratio of 2.0 or more, below bypass ratios of about 8, because the core jet velocity is larger than the fan jet velocity in this range of bypass ratios, with the fan pressure ratio of 1.6. The lowest fan pressure ratio shown, 1.2, results in too little energy extraction from the core flow up to a bypass ratio of about 14.

The effects of variations from the base values of compressor pressure ratio, compressor and fan polytropic efficiency, and turbine efficiency are shown in figures 3.7, 3.8, and 3.9 for the base case of $\alpha = 5$, $\pi_f(\text{ideal}) = 1.6$, $\theta_t = 7.5$. Note that there is considerable benefit to be had from pressure

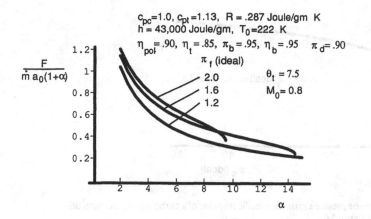

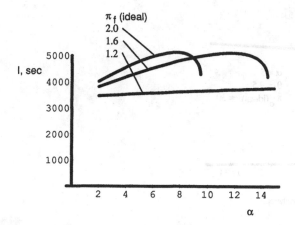

Figure 3.6
Thrust and specific impulse of turbofan as a function of bypass ratio, for fixed nominal fan
pressure ratios of 1.2, 1.6, and 2.0. Other parameters are as shown in the figure.

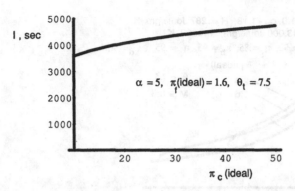

Figure 3.7
Effect of compressor pressure ratio on specific impulse of a turbofan engine, with other
parameters as in figure 3.6.

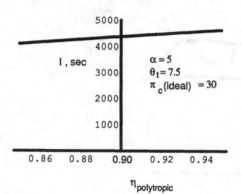

Figure 3.8
Effect of compressor polytropic efficiency on specific impulse of a turbofan engine, with other
parameters as in figure 3.6.

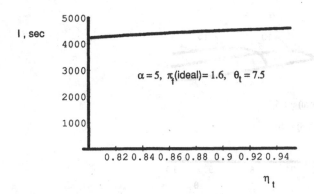

$\alpha = 5,\ \pi_f(\text{ideal}) = 1.6,\ \theta_t = 7.5$

Figure 3.9
Effect of turbine efficiency on specific impulse of a turbofan engine, with other parameters as in figure 3.6.

ratios higher than 30. Also, for these high-bypass engines, the benefit of increased turbine efficiency is about the same as that of increased compressor efficiency.

From figure 3.10 we see the effect of turbine inlet temperature for fixed fan pressure ratio and compressor pressure ratio. Within the range of $5 < \theta_t < 10$ shown, there is an optimum for each of the bypass ratios of 5, 10, and 15. The decline beyond the peak value is in each case to be attributed to reduced propulsive efficiency, since the thermal efficiency continues to rise as the turbine inlet temperature is increased. Larger increases in specific impulse could be had by increasing the compression ratio as the turbine inlet temperature is increased.

Figure 3.10 also shows the minimum turbine inlet temperatures required for the three bypass ratios of 5, 10, and 15 at the assumed fan pressure ratio of 1.6. For example, this value is about 6.1 for a bypass ratio of 10. Below this value the core provides too little power to drive the fan.

Figure 3.11 shows the effects on thrust and specific impulse of the various losses that have been included in the foregoing calculations. The top curves show the thrust and specific impulse given by the ideal cycle analysis, with equal core and fan jet velocities, i.e., the values given by equations 2.22 and 2.24. The next lowest curve is for an ideal cycle with, however, fixed compressor and fan pressure ratios of 30 and 1.6, and with specific heats as in figure 3.5. Comparing the top and bottom figures, we see that propulsion inefficiency affects specific impulse more than it affects thrust, which is offset somewhat by the difference in specific heats.

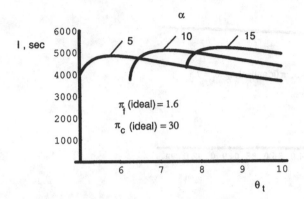

Figure 3.10
Effect of turbine inlet temperature on specific impulse of a turbofan engine for three bypass ratios, with other parameters as in figure 3.6.

The effect of inlet and burner pressure losses is largest at high bypass ratios, where the thrust per unit of airflow is smaller, so that the pressure loss is proportionately larger relative to the core nozzle pressure ratio.

The loss due to the choked nozzle is quite small even for low bypass ratios, where the core nozzle is significantly underexpanded.

3.9   Regenerated Gas Turbine with Losses

Shaft power P is the useful output of the regenerative gas turbine. From the compressor-turbine power balance, we get

$$P = \dot{m}(1 + f_b)c_{pt}T_{t4}(1 - \tau_t) - \dot{m}c_{pc}T_0(\tau_c - 1),$$

where the notation is as in figure 1.8. In dimensionless form,

$$W = \frac{P}{\dot{m}c_{pc}T_0} = (1 + f_b)\frac{c_{pt}}{c_{pc}}\theta_t(1 - \tau_t) - (\tau_c - 1).$$

Since τ_c and θ_t are to be regarded as design parameters, we have only to determine τ_t (and f_b) in order to compute $P/\dot{m}c_p T_0$. Following the changes in stagnation pressure through the engine, as in equation 3.9, we find

$$p_{t6'} = p_0 \pi_c \pi_b \pi_t \pi_r,$$

where $\pi_r = (p_{t3'}/p_{t3})(p_{t6'}/p_{t6})$ includes the pressure losses in both sides of the regenerator.

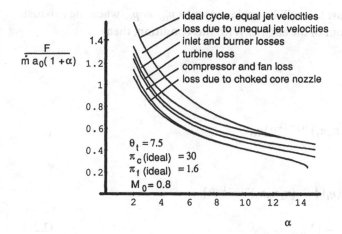

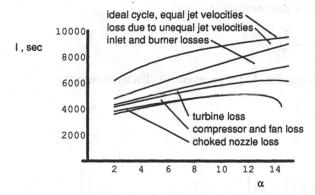

Figure 3.11
Effects of the several loss mechanisms on thrust and specific impulse of turbofan engines, for a range of bypass ratios.

The least exhaust loss will be achieved when $p_{16'} = p_0$, when the exhaust gas can just flow out of the engine. With this condition, then,

$$\pi_t = \frac{1}{\pi_c \pi_b \pi_r}$$

and

$$\tau_t = 1 - [1 - (\pi_c \pi_b \pi_r)^{-(\gamma_t - 1)/\gamma_t}]\eta_t,$$

so that finally

$$W = (1 + f_b)\frac{c_{pt}}{c_{pc}}\theta_t \eta_t [1 - (\pi_c \pi_b \pi_r)^{-(\gamma_t - 1)/\gamma_t}]$$

$$- [\pi_c^{(\gamma_c - 1)/\gamma_c} - 1]/\eta_c. \qquad (3.29)$$

The energy balance for the combustor gives

$$f_b = \frac{\bar{c}_p T_0}{\eta_b h}\left((1 + f_b)\theta_t - \frac{T_{3'}}{T_0}\right);$$

here the definition (equation 3.5) of regenerator effectiveness gives

$$\frac{T_{3'}}{T_0} = \varepsilon(\theta_t \tau_t - \tau_c) + \tau_c,$$

so that

$$\frac{hf_b}{c_{pc} T_0} = \frac{\bar{c}_p}{\eta_b c_{pc}}\left\{(1 + f_b)\theta_t - \varepsilon\theta_t\left[1 - \eta_t\left(1 - \frac{1}{(\pi_c \pi_b \pi_r)^{(\gamma_t - 1)/\gamma_t}}\right)\right]\right.$$

$$\left. - (1 - \varepsilon)\left(1 + \frac{\pi_c^{(\gamma_c - 1)/\gamma_c} - 1}{\eta_c}\right)\right\}. \qquad (3.30)$$

The thermal efficiency is now

$$\eta_{\text{thermal}} = \frac{P}{hf_b \dot{m}} = \frac{P/\dot{m}c_{pc} T_0}{hf_b/c_{pc} T_0}; \qquad (3.31)$$

the numerator is given by equation 3.29 and the denominator by equation 3.30.

As for the turbojet, not much is gained by further manipulation of equations 3.29 and 3.30. Computation is necessary to show the effects of the

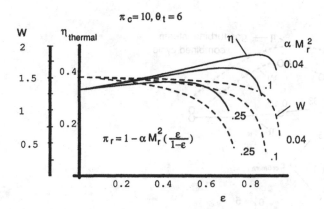

Figure 3.12
Efficiency η_{thermal} and specific work W of regenerated gas turbine with losses, as functions of regenerator effectiveness, ε, for various regenerator Mach numbers M_r.

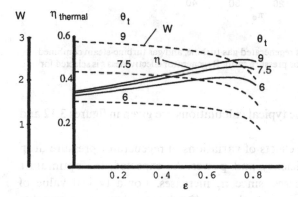

Figure 3.13
Effects of turbine inlet temperature ratio θ_t and of regenerator effectiveness on the efficiency and specific power of a regenerated gas turbine.

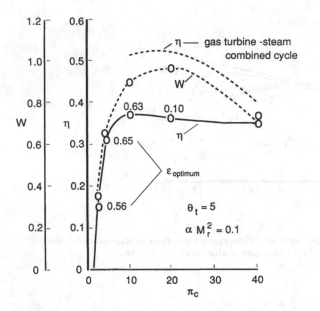

Figure 3.14
Efficiency and specific power of regenerated gas turbine, and gas turbine-steam combined cycle, as functions of compressor pressure ratio. Regenerator effectiveness ε is selected for maximum efficiency.

losses. The results of some typical calculations are given in figures 3.12 and 3.13.

Figure 3.12 shows the effects of variations of regenerator pressure drop and effectiveness on efficiency and power. As expected, the optimum ε increases as αM_r^2 decreases, since π_r increases. For a typical value of $\alpha M_r^2 = 0.1$, the peak efficiency is about 0.42, while the power is reasonably close to the maximum of $P/\dot{m}c_{pc}T_0 = 1.50$.

The effect of increasing θ_t, with fixed $\pi_c = 10$, is shown in figure 3.13. We see that the power increases dramatically, but the efficiency less so.

From a series of calculations such as those shown in figure 3.13, for different π_c the points of peak η can be selected, with the results summarized in figure 3.14. Here θ_t has been chosen as 5, a value representative of advanced, long-life, ground-based gas turbines, to facilitate comparison with the gas turbine-steam cycle in the next section. Note that for $\theta_t = 5$ a π_c of 10 maximizes η, and that the optimum ε at that point is 0.63. The power is somewhat below its maximum reached at $\pi_c \approx 20$, as would be expected.

3.10   Combined Gas Turbine-Steam Cycles with Losses

An assessment of this cycle, diagrammed in figure 1.10, necessitates a parametric study of both the steam "bottoming" cycle (a, b, c, d) and the gas turbine "topping" cycle $(0, 3, 4, 5)$. The requirement that the steam cycle draw its heat from the gas turbine exhaust implies that the gas temperature along the line $5 \rightarrow 0$ is larger than the steam (or water) temperature along $a \rightarrow c$ at each set of points between which the fluids are exchanging energy. Some reflection will show that this implies

$$T_6 - T_b = \Delta T_b > 0$$

and

$$T_5 - T_c = \Delta T_s > 0,$$

where ΔT_b and ΔT_s are the gas-water temperature drops in the boiler and the superheater, respectively. A further condition is

$$\dot{m}_{H_2O}(h_c - h_b) = \dot{m}_{gas} c_p (T_5 - T_6), \tag{3.32}$$

where h is the enthalpy of the steam.

Within these constraints lie wide choices of possible steam cycles and gas cycles. Thus, given a maximum feasible value of T_4 and a heat-rejection temperature of $T_a = T_d$, we can choose a high steam pressure and temperature T_b with a correspondingly high value of T_6 or a relatively low T_b and T_6. The first choice would result in an efficient steam cycle and a relatively inefficient gas cycle; the second choice would reverse the situation.

To analyze this cycle, we begin by prescribing the two most important parameters of the steam cycle: the boiling temperature T_b (and hence the steam pressure p_b) and the superheat temperature T_c. These values determine the enthalpies h_c, h_b, and h_a. If we specify the steam turbine's efficiency, the steam turbine's work is given by

$$h_c - h_{d'} = \eta_{t(steam)}(h_c - h_d),$$

where h_d is the ideal value of enthalpy at the steam turbine exit (that is, the value for which the entropy at d equals that at c), and $\eta_{t(steam)}$ is the efficiency of the steam turbine. An example will clarify this calculation. We choose

$p_a = p_d = p_{d'} = 2 \text{ mm Hg}$,

$\bar{p}_b = 2400 \text{ psia}$,

and

$T_c = 1000°F$.

Then, from *Thermodynamic Properties of Steam*, by J. H. Keenan and F. G. Keyes (Wiley, 1936), we find the following:

$T_b = 62°F$, $h_b = 718 \text{ BTU/lb}$,

$h_c = 1462 \text{ BTU/lb}$, $s_c = 1.534 \text{ BTU/lb °F}$,

$T_a = 101°F$, $h_{af} = 69.1, h_{afg} = 1037 \text{ BTU/lb}$,

$s_{af} = 0.132$, $s_{afg} = 1.848 \text{ BTU/lb °F}$,

where the subscripts f and fg refer to the properties for the liquid and for vaporization, respectively. Equating the entropies at c and d for ideal expansion gives

$s_c = 1.534 = s_{af} + x_d s_{afg} = 0.132 + 1.848\, x_d$,

or

$x_d = 0.759$.

Then

$h_d = 69.1 + 0.759(1037) = 856$,

and if $\eta_{t(steam)} = 0.85$, then

$h_c - h_d = 0.85(1462 - 856) = 515 \text{ BTU/lb}$.

The ratio of steam mass flow to gas turbine mass flow follows from equation 3.32:

$$\frac{\dot{m}_{H_2O}}{\dot{m}_{gas}} = \frac{c_{pt}(T_5 - T_6)}{h_c - h_b} = \frac{c_{pt}(T_c + \Delta T_s - T_b - \Delta T_b)}{h_c - h_b}.$$

For $\Delta T_s = 100°F$ and $\Delta T_b = 50°F$,

$$\frac{\dot{m}_{H_2O}}{\dot{m}_{gas}} = \frac{0.27(1000 + 100 - 662 - 50)}{1462 - 718} = 0.141.$$

The work produced by the steam turbine per unit of gas turbine mass flow is

$$(h_c - h_{d'})(\dot{m}_{H_2O}/\dot{m}_{gas}).$$

In the gas cycle, for a fixed T_{t4} the turbine temperature ratio is known, since $T_{t5} = T_c + \Delta T_s$ is fixed by the steam cycle; that is,

$$\tau_t = T_{t5}/T_{t4} = (T_c + \Delta T_s)/T_{t4}.$$

The turbine pressure ratio follows from $\pi_t = [1 - (1 - \tau_t)/\eta_t]^{\gamma_t/(\gamma_t-1)}$. Tracing the stagnation pressure ratios through the engine, we get

$$\pi_c \pi_b \pi_t = 1, \tag{3.33}$$

where π_b represents the pressure ratio of the combustor and the boiler together. Thus π_c is set, and

$$\tau_c = 1 + \frac{\pi_c^{(\gamma_c-1)/\gamma_c} - 1}{\eta_c}.$$

The work of the gas turbine is then

$$\dot{m}_{gas}[c_{pt} T_{t4}(1 - \tau_t) - c_{pc} T_0(\tau_c - 1)],$$

and the heat input to the combustor is

$$\bar{c}_p(T_{t4} - T_0 \tau_c).$$

Finally, $\eta_{combined}$, the efficiency of the combined cycle, is the sum of the gas turbine and steam turbine works divided by the heat input:

$$\frac{\dfrac{c_{pt}}{\bar{c}_p}\left[1 - \dfrac{T_c + \Delta T_s}{T_0 \theta_t} + \left(\dfrac{T_c + \Delta T_s}{T_0} - \dfrac{T_b + \Delta T_b}{T_0}\right)\dfrac{h_c - h_{d'}}{h_c - h_b} - \dfrac{\pi_c^{(\gamma_c-1)/\gamma_c} - 1}{h_c}\dfrac{c_{pc}}{c_{pt}}\right]}{\theta_t - 1 - \dfrac{\pi_c^{(\gamma_c-1)/\gamma_c} - 1}{\eta_c}}.$$

Continuing the above example for $\theta_t = 5$, $T_0 = 500°R$, we find $\tau_t = 1560/2500 = 0.624$, and $\pi_t = 0.1002$ if $\eta_t = 0.85$. Then for $\pi_b = 0.95$ we have $\pi_c = 10.5$, and from the last equation we have $\eta = 0.529$.

The results of a series of such calculations, for different steam pressures and maximum steam temperatures, are summarized in figure 3.15, where the lowest curve is for saturated steam (no superheat) and the upper two are for different superheat temperatures. The highest efficiencies are attained

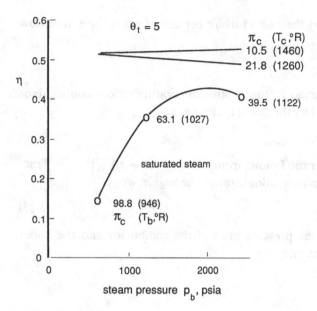

Figure 3.15
Efficiency of the gas turbine-steam combined cycle as a function of steam pressure for superheated (upper curves) and saturated (lower curve) steam cycles.

with superheated steam cycles, which they provide a closer match between the heat-rejection curve of the gas turbine and the heat-absorption curve of the steam. The efficiency is insensitive to steam pressure, with a slight advantage for the higher pressures. Note that peak efficiencies near 0.52 are predicted for this cycle.

With π_c used as a parameter, the efficiency of the gas turbine steam cycle is compared with that of the regenerative gas turbine in figure 3.14. It offers about a 0.15 improvement, a large margin for ground-based power plants.

3.11   Concluding Comments

Variations on the theme of this chapter could fill a much larger book. The possible range of parametric variations even for the simple turbojet is such that there is little sense in attempting to produce a complete set of characteristics. Rather, the characteristics of a particular engine cycle must be calculated when the need arises. If these examples aid the reader in this process, they will have served their purpose.

A great many variations of the basic turbojet, turbofan, turboshaft, and other cycles can be treated by slight modifications of this approach. These include turbine cooling and compressor bleed. Suggestions for such analyses will be found in the exercises at the end of this chapter, and some of their ramifications will be discussed in later chapters.

Problems

3.1   Write expressions for $F/\dot{m}a_0$ and I for a ramjet; include all appropriate losses (π_d, π_b, and so on), and account for changes in c_p with temperature. Assume that the nozzle is ideally expanded.

3.2   Derive expressions for $F/\dot{m}a_0$ and I for a turbojet that is ideal except that it has a choked convergent nozzle so that $M_6 = 1$. Compare your results numerically against those of figure 2.3.

3.3   Write expressions for $F/\dot{m}(1 + \alpha)$ and for I for a single-nozzle turbofan (figure 2.9), including the effects of inefficiency in the compression and expansion processes and those of changes in c_p with temperature. Assume that the fan airstream and the core stream, which may have different stagnation pressures and temperatures, mix at equal static pressures in a constant flow area and exit through a common nozzle. Compare your results to those of figure 3.6 for $\alpha = 5$ and $\pi_f = 1.6$.

4 Nonrotating Components

In chapters 2 and 3, engine performance was related to the performance of the several major components of the engine (inlet, compressor, burner, turbine, and nozzle); each component was described by the parameters through which it influences the thermodynamic cycle. Chapters 4–8 show how the behavior of each of the components is determined by its shape and other mechanical characteristics and what factors limit the performance of the components, and hence that of the engine.

Some of these limiting factors stem from fluid-mechanical phenomena, some from thermal effects, and some from mechanical or structural limitations. Even if they could all be quantified, a complete, quantitative treatment of all of them would require many volumes, and no such treatment will be attempted here. Consistent with the overall objective of this text, which is to convey an understanding of the behavior of aircraft engines, the phenomena will be discussed in physical terms, with as much analysis as is required to illustrate the important compromises controlling the design of any engine.

This chapter begins the process with a discussion of those components in which there is no dynamic energy exchange such as occurs between flowing gas and moving blades in the compressor and the turbine. Understanding the behavior of these components requires some background in gas dynamics and solid mechanics. Particularly essential is an understanding of compressible channel flow, shock waves, and the rudiments of boundary layer flows. Since these phenomena will play a central role in much of the discussion to follow, a brief summary of each will be given. The reader with no prior exposure to compressible flow would do well to study the relevent sections of one of the excellent texts, such as references 4.1 and 4.2.

4.1 Topics in Gas Dynamics

Three topics from gas dynamics will be discussed here: channel flow, because it provides an intuitive understanding of the way flow through the internal passages of engines is controlled by the shape of the passage walls; shock waves, because they lead to large qualitative differences between subsonic and supersonic flows; and boundary layers, because they lead to important limitations on the performance of the engine components.

4.1.1 Channel Flow

By *channel flow* we mean flow through a closed passage of gently varying cross-sectional area, such that the velocity component along the axis of the channel is much larger than the components perpendicular to the axis. When this condition is met, the kinetic energy of the flow and its momentum may be assumed to be those associated with only the velocity component along the axis. In this sense the flow is one-dimensional, but the variations in pressure, velocity, and temperature all result from the small velocities perpendicular to the axis, which result from the shape of the channel wall. These effects are brought into the analysis by the mass-flow continuity equation in the form

$$\rho u A = \dot{m} = \text{const},\tag{4.1}$$

where A is the cross-sectional flow area of the duct.

Combining the first law of thermodynamics and the axial momentum equation gives an equation for the total energy, or the stagnation enthalpy h_t, in terms of the thermal enthalpy h and kinetic energy:

$$h + \frac{u^2}{2} = h_t = \text{const},\tag{4.2}$$

where the subscript t denotes the stagnation state.

If the flow is isentropic, the entropy is constant:

$$s = s_t = \text{const}.\tag{4.3}$$

If the gas is *thermally* perfect,

$$p = \rho R T.\tag{4.4}$$

In most of the following discussion the gas will also be assumed to be *calorically* perfect, so that $h = c_p T$, where c_p is constant. Then, from equation 4.2,

$$c_p T \left(1 + \frac{u^2}{2c_p T}\right) = c_p T \left(1 + \frac{\gamma - 1}{2} M^2\right) = c_p T_t,$$

or   (4.5)

$$\frac{T_t}{T} = 1 + \frac{\gamma - 1}{2} M^2.$$

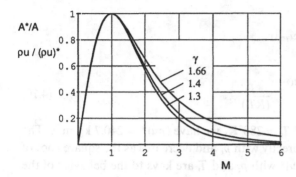

Figure 4.1
Ratio of mass flow density to that at $M = 1$ as a function of M for isentropic channel flow.

Since $s = s_t + c_p \ln(T/T_t) - R \ln(p/p_t)$, we have, from equation 4.3,

$$\frac{p_t}{p} = \left(\frac{T_t}{T}\right)^{\gamma/(\gamma-1)} = \left(1 + \frac{\gamma - 1}{2}M^2\right)^{\gamma/(\gamma-1)}. \tag{4.6}$$

The mass flow per unit of flow area is $\rho u = (p/RT)M(\gamma RT)^{1/2}$, so that

$$\frac{\rho u}{\rho_t a_t} = \frac{M}{\{1 + [(\gamma - 1)/2]M^2\}^{(\gamma+1)/2(\gamma-1)}}, \tag{4.7}$$

where a_t is the speed of sound at stagnation conditions. For given ρ_t and a_t, ρu is a maximum for $M = 1$. This condition is denoted by an asterisk:

$$\frac{(\rho u)^*}{\rho_t a_t} = \left(\frac{2}{\gamma + 1}\right)^{(\gamma+1)/2(\gamma-1)}. \tag{4.8}$$

Then, from equation 4.1,

$$\frac{A}{A^*} = \frac{(\rho u)^*}{\rho u} = \frac{1}{M}\left(\frac{1 + [(\gamma - 1)/2]M^2}{(\gamma + 1)/2}\right)^{(\gamma+1)/2(\gamma-1)}. \tag{4.9}$$

This relation connects the various flow properties to A, through M. It is plotted in figure 4.1. For the present, we simply note that ρu peaks at $M = 1$ and drops off by about a factor of 5 from this peak value at $M = 3$. This large range of variation greatly complicates the design of supersonic inlets. The significance of these relations will be further explored in the context of discussions of the behavior of the various components.

From equation 4.8,

$$(\rho u)^* = \rho_t a_t \left(\frac{2}{\gamma+1}\right)^{(\gamma+1)/2(\gamma-1)}$$

$$= \gamma^{1/2} \left(\frac{2}{\gamma+1}\right)^{(\gamma+1)/2(\gamma-1)} \frac{p_t}{(RT_t)^{1/2}}. \tag{4.10}$$

For air at $p_t = 1$ atm and $T_t = 288°$K, we have $(\rho u)^* = 240.7$ kg/m^{2}s. The value of $(\rho u)^*$ increases directly with p_t and decreases as the square root of T_t. These variations of $(\rho u)^*$ with p_t and T_t are keys to the behavior of the flow in the engine.

4.1.2 Shock Waves

By introducing jumps in pressure, temperature, and velocity, and by increasing the entropy and hence reducing the stagnation pressure of the bulk flow away from solid surfaces, shock waves exert a strong influence on the behavior of supersonic flows. Their existence is explained by the fact that the equations of conservation of momentum, mass, and energy have not a unique solution but rather two solutions, one corresponding to $M > 1$ and one to $M < 1$, and by the fact that the subsonic state is the one of higher entropy. Writing these relations across a discontinuity we have

$$\rho_1 u_1 = \rho_2 u_2,$$

$$p_1 + \rho_1 u_1^2 = p_2 + \rho_2 u_2^2,$$

$$c_p T_1 + \frac{u_1^2}{2} = c_p T_2 + \frac{u_2^2}{2}, \quad \text{or } T_{t1} = T_{t2},$$

where u is the velocity *perpendicular to the discontinuity*. These equations can be manipulated to give

$$u_1 u_2 = (a^*)^2,$$

$$M_2^2 = \frac{1 + [(\gamma-1)/2]M_1^2}{\gamma M_1^2 - (\gamma-1)/2},$$

$$\frac{p_2}{p_1} = \frac{2\gamma}{\gamma+1} M_1^2 - \frac{\gamma-1}{\gamma+1},$$

$$T_{t2} = T_{t1},$$

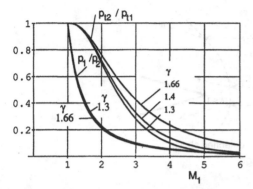

Figure 4.2
Stagnation pressure ratio and static pressure ratio across shock as function of normal upstream Mach number.

and

$$\frac{p_{t1}}{p_{t2}} = \left(\frac{2\gamma M_1{}^2}{\gamma + 1} - \frac{\gamma - 1}{\gamma + 1}\right)^{1/(\gamma-1)} \left(\frac{(\gamma - 1)M_1{}^2 + 2}{(\gamma + 1)M_1{}^2}\right)^{\gamma/(\gamma-1)}. \qquad (4.11)$$

This last equation, giving the stagnation pressure ratio across the shock as a function of the normal upstream Mach number, is particularly important for the following arguments. It is plotted in figure 4.2 with p_1/p_2, the static pressure ratio across the shock. Note that for M_1 slightly larger than unity the shock produces a significant static pressure ratio with little stagnation pressure loss, but as the Mach number increases above about 1.5 the loss increases rapidly. As we shall see, this implies that weak shocks are efficient means for diffusion of supersonic flows, but strong shocks introduce large stagnation pressure losses.

Since $(\rho u)^* \propto p_t$, and since T_t is constant across the shock,

$$\frac{(\rho u)_2^*}{(\rho u)_1^*} = \frac{p_{t2}}{p_{t1}}. \qquad (4.12)$$

4.1.2.1 Oblique Shock Waves   The above relations apply equally to all shock waves, but it is useful to distinguish shocks by their orientation in the flow. The convention is to refer to the shock as "normal" if it stands perpendicular to the total flow velocity, so that M_1 is the Mach number of that total velocity. If the shock stands at an angle to the total flow, it is termed an "oblique" shock, and $M_1 = M_{1n}$ is then to be interpreted as the Mach number associated with the velocity perpendicular to the shock sur-

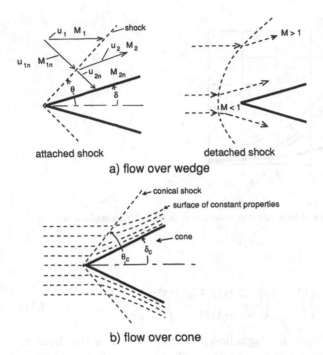

a) flow over wedge

b) flow over cone

Figure 4.3
(a) Supersonic flow over two-dimensional wedge with attached flow at left and detached flow at right, showing definitions of wave angle θ and deflection (wedge half) angle δ.
(b) Supersonic flow over axisymmetric cone, showing conical shock, and continued compression of flow over surface of cone.

face. Such an oblique shock can be generated by a wedge, as shown in figure 4.3a, or by a cone, as shown in figure 4.3b.

For the wedge, the oblique shock is completely determined by M_1 and the wave angle θ, but it is convenient to relate the wave angle θ (and therefore all properties of the shock) to the deflection angle δ, since δ specifies the shape of the wedge. When this is done, the results shown in figures 4.4 and 4.5 are found (reference 4.3). For a given δ, there is a minimum M_1 below which the oblique shock will not turn the flow sufficiently. For lower values of M_1 the shock "detaches," forming a normal shock in front of the nose of the wedge as in the right diagram of figure 4.3a. The subsonic flow behind this normal shock can turn through the required deflection angle. The flow behind the curved shock is much more complex than that behind a simple oblique shock because it is partially subsonic and partially supersonic.

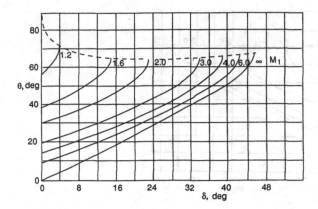

Figure 4.4
Wave angle θ as a function of deflection angle δ and initial Mach number M_1 for oblique shock on two-dimensional wedge.

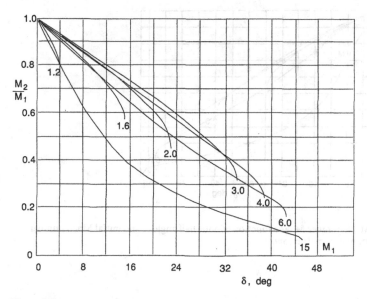

Figure 4.5
Ratio of final to initial Mach number, M_2/M_1, as a function of deflection angle δ and initial Mach number M_1 for oblique shock on two-dimensional wedge.

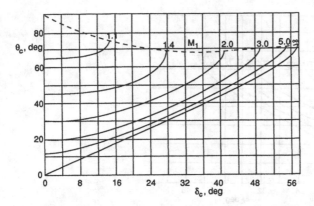

Figure 4.6
Wave angle as a function of cone half angle and initial Mach number for flow over axisymmetric cone.

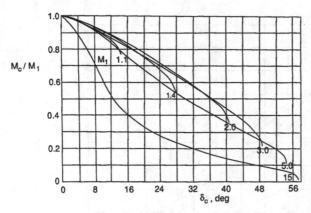

Figure 4.7
Cone surface Mach number M_c as a function of cone half angle δ_c and initial Mach number M_1 for flow over axisymmetric cone.

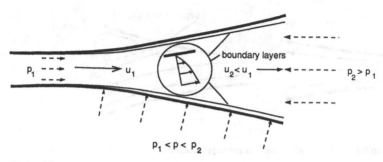

Figure 4.8
Diffusion in a subsonic channel flow.

For the two-dimensional wedge with attached shock, the flow behind the shock is uniform and parallel to the surface of the wedge. The analogous situation for an axisymmetric cone is that the flow has constant properties along lines radiating from the tip of the cone. Because the shock surface is made up of such lines, it has constant normal Mach number, pressure ratio, and stagnation pressure loss. The surface of the cone is another surface with constant properties different from those just behind the shock, and the flow field outside the cone has curved streamlines, as sketched in figure 4.3b. The variations of shock angle and Mach number at the cone surface are given for such flows in figures 4.6 and 4.7 (reference 4.3); they are useful for estimating the performance of axisymmetric inlets.

4.1.3   Boundary Layers and Separation

Blading and passages are designed to control the air movement in an engine to produce the directions and velocities required for energy exchange with the fluid, or between its thermal and kinetic energies, with the minimum entropy production (losses). In the main the fluid is controlled by pressure gradients, which result from changes in velocity traceable to changes in flow area through the mass flow continuity condition. Thus, in the simple divergent passage sketched in figure 4.8, the (subsonic) flow is decelerated (diffused) primarily by the pressure gradient in the flow direction, and there is an exchange between momentum and pressure as the velocity decreases. In the main part of the flow, viscous effects are small because the shear forces are small.

But in the immediate neighborhood of the walls, the velocity must change rapidly from the bulk value to zero to satisfy the no-slip condition

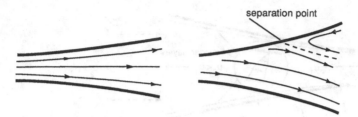

Figure 4.9
Unseparated (left) and separated flow in a subsonic channel.

at the surface. As a result, in a region near the wall, the behavior of the fluid is controlled by pressure and viscous shear forces, the momentum being very small. If this region is very thin relative to the dimensions of the channel (or, in the case of a blade, relative to its chord), the overall flow may be well represented by that in a passage made slightly smaller to allow for the viscous region (boundary layer) but of essentially the "inviscid" shape. One of the uses of boundary layer theory is to define the correction to passage shape, termed the "displacement thickness" of the boundary layer. A more difficult and more important use of the theory is to determine when the flow will fail to follow the surfaces of the passage because of the presence of the viscous region. This failure is termed *separation*.

To appreciate the importance of separation, we must recognize that for almost all gas-dynamic devices, the viscous shear forces acting on the surfaces are small relative to the pressure forces, which largely govern the behavior of the flow. Thus, in the channel in figure 4.9, if the flow remains attached to the walls, as in the left diagram, the loss associated with viscous effects is of the order of the ratio of boundary layer thickness to channel dimension—ordinarily a very small number. On the other hand, if the flow separates from the wall, as in the right diagram, the diffusing effect of the downstream portion of the passage is lost because the walls no longer control that part of the flow, and the diffuser's loss in performance can be very large.

Another example is provided by figure 4.10, where attached flow is shown at the top and separated flow at the bottom. Potential flow theory tells us that if the streamlines close smoothly at the trailing edge of the blades, the pressure drag of the airfoil (the net force in the direction of flow due to normal forces on the airfoil surface) is zero in two-dimensional flow. All the drag is then due to viscous shear on the surfaces. If the flow sepa-

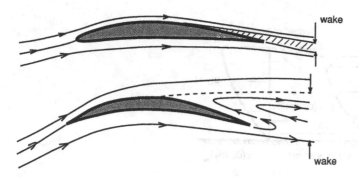

Figure 4.10
Unseparated (top) and separated flow on the suction surface of a blade, showing increase in
wake width.

rates, as in the lower diagram, the viscous drag is actually lower than for
attached flow because the velocity gradient is reduced on the downstream
portion of the airfoil, but the pressure drag increases greatly because the
pressure over the downstream portion of the suction side of the airfoil is
lower than it would be if the flow were smoothly diffused as in the attached
case. The increased drag generates a broader wake of low-momentum fluid
than for the case of attached flow, and the mixing of this wake with the
inviscid outer flow ultimately leads to an entropy increase in the flow.

Boundary-layer theory is one of the more difficult branches of fluid
mechanics, and the prediction of separation is one of the most difficult
aspects of boundary-layer theory; thus, a quantitative treatment of this
controlling phenomenon is beyond the scope of this book. The discussion
will be limited to qualitative descriptions of the important phenomena.
For quantitative treatments, see reference 4.4.

Consider the region of viscous flow near the wall, as sketched in figure
4.11. The flow velocity u along the surface increases from zero at the wall to
the free-stream value at the edge of the viscous region. The viscous shear
force at the wall is transferred outward by shear forces in the fluid, and
decelerates the flow in the viscous region. In the absence of pressure gradients along the flow direction, this deceleration causes the thickness of the
viscous region to grow in the flow direction; that is, the boundary layer
thickens as it entrains more fluid from the flow outside the boundary layer.

If there is a pressure gradient in the flow direction, the rate of change of
the boundary-layer thickness is altered. The reason is that any particular

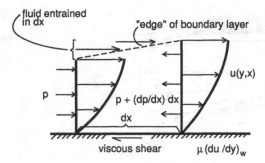

Figure 4.11
The forces acting on an element of boundary-layer flow.

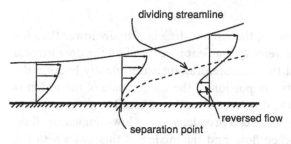

Figure 4.12
Flow at a two-dimensional separation point in a boundary layer.

pressure change requires a larger fractional change of the flow velocity in the boundary layer than is required in the inviscid flow, because the velocity is lower in the boundary layer. Thus, if the free-stream pressure decreases along the flow, the boundary-layer fluid is *accelerated* relative to the inviscid flow, the boundary layer is thinned, and we say that the pressure gradient is favorable. If the pressure increases in the flow direction, the boundary-layer fluid is *decelerated* relative to the inviscid flow, and the boundary layer is thickened. The greater thickness implies smaller viscous shear than in the case of zero pressure gradient, and if the pressure gradient is too large it will overcome the shear forces that transfer momentum toward the wall. In this case, the fluid near the wall will stagnate and, in fact, reverse, as sketched in figure 4.12. This is what is meant by "separation" in two-dimensional flow.

Laminar boundary layers can support only a very small amount of diffusion before separation—on the order of a 10 percent reduction in flow velocity if the velocity decreases linearly. Turbulent boundary layers do better, and one method of preventing separation is to induce transition to turbulent flow. Most aircraft engine components operate at Reynolds numbers such that transition to turbulent flow occurs somewhere in the passages, or on the blades; if it can be arranged that the pressure rise occurs mainly after transition, separation problems can be made less serious than if it occurs where the layers are laminar.

The location of the transition point can be a critical factor in controlling performance. This is the case for cooled turbines (chapter 6) and for prospective hypersonic vehicles (chapter 10).

Shock waves are an additional complicating factor, as they can induce severe separation problems when they impinge on walls, by imposing sudden pressure changes on the boundary layers.

Although much effort has been devoted to developing methods for predicting separation of turbulent boundary layers, methods are not available at present to reliably predict separation on the blades or casings of turbomachines. Unsteadiness and three-dimensional effects make the situation so complex that separation is even hard to define. The usual approach has been to use data correlations to extend the base of experience to new designs.

For the stationary components, such as the subsonic portion of the inlet and the diffuser between the compressor exit and the combustor, some guidance can be obtained from systematic experimental studies of diffusers. Such studies are reported in reference 4.5 for straight-walled planar (i.e., two-dimensional) diffusers, in reference 4.6 for straight-walled conical diffusers, and in reference 4.7 for annular diffusers as well as the other two types. In these studies the performance of the diffusers is characterized by a pressure coefficient, defined as

$$c_p = \frac{p_e - p_i}{p_{ti} - p_i},$$

where p_e and p_i are the exit and inlet static pressures and p_{ti} is the inlet stagnation pressure on the central streamline (i.e., in the inviscid flow). With this definition it is found that pressure coefficients up to 0.8 can be realized when the blockage of the inlet by viscous layers is less than about 0.02. As the blockage increases, the pressure coefficient decreases, to about 0.6 at a blockage of 0.12.

E. E. Zukoski (reference 4.8) has pointed out that these results can be generalized somewhat by defining a mean pressure coefficient

$$\bar{c}_p = \frac{p_e - p_i}{\bar{p}_{ti} - p_i},$$

where $\bar{p}_{ti}$ is a mean inlet stagnation pressure defined through equation 4.6 in terms of a mean inlet Mach number $\bar{M}$. This mean inlet Mach number is related to the true average mass flow density through equation 4.7. With this definition, a value of $\bar{c}_p = 0.85$ correlates the data of reference 4.6 for a range of blockages from 0.02 to 0.12, and for values of $\bar{M}$ from 0.2 to approximately 1.0.

Very recently, V. Filipenko (reference 4.9) has found in studies of a radial diffuser for a centrifugal compressor that the stalling incidence and the pressure recovery of the diffuser can be correlated over a wide range of inlet conditions, equivalent to the blockage changes discussed above, by referring the performance to an "availability averaged" inlet stagnation pressure. Physically this reference pressure is the one that would result if all the entering fluid were brought reversibly to a uniform state while conserving energy and entropy. Although this correlation has been experimentally confirmed only in the context of a radial diffuser, it seems probable that it applies more generally. In a form appropriate to the more general situation, the availability-averaged stagnation pressure would be given by

$$(\overline{p_{t1}})_{av} = \exp\left(\frac{\int \ln(p_{t1})\rho_1 u_1 \, dA}{\int \rho_1 u_1 \, dA}\right),$$

where the integral is over the inlet flow area, and $\rho_1 u_1$ is the inlet mass flux. The pressure coefficient then would be defined as

$$(\overline{c_p})_{av} \equiv \frac{p_2 - p_1}{(\overline{p_{t1}})_{av} - p_1}.$$

4.1.4 Gas-Solid Heat Transfer

Wherever the fluid exerts viscous shear stresses on the passage walls, there is also the possibility of thermal energy exchange between the fluid and the wall. In gases, momentum transport and thermal transport occur by essentially the same mechanisms. For laminar flows, the mechanism is the random motion of molecules, which leads molecules originating in hotter regions of the fluid to transfer part of their kinetic energy to molecules in

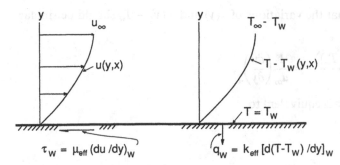

Figure 4.13
Illustration of the analogy between momentum transport in a velocity gradient (left) and heat transfer in a temperature gradient.

cooler regions by collisional interaction. For turbulent flows this mechanism is greatly augmented by more-or-less random motion of fluid eddies. This equivalence of transport mechanisms does not exist for all fluids. In liquid metals, for example, electronic transport is a major factor in thermal conduction, but not in momentum transport.

Restricting ourselves to ordinary gases at low Mach numbers, we can construct an analogy between heat transfer and shear along the lines indicated in figure 4.13. The shear stress at the wall τ_w is related to the velocity gradient by an effective viscosity μ_{eff} equal to the fluid viscosity for laminar flow and larger for turbulent flow. For low Mach numbers, the driving force for heat transport to the wall is the difference between fluid and wall temperatures; thus, if the momentum and the energy are transported by the same mechanism, we would expect the variations of $u(y)$ and $T(y) - T_w$ to be similar. To make this statement quantitative, we must nondimensionalize the shear and the heat transfer. Customarily, we nondimensionalize the shear by the free-stream dynamic pressure to yield a friction coefficient:

$$c_f \equiv \frac{2\tau_w}{\rho u_\infty^2} = \frac{2\mu_{\text{eff}}(\partial u/\partial y)_w}{\rho_\infty u_\infty^2}. \tag{4.13}$$

Similarly, we divide the heat flux by the rate at which thermal energy (referenced to the surface temperature) is convected along the surface to yield a Stanton number:

$$\text{St} \equiv \frac{q_w}{\rho_\infty u_\infty c_p(T_\infty - T_w)} = \frac{k_{\text{eff}}(\partial(T - T_w)/\partial y)_w}{\rho_\infty u_\infty c_p(T_\infty - T_w)}. \tag{4.14}$$

The statement that the variations of $u(y)$ and $T(y) - T_w$ should be similar implies that

$$\frac{1}{T_\infty - T_w}\left(\frac{\partial(T - T_w)}{\partial y}\right)_w = \frac{1}{u_\infty}\left(\frac{\partial u}{\partial y}\right)_w,$$

so we see that this is equivalent to

$$St = \frac{k_{eff}}{c_p \mu_{eff}}\frac{c_f}{2}.$$

The dimensionless group $c_p \mu/k$ is called the *Prandtl number* (denoted Pr), and is near unity for most gases (0.7 for air). For turbulent flows, where k_{eff} and μ_{eff} are dominated by eddy transport, $c_p \mu_{eff}/k_{eff}$ is quite close to unity. Thus we arrive at a form of Reynolds analogy,

$$St \approx \frac{c_f}{2}, \tag{4.15}$$

where St and c_f are defined by equations 4.14 and 4.13. In physical terms, this relation states that

$$\frac{\text{Heat flux to wall}}{\text{Convected heat flux}} = \frac{\text{Momentum flux to wall}}{\text{Convected momentum flux}},$$

or

$$\frac{q_w}{(\rho_\infty u_\infty)c_p(T_\infty - T_w)} = \frac{\tau_w}{(\rho_\infty u_\infty)u_\infty}.$$

We will use this relationship to understand some important characteristics of cooled turbines in section 6.3.

4.2 Diffusers

The function of the diffuser (or inlet) is to bring the air from ambient conditions to the conditions required at the inlet to the engine. As was noted in section 1.12 and indicated schematically in figures 3.1, 1.13, and 1.14, diffusers designed for purely subsonic flight differ greatly from those designed for supersonic flight, which must meet the requirements for decelerating the flow from supersonic speeds in addition to the requirements the subsonic diffuser must meet.

In this section it will be necessary to refer to some of the mass flow characteristics of compressors that have not been described yet. A preliminary look at chapter 5, where these characteristics are discussed, may be helpful.

4.2.1 Subsonic Diffusers

The two principal constraints that must be met by the subsonic diffuser are that it must supply air to the engine at the axial Mach number M_2 which the compressor or fan demands and that it must efficiently capture the entering streamtube over a wide range of free-stream Mach numbers.

As we shall see, M_2 is mainly determined by the rotational Mach number of the compressor or fan. It therefore depends on rotational speed and inlet air temperature and is largest at conditions of high altitude (low T_0) and full engine speed. M_2 is smallest for conditions of low altitude (high T_0) and low engine speed. For a typical subsonic transport application, the most important requirements are for takeoff at full engine speed and high T_0, and for cruise at lower T_0 (but at $M_0 \approx 0.8$) and perhaps reduced engine speed. The decrease in T_0 tends to be offset by the increase in M_0 and the reduction in engine speed, so the variation in M_2 is not large. A decrease of about 20 percent from takeoff to high subsonic cruise is typical.

A more demanding requirement stems from the change in M_0 from zero at takeoff to about 0.8 at cruise. With M_2 nearly fixed, this imposes large changes in the geometry of the streamlines entering the diffuser, with attendant problems of boundary-layer separation when the inlet surfaces must turn the flow through large angles. From the curve of mass flow density ρu versus M of figure 4.1, the ratio of inlet streamtube area to engine inlet area A_0/A_2 can be computed:

$$\frac{\rho u(M_2)}{\rho u(M_0)} = \frac{A^*/A(M_2)}{A^*/A(M_0)}.$$

If $M_2 = 0.5$, for example, then A_0/A_2 varies from ∞ at $M_0 = 0$ to 0.78 at $M_0 = 0.8$. The shape of the captured streamtube therefore varies somewhat as shown in figure 4.14, and the problem is to design an inlet that will accept the flow with the indicated large changes in direction without excessive losses due to separation.

If optimized for the $M_0 = 0.8$ cruise, the inlet might have the contour in the upper half of figure 4.14 with a thin lip to minimize the increase in Mach number in the external flow over the lip. But this inlet would sepa-

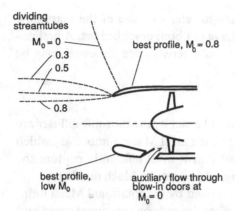

Figure 4.14
Schematic diagrams of a subsonic diffuser. The upper diagram shows the streamtube shapes
for various flight Mach numbers and an optimum shape for cruise; the lower one shows the
shape dictated by compromises for takeoff and low-speed flight.

rate badly on the inside at takeoff and low subsonic conditions because the
turn around the sharp lip would impose severe pressure gradients. The
compromise is to round the lip somewhat, making it less sensitive to flow
angle, but incurring some loss in the exterior flow, which accelerates to
supersonic speeds as it turns over the outer contour of the inlet. The Mach
number at this point at the cruise condition may be 1.3 or even higher, with
attendant shock losses in the external flow.

To minimize these problems, it is desirable to make the minimum flow
area of the inlet no larger than is necessary to pass the required engine flow
at the highest corrected speed (or M_2), which will usually occur at takeoff
conditions.

In addition, "blow-in doors" are sometimes added to allow some flow to
bypass the minimum area of the inlet and so reduce separation problems at
takeoff. These doors normally are open only at full-throttle conditions
during takeoff. They were used in the turbofan-powered Boeing 707, and
in an early model of the 747. It is interesting to note that the blow-in
doors have been deleted from later 747s, because the nonuniform flow
into the fan, which resulted from their opening on takeoff, led to unacceptable increases in fan noise. Eliminating the doors required an increase in
minimum area, a more rounded lip, and somewhat more drag at cruise
conditions—a price to be paid for lower takeoff noise.

The details of the design of a subsonic inlet depend on the way these compromises are struck. Modern computational tools make possible detailed calculations of the boundary layer behavior at all operating conditions, so that the compromises can be made quite rationally. With wing-mounted engines, they also deal quite effectively with the interaction between the external flow over the nacelle and the flow over the wing. An example of great success in this area is the nacelle installation for the Boeing 737-300. For this version of the 737, the original low-bypass JT8D engines were replaced by high-bypass CFM-56 engines of much larger airflow, consequently requiring larger-diameter nacelles. To avoid lengthening the landing gear it was necessary to mount the nacelles higher on the wing than early practice would have allowed without serious interference drag arising. Through the extensive use of computational fluid dynamics, a design was evolved with less interference drag than the original low-bypass installation.

When fully developed, a good inlet will produce a pressure recovery π_d between 0.95 and 0.97 at its optimum condition. For a more detailed discussion of subsonic inlet design, see reference 4.10.

4.2.2 Supersonic Diffusers

Flight at supersonic speeds complicates the design of the diffuser for three reasons. The most fundamental is the existence of shock waves, which introduce a wholly new loss mechanism that can lead to large decreases in stagnation pressure, even in the absence of viscous effects, and to bistable operation, with large changes in losses and in mass flow between the two modes. Much of the emphasis in discussions of supersonic diffusers has been on this aspect of their behavior. A second reason is that the variations in capture streamtube diameter between subsonic and supersonic flight for a given engine are very large (as much as a factor of 4 between $M_0 = 1.0$ and $M_0 = 3.0$), and an aircraft that is to fly at $M_0 = 3.0$ must also operate at $M_0 = 1$! Finally, as M_0 increases, the inlet compression becomes a larger fraction of the overall cycle compression ratio; as a result, the specific impulse and thrust per unit of mass flow become more sensitive to diffuser pressure recovery. This is especially evident for hypersonic air-breathing propulsion systems.

A typical diffuser, such as the one at the right in figure 3.1, is made up of a supersonic diffuser, in which the flow is decelerated by a combination of shocks and diffuse compression, and a subsonic diffuser, which reduces the

Mach number from the high subsonic value after the last shock to the value acceptable to the engine.

Focusing on the supersonic diffusion, we may divide the compression process into two types: *external compression*, in which the streamtube is bounded only on one side by solid surfaces, and *internal compression*, in which the flow is through a passage bounded on both sides (that is, a channel). Modern diffusers generally use a combination of external compression and internal compression. We will begin with a description of internal compression as an introduction to many of the phenomena that characterize supersonic diffusers.

4.2.2.1 Internal Compression Conceptually, an internal compression diffuser may be thought of as a convergent-divergent channel in which the supersonic flow is decelerated by a diffuse series of weak compression waves to sonic velocity, then diffused subsonically. In the context of channel flow the exact shape of the passage is immaterial; only the variation of flow area along the streamwise coordinate enters into the representation of the geometry, and the diffuser may be thought of schematically as the analogue of a convergent-divergent, or Laval, nozzle. In a real diffuser, however, the shape is of crucial importance, because it controls the strength and location of the compression waves that carry out the free-stream compression and the pressure gradient to which the boundary layer is subjected. It is this pressure gradient that determines whether the boundary layer separates, and hence whether the shape of the passage controls the flow behavior. While a simple convergent-divergent passage could function as a nearly isentropic diffuser in principle, it is impractical for at least two reasons. One is the starting problem. The ideal shock-free diffusion cannot be attained by increasing the flight Mach number to the final value with fixed geometry of the diffuser, because the existence of a shock ahead of the inlet prevents it from passing the design mass flow. This will be explained in detail below. The second is that boundary-layer growth would prevent smooth diffusion to $M = 1$ in a convergent passage bounded on both sides. Recall (figure 4.1) that the passage area passes through a minimum at $M = 1$; thus, for a given Mach number at entrance to the diffuser, the Mach number at the point of minimum area (throat) would be very sensitive to the thickness of the boundary layer there. Excess boundary-layer growth would tend to cause the flow to shock, and the shock would be expelled to a position ahead of the diffuser entrance, because it is unstable in the convergent passage. These points will also be elaborated.

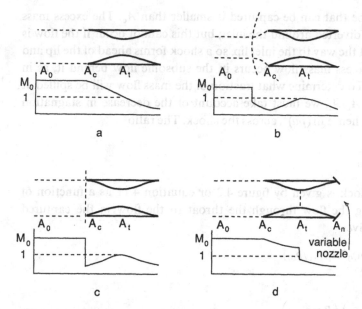

Figure 4.15
Schematics of internal-compression diffuser, showing (a) ideal isentropic diffusion from M_0 through unity to $M < 1$, (b) operation below the critical (starting) Mach number, (c) operation at the critical M_{0c}, but not started, and (d) operation at the critical M_{0c} and started, with the shock positioned at the throat.

Addressing first the "starting" problem, we consider the development of the flow with increasing flight Mach number for a diffuser of fixed area ratio, A_c/A_t, between capture and throat, as shown schematically in figure 4.15. The ideal, shock-free operation (figure 4.15a) would require a value of M_0 such that

$$\frac{A(M_0)}{A^*} = \frac{A_c}{A_t}.$$

For any smaller M_0,

$$\frac{A(M_0)}{A^*} < \frac{A_c}{A_t},$$

and if the flow is choked at the throat, so that $A^* = A_t$, then

$$A(M_0) < A_c;$$

the streamtube that can be captured is smaller than A_c. The excess mass flow must be diverted around the inlet; but this cannot occur if the flow is supersonic all the way to the inlet lip, so a shock forms ahead of the lip and the spill of excess mass flow occurs in the subsonic flow behind it, as in figure 4.15b. To determine what fraction of the mass flow will be spilled at any M_0 and A_c/A_t, we must take account of the decrease in stagnation pressure, and hence in $(\rho u)^*$, across the shock. The ratio

$$\frac{(\rho u)_1{}^*}{(\rho u)_0{}^*} = \frac{p_{t1}}{p_{t0}}$$

across the shock is given by figure 4.2 or equation 4.11 as a function of M_0. Equating the flow through the throat to the flow in the captured streamtube gives

$$\rho_0 u_0 A_0 = (\rho u)_0{}^* A_t (p_{t1}/p_{t0}),$$

or

$$\frac{A_0}{A_t} = \left(\frac{A}{A^*}(M_0)\right)\left(\frac{p_{t1}}{p_{t0}}(M_0)\right). \tag{4.16}$$

This ratio of actual captured streamtube area to throat area is plotted in figure 4.16 as a function of M_0, along with the corresponding ratio (simply A_0/A^*) for the shock-free diffuser. When the flight Mach number M_0 is increased to the critical value M_{0c} such that $A_0/A_t = A_c/A_t$, the normal shock will stand just at the lip, as in figure 4.15c. But in this position it is unstable and will move downstream if perturbed.

This instability is easily understood by imagining that the shock is moved slightly downstream by some disturbance. Since the Mach number decreases downstream with supersonic flow in the converging passage, the perturbed shock will stand at a lower Mach number and will cause less stagnation pressure loss. The choked throat will then pass more mass flow than is captured by the lip, with a consequent net outflow of mass from the convergent section of the diffuser. This requires a reduction in average density that can be effected only by discharging the shock downstream, thus "starting" the diffuser.

To achieve the best possible pressure recovery with the diffuser, once it is started, the back pressure would be adjusted so that the shock stands at the throat, where the Mach number is smallest, as in figure 4.15d.

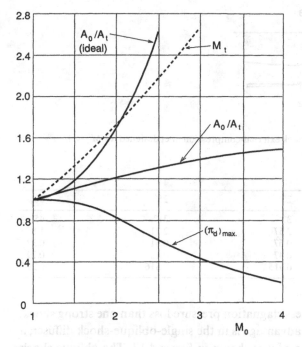

Figure 4.16
Characteristics of a fixed-geometry internal-compression inlet, including best pressure
recovery, capture-to-throat area ratio, and throat Mach number.

The Mach number M_t ahead of the shock at the throat is readily found
by noting that it corresponds to a value of A/A^* given by

$$\frac{A(M_t)}{A^*} = \frac{A(M_0)}{A^*} \frac{A_t}{A_c}. \tag{4.17}$$

The best pressure recovery this inlet can achieve is that for a normal shock
at M_t.

Values of M_t and $(\pi_d)_{max}$ for such inlets are given in figure 4.16. The
pressure recovery is good up to $M_0 \approx 1.5$, but it drops off rapidly at higher
Mach numbers because the starting requirement dictates an area ratio that
leads to a strong normal shock at the throat.

4.2.2.2 External Compression   Much better pressure recovery can be
had at high M_0 by taking advantage of the characteristic of shock waves
that, for a given overall pressure (or Mach number) ratio, a series of weak

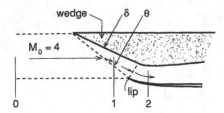

Figure 4.17
Schematic of single-oblique-shock external-compression inlet designed for $M_0 = 4$.

Table 4.1

$\delta =$	20°	25°	30°
$\theta =$	32°	38°	45°
$M_1 =$	2.57	2.21	1.85
$p_{t1}/p_{t0} =$	0.67	0.51	0.37
$p_{t2}/p_{t1} =$	0.47	0.62	0.79
$\pi_d =$	0.315	0.316	0.295

shocks produces much less stagnation pressure loss than one strong shock. The simplest use of this advantage is in the single-oblique-shock diffuser; a two-dimensional version of it is shown in figure 4.17. The oblique shock decelerates the flow to some M_1 between M_0 and 1, and it is again decelerated through a nearly normal shock at the throat.

To illustrate the characteristics of such a diffuser, consider a design for $M_0 = 4$. For a set of three shock deflection angles δ, one finds the shock angles θ from figure 4.4 and M_1 from figure 4.5 which are tabulated in table 4.1. The stagnation pressure ratio across the oblique shock follows from figure 4.2 with $M_{0n} = M_0 \sin\theta$, and that across the normal shock follows directly from figure 4.2 with $M_{1n} = M_1$.

Evidently the best wedge angle for $M_0 = 4$ is about 22.5°. Comparing the pressure recovery of 0.32 to the value of 0.2 attainable with a simple internal compression inlet (see figure 4.16), we see a marked improvement.

Further gains can be made by introducing more oblique shocks. Three, as shown schematically at the top of figure 4.18, would give $\pi_d = 0.63$ at $M_0 = 4$. The limit in this progression is the isentropic wedge diffuser at the bottom of figure 4.18, in which compression takes place to $M = 1$ through a series of very weak compression waves. There are some serious difficulties with such nearly isentropic diffusers, however. One is that the

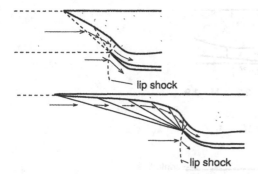

Figure 4.18
Schematics of (above) three-oblique-shock external-compression inlet and (below) isentropic spike inlet, showing how air turning angle increases with approach to isentropy.

shocks originating on the wedge in the multi-wedge diffuser will tend to separate the boundary layer there. The streamwise pressure increase along the spike has the same effect in the isentropic diffuser. Another disadvantage is that, as the compression approaches isentropy, the flow deflection angle increases from zero for a normal shock to 68° at $M_0 = 4$ for an isentropic diffuser. The near-sonic flow at this angle must be turned back to the axial direction; this leads to a large angle on the lip, which will form a shock in the external flow and produce considerable "cowl drag." It is clear that a compromise must be made between pressure recovery and drag for such external-compression inlets.

4.2.2.3 Mixed Compression By combining an initial external compression through an oblique shock with internal compression inside the lip, one can approach the high-pressure recovery of the multi-oblique-shock diffuser without incurring as much cowl drag as with a pure external-compression inlet. The scheme is illustrated in two dimensions in figure 4.19. The three-shock diffuser shown would have an ideal $\pi_d = 0.55$ at $M_0 = 4$. This is considerably better than the 0.32 of the two-shock external-compression inlet of figure 4.17, and the diffuser as shown would have low cowl drag. The main disadvantage is that the internal contraction reintroduces the starting problem, as discussed in subsection 4.2.2.1, but this difficulty can be overcome by reasonable variations in geometry in flight. An axisymmetric version of this type of inlet is used on the SR-71 Mach 3 + reconnaissance aircraft. This inlet will be discussed in subsection 4.2.2.7.

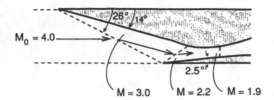

Figure 4.19
Mixed-compression inlet configured for $M_0 = 4$.

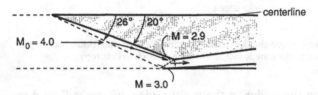

Figure 4.20
Axisymmetric mixed-compression inlet configured for $M_0 = 4$; compare figure 4.19.

4.2.2.4 Axisymmetric Diffusers Thus far the quantitative discussion of supersonic diffusers has dealt with two-dimensional devices. Although many practical inlets are essentially two-dimensional, in some applications an axisymmetric inlet is desirable, as in pod-mounted engine installations such as that on the SR-71. Indeed, the popular conception of a supersonic inlet is axisymmetric.

The flow is more complex in such an inlet than in two-dimensional ones because the shocks are curved rather than plane surfaces and in general are of varying strength over their surface. Fortunately, some simplification results from the characteristics of conical flow fields (subsection 4.1.2). The shock generated by a cone, as indicated in the cross-sectional schematic of figure 4.20, has constant properties, pressure ratio and turning, but the flow behind it undergoes a continued isentropic compression, as discussed in subsection 4.1.2.1. This is an advantage because the flow between the shock and the cone surface is compressed isentropically, but as a result the flow at the lip, just behind the shock, has a higher Mach number than that at the cone surface. In fact, this Mach number is the same as for a two-dimensional inlet with the same shock angle. Both effects are rather small in practice, however, as we shall see from an example.

A direct comparison of the two types of inlet can be made by choosing the cone angle so the initial oblique shock is the same strength (lies at the same angle) as in the two-dimensional diffuser. Compare figure 4.20 and figure 4.19; the cone-angle and the cone-surface Mach numbers are taken from figures 4.6 and 4.7. Note that the cone-surface Mach number differs only slightly from that at the lip, so that the second shock would be of nearly uniform strength in this example, and the pressure recovery would be almost the same as for the two-dimensional inlet.

The axisymmetric inlet offers somewhat better boundary-layer behavior on its spike than on the first ramp of the two-dimensional inlet, because of the lateral stretching effect on the boundary layer. As the circumference of the cone increases in the flow direction, the boundary layer is thinned relative to a layer in the same pressure gradient on a two-dimensional ramp. On the other hand, the axisymmetric inlet suffers from cross-flow effects that occur at angle of attack, the boundary layer tending to flow around the spike from the upwind side to the downwind side.

4.2.2.5 Boundary-Layer Control Each of these external compression inlets has at least one point where a shock impinges on a surface. As was noted in subsection 4.1.3, this is almost sure to result in separation of the boundary layer unless some preventive action is taken. Probably the most effective remedy is to bleed off the boundary layer just upstream of the point of shock impingement. The shock then effectively stands on a surface free of boundary layer. Bleeding can also be used to control the boundary layer in the subsonic portion of the diffuser downstream of the shock system. The amounts of flow that must be taken through the suction slot or through porous walls are very small relative to the throughflow of the diffuser—on the order of a few percent. Normally the bled flow is dumped overboard, but in some cases it has been used for afterburner or nozzle cooling. These techniques for boundary-layer control will be more fully discussed in subsection 4.2.2.7, where an actual supersonic diffuser will be described.

4.2.2.6 Off-Design Behavior A fixed-geometry inlet has at least three important operating variables: flight Mach number, angle of attack, and mass flow (or the backpressure imposed by the engine). Leaving the effects of angle of attack aside for the moment, we consider the effects of varying Mach number and mass flow on the two simple types of diffuser: internal-compression and external-compression.

As the flight Mach number is increased for a simple diffuser of the type shown in figure 4.15, a normal shock will form at $M_0 = 1$ and will gradually move toward the lip as M_0 approaches M_{0c}. At M_{0c} it will pop through the convergent section, provided the downstream condition is such that the inlet can pass the full mass flow. If we imagine this downstream condition to be imposed by a variable-area nozzle, as in figure 4.15d, and if it is choked, the requirement on the nozzle area is that

$$A_n(\rho u)_n^* = A_c(\rho u)_0^* \frac{A^*}{A(M_0)}. \qquad (4.18)$$

Now if the shock moves downstream in the divergent passage, it occurs at a higher M and so produces a larger stagnation pressure loss, reducing $(\rho u)_n^*/(\rho u)_0^*$ and requiring larger A_n. By adjusting A_n to the above value, the shock can be put near the throat, and the diffuser then has the pressure recovery given in figure 4.16. Increasing M_0 beyond the starting value M_{0c} while holding the shock at the throat results in somewhat poorer pressure recovery than that in figure 4.16 because the contraction ratio A_0/A_t is smaller than it should be for best π_d. For this off-design condition, π_d is the stagnation pressure ratio for a normal shock at the throat Mach number M, found from

$$\frac{A(M_t)}{A^*} = \frac{A(M_0)}{A^*} \frac{A_t}{A_c}. \qquad (4.19)$$

As an example, consider an inlet with $M_{0c} = 3$. From figure 4.16, A_c/A_t should be 1.38. Using this relation and figures 4.1 and 4.2 we find π_d to be as shown in figure 4.21. As M_0 increases from 1, π_d corresponds to a normal shock until the inlet "starts" at M_{0c}, when it is given by a normal shock at M_t. If M_0 is reduced while the shock is kept at the throat by adjusting A_n, the shock weakens until $M_t = 1$. Below this value of M_0 (1.72 for this case) the throat will not pass the full capture mass flow, so a spill shock forms ahead of the inlet, and the pressure recovery drops to that for a normal shock at M_0.

With the shock at the throat, the inlet is unstable in the sense that any perturbation moving the shock slightly upstream will cause it to "pop" forward, leading to unstart. It is therefore necessary to operate with the shock a bit downstream of the throat to ensure stability, and this lowers π_d somewhat.

Unstart can be an unsettling experience for the pilot or the passengers of a supersonic aircraft. It is attended by a significant loss of thrust, which

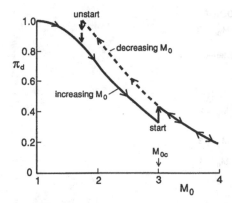

Figure 4.21
Off-design behavior (with Mach number) of the simple internal-compression diffuser of figure 4.15, showing hysteresis connected with starting.

results from the combination of reduced stagnation pressure recovery and reduced mass flow. Occurring almost instantaneously to a human observer, the thrust loss is apparent as a rapid deceleration of the aircraft, and may also lead to a strong yawing moment in multi-engine aircraft.

Mass flow variations at fixed M_0, as might be caused (schematically) by variation of A_n, also cause changes in the shock system and variations in π_d. Consider first operation at $M_0 = M_{0c}$, so that the inlet can swallow the shock if A_n is increased to a large enough value. If A_n is reduced below the value that places the shock at the throat, the shock will pop in front of the inlet, and π_d will be reduced to that for a normal shock at M_0. It will retain this value as A_n is further reduced, with a corresponding reduction in mass flow. The excess mass flow is spilled over the lip behind the normal shock. This mode of operation is termed *subcritical*.

If A_n is increased from the critical value, the mass flow cannot increase, because the flow is supersonic up to the throat. To accommodate the increased A_n, the shock will move downstream into the divergent portion of the diffuser, where the flow is now supersonic, until it stands at a high enough M to produce, through shock losses and viscous dissipation, the reduction in $(\rho u)^*$ required to match the larger A_n. The pressure recovery will then be given by

$$\pi_d = \frac{(\pi_d)_c A_{nc}}{A_n},$$ (4.20)

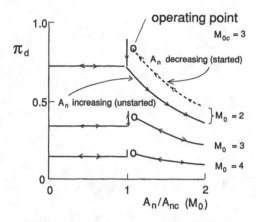

Figure 4.22
Off-design behavior with changing mass flow for the simple internal-compression inlet of figure 4.15, showing double-valued (started and unstarted) operation for $M_0 < M_{0c}$.

where the subscript c denotes the condition with the shock at the throat. This mode of operation is termed *supercritical*. Figure 4.22 shows this behavior at $M_0 = 3$ for an inlet with $M_{0c} = 3$. At higher M_0 the behavior is similar because the inlet can start as A_n is increased through the value $A_{nc}(M_0)$ that puts the shock at the throat.

For $M_0 < M_{0c}$, on the other hand, there is hysteresis, because the inlet will not start if A_n is increased continuously from a small value. The bow shock will remain in front of the inlet, and when A_n reaches A_{nc} a weak shock will form at the (now sonic) throat and will increase in strength as it moves downstream with further increasing A_n. If the inlet has been started as in figure 4.21, by going to higher M_0 for example, it operates on the upper curve; if An is reduced below A_{nc}, it unstarts and reverts to the lower curve. The most desirable operating point for a diffuser of this type is just above critical, as marked by the circles on figure 4.22.

External-compression inlets do not suffer the complications of starting and unstarting; thus their behavior is somewhat simpler. Consider, for example, the behavior of the diffuser of figure 4.17 as M_0 is reduced from the design value of 4. Reference to figure 4.4 shows that with fixed δ the shock angle θ increases as M_0 decreases, so the shock stands off the lip. Because the flow is deflected by the wedge behind this shock, the captured streamtube becomes smaller than that which intersects the lip, as indicated at the left in figure 4.23. The excess mass flow is spilled by supersonic

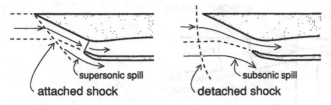

Figure 4.23
External-compression diffuser at flight Mach number below design value, with attached shock and supersonic spill (left) and with detached shock and subsonic spill (right).

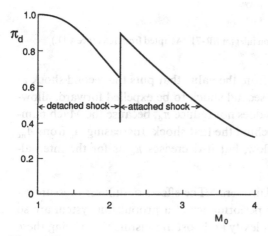

Figure 4.24
Off-design performance of external-compression inlet: variation of π_d with M_0.

turning over the lip. With the proper backpressure, the second shock can be positioned at the lip, which we take here to be coincident with the throat.

To estimate the pressure recovery, the shock angle is found from figure 4.4 and M_t from figure 4.5. The stagnation pressure ratio for each shock is taken from figure 4.2, and π_d is the product of the two stagnation pressure ratios. The result is shown in figure 4.24.

As M_0 decreases from the design value, the pressure recovery increases continuously to the value of M_0 at which the oblique shock detaches from the wedge. Below this value π_d is essentially that for a normal shock inlet. Note that although the detachment causes a discontinuity in π_d, there is no hysteresis as in the internal-compression inlet.

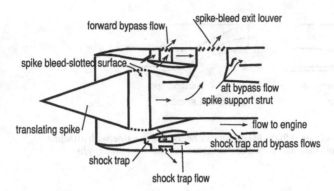

Figure 4.25
Variable-geometry mixed-compression inlet for SR-71. (Adapted from reference 4.11.)

A reduction in nozzle area from the value that puts the second shock at the lip, called A_{nc}, causes the second shock to be expelled forward, allowing subsonic spill. Ideally this does not reduce π_d, because the Mach number is uniform in the region behind the first shock. Increasing A_n from A_{nc} does not increase the mass flow, but it decreases π_d as for the internal-compression inlets.

4.2.2.7 Variable-Geometry Diffusers The effects of diffuser pressure recovery and mass flow on the performance of a propulsion system are so strong that considerable complexity and cost are justifiable to bring these parameters close to the optimum in modern high-performance aircraft. The requirements vary in detail with the mission of the aircraft. Some examples will be presented here.

The type of inlet used on the SR-71 Mach 3 + reconnaissance aircraft is shown schematically in figure 4.25. A similar inlet was under development for the Boeing 2707-300 SST. The SR-71 inlet has a conical spike which translates fore and aft to vary the effective throat area and also to help position the shock on the lip as the Mach number changes. The internal lip shock falls on a slotted bleed surface, which prevents boundary-layer separation due to the shock impingement. Further aft is a "shock trap" which serves to locate the final normal shock. It stabilizes the shock by providing a position-sensitive bleed. If the shock moves ahead of the lip of the trap, the high-pressure air behind the shock passes out through the trap, tending to move the shock downstream. Thus, the shock tends to sit just at the lip of the trap. There are two controlled bypasses: one forward, which dumps

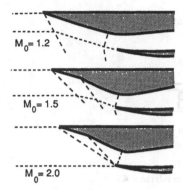

Figure 4.26
Schematic of variable-geometry external-compression inlet with sliding wedge, and with variable angle on second half of wedge.

overboard, and one just ahead of the compressor, which bypasses flow to the secondary passages of the nozzle. The flow bled from the spike is dumped overboard through louvers. Each of these devices has a key role in maintaining stable flow or in matching the inlet to the engine at one or more flight conditions. At takeoff and low speeds, the spike is forward, to maximize the flow area. Some additional flow is drawn in through the spike bleed and the forward bypass as well, to meet the engine's airflow demand at takeoff power. When sonic speed is passed and a shock system forms on the spike, the spike is retracted to minimize the supersonic spill behind this shock system. At some flight Mach number the forward bypass is opened to pass the normal shock through the throat and "start" the inlet. The shock is then positioned by the shock trap. When the vehicle reaches cruise speed, the spike shock is positioned at the lip, and the internal or lip shock impinges on the slotted surface of the spike, as noted above. Depending on the engine's airflow needs, some air will pass around the engine to the nozzle, through the aft bypass. The control of these complex devices is integrated with the engine control.

A second type of variable-geometry inlet, using external compression, is shown configured for three values of M_0 in figure 4.26. The lower sketch depicts the design condition at $M_0 \approx 2$ with three shocks converging on the lip. At lower M_0 the spike is slid forward and the second wedge is rotated to a lower angle; at the slightly supersonic condition $M_0 \approx 1.2$ it is straight, at the low angle of the front wedge. This prevents shock detach-

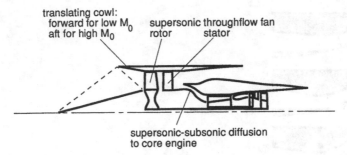

Figure 4.27
Schematic of a supersonic throughflow fan. (Adapted from reference 4.13.)

ment at this low M_0. The Concorde uses an inlet of this type, as does the
F-15, both with bleeds and shock positioning systems of the sort described
above.

4.2.2.8 Supersonic Throughflow Fan It will be clear from the above discussion that inlets for supersonic propulsion systems can be quite complex.
They are also heavy, and they introduce complicated control requirements.
The concept of the supersonic throughflow fan was introduced by Dr.
Antonio Ferri, and was apparently first documented in reference 4.12. It
was proposed for use in supersonic transports in reference 4.13; a later
study is reported in reference 4.14. In this concept, shown schematically in
figure 4.27, the axial velocity in the fan is supersonic at the design point, in
contrast to the subsonic axial velocity in all compressors and fans in present operation. When the flight Mach number is sufficiently above unity,
the flow remains supersonic from the free stream through the rotor and
stator of the fan and out the nozzle, so that the losses and the complexity
associated with diffusion to subsonic speed ahead of the fan are reduced.

The flow in the fan itself will be discussed in chapter 5. Here it will simply
be noted that, at least for the concept discussed in references 4.13 and 4.14,
the flow through the fan would be sonic at takeoff and at low flight Mach
numbers, increasing to a value just a bit below the flight Mach number
at supersonic cruise, as determined by an internal-compression diffuser
with a conical spike as shown in figure 4.27. At subsonic flight conditions
the flow would accelerate to sonic velocity in this convergent passage; in
supersonic flight it would be diffused from the free-stream Mach number
to a value at the fan face on the order of 2 (for a flight Mach number of
2.7). It would still be necessary to diffuse the flow into the core engine to

subsonic speed; this would occur downstream of the fan. Though still in the research phase, the supersonic throughflow fan offers substantial performance benefits to supersonic cruise aircraft, such as advanced SSTs.

4.3 Exhaust Nozzles

Like the inlet, the exhaust nozzle increases in complexity as the maximum flight Mach number of the aircraft engine increases, with resultant increases in nozzle pressure ratio and nozzle exit Mach number. The nozzle pressure ratio for a simple turbojet varies from a value between 2 and 3 at takeoff to as large as 40 at $M_0 = 3$. To achieve perfect expansion at all flight conditions, the nozzle would have to be convergent-divergent and capable of a wide range of variations of the ratio of exit area to throat area. In addition, the throat area required for best engine performance changes somewhat with flight Mach number. It increases by a factor of 1.4 or so when the engine changes from dry to afterburning operation.

These trends are easily seen from the ideal cycle analysis of the simple turbojet. The ratio of exit to throat area of the nozzle required for perfect expansion is determined by the nozzle exit Mach number M_7, given by (see section 2.3)

$$M_7{}^2 = \frac{2}{\gamma - 1}\left[\theta_0\tau_c\left(1 - \frac{\theta_0}{\theta_t}(\tau_c - 1)\right) - \right].$$ (4.21)

The area ratio is then $A(M_7)/A^*$. If the nozzle throat is choked, as is almost always the case for turbojets, the ratio of nozzle throat area to compressor inlet area is

$$\frac{A_n}{A_2} = \frac{(\rho u)_2^*}{(\rho u)_6^*}\frac{A^*}{A(M_2)}.$$

Noting that $(\rho u)^* \propto p_t/\sqrt{T_t}$, and taking the pressure and temperature ratios from section 2.3, we get

$$\frac{A_n}{A_2} = \sqrt{\frac{\theta_t}{\theta_0\tau_c}}\left[\tau_c\left(1 - \frac{\theta_0}{\theta_t}(\tau_c - 1)\right)\right]^{-(\gamma+1)/2(\gamma-1)}\frac{A^*}{A(M_2)}\quad\text{(nonafterburning)}$$

and   (4.22)

$$\frac{A_n}{A_2} = \sqrt{\frac{\theta_a}{\theta_0}}\left[\tau_c\left(1 - \frac{\theta_0}{\theta_t}(\tau_c - 1)\right)\right]^{-\gamma/(\gamma-1)}\frac{A^*}{A(M_2)}\quad\text{(afterburning)}.$$

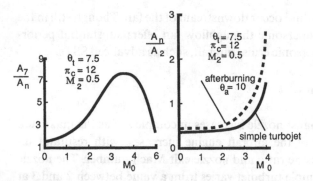

Figure 4.28
Trends in nozzle throat area and area ratio with flight Mach number for turbojets.

These results are plotted in figure 4.28 for a turbojet with $\pi_c = 12$ and $\theta_t = 7.5$. From the right graph, the ratio of nozzle throat area A_n to compressor inlet area A_2 is nearly constant up to $M_0 = 2$; it then rises sharply. Assuming that M_2 is constant with varying M_0 is equivalent to assuming that the corrected weight flow $W_2\sqrt{\theta_2}/\delta_2$ of the engine is constant, which in turn implies that $N/\sqrt{\theta_2}$ is constant (see chapter 5). Ordinarily engines are operated at more nearly fixed N, so as θ_2 increases with increasing M_0, $N/\sqrt{\theta_2}$ decreases and so does M_2, tending to decrease the required A_n/A_2 somewhat.

The ideal nozzle area ratio, A_7/A_n, changes markedly with M_0, becoming as large as 7 at $M_0 = 3$. As is indicated in figure 3.3, the loss associated with the use of a simple convergent nozzle ($A_7/A_n = 1$) is fairly small for M_0 up to 1, but is 14 percent at $M_0 = 1.5$ and more than 50 percent at $M_0 = 3$; thus it is essential to use a convergent-divergent nozzle on supersonic aircraft.

4.3.1 Fixed Nozzles

A simple convergent-divergent internal-expansion nozzle is shown schematically in figure 4.29. Having an area ratio of 4, this nozzle might operate ideally on a turbojet at $M_0 \approx 2$, as in diagram c. At higher M_0, it would have too low an area ratio, so that the flow would expand outside the nozzle through an expansion fan centered on the nozzle lip, as shown in diagram d. For values of M_0 less than 2 the nozzle would be overexpanded; that is, the pressure at the nozzle exit plane would be less than the ambient

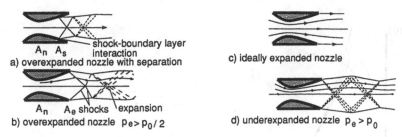

Figure 4.29
Design point (case c) and off-design flow in a convergent-divergent nozzle.

pressure if the nozzle were to flow full. If the pressure at the exit plane were greater than about half the ambient pressure, compression through an oblique shock would adjust the pressure, as indicated in diagram b. At yet lower M_0 where the pressure ratio required of the exit shocks in configuration b exceeds that which would cause the boundary layers to separate from the nozzle walls as a result of the sharp pressure rise across the shock, the flow would no longer fill the nozzle (diagram a). The pressure at the wall in the portion of the nozzle downstream of the separation point is only slightly below ambient, so the thrust of the nozzle in this flow configuration is nearly that of a nozzle of area ratio A_s/A_n, where A_s is the flow area at the separation point.

The thrust of the nozzle can be estimated very well for any of these operating conditions from equation 1.12, provided A_e is interpreted as A_s and p_e as the pressure at A_s in case a and provided the external flow is known, so that the effective ambient pressure can be determined. However, the external flow interacts with the internal flow. If this interaction influences either the internal flow or the external flow upstream of the nozzle exit plane as in case a, then neither the drag nor the thrust can be determined without understanding the interaction.

For the overexpanded nozzle (case a), the pressure at separation A_s is controlled by the static pressure rise in the shock-boundary layer interaction. According to E. E. Zukoski (reference 4.8), this separation occurs at a point in the nozzle such that the pressure at the separation point, p_s, is given by

$$p_s = \frac{p_0}{1 + M_s/2}, \tag{4.23}$$

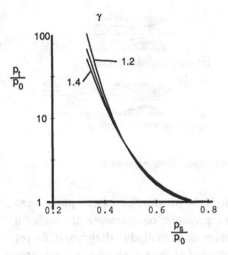

Figure 4.30
Separation pressure ratio as a function of stagnation pressure ratio for overexpanded nozzles.

where p_0 is the ambient or discharge static pressure and M_s is the flow
Mach number ahead of the separation. Since p_s and M_s are related by

$$p_t = p_s\left(1 + \frac{\gamma - 1}{2}M_s^2\right)^{\gamma/(\gamma-1)},$$

where p_t is the stagnation pressure, we can obtain a relationship between
p_t/p_0 and p_s/p_0:

$$\frac{p_t}{p_0} = \frac{p_s}{p_0}\left[1 + 2(\gamma - 1)\left(\frac{p_0}{p_s} - 1\right)^2\right]^{\gamma/(\gamma-1)}. \tag{4.24}$$

This expression is plotted in figure 4.30, from which we see that, for pressure ratios in the range from 2 to 50, p_0/p_s ranges from about 1.6 to 2.5.

Just as the off-design problems of supersonic inlets can be eased somewhat by utilizing external compression, the nozzle problem can be reduced
by incorporating external expansion in the design. The plug nozzle shown
in figure 4.31 is the idealized example of this. It is the exact analogue of the
isentropic spike inlet of figure 4.18. The improvement in off-design performance results from the flow remaining attached to the spike at pressure
ratios below design, while the streamtube leaving the nozzle contracts to
satisfy the requirements for lower expansion ratio. This mode of operation
is shown in figure 4.31b.

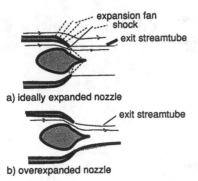

a) ideally expanded nozzle

b) overexpanded nozzle

Figure 4.31
Design point (case a) and overexpanded (case b) operation of an external-expansion "plug" nozzle.

Because the flow is expanding rather than compressing, less care need be given to the form of the centerbody than in the case of the inlet; however, for the same pressure ratio the turning must be the same in the two cases, and just as the inlet had a serious lip drag problem, the nozzle has a serious base drag problem due to the expansion around the nozzle shroud.

Cooling the plug is a serious problem when the plug nozzle is used on an afterburning engine. It has not been much used in the simple form of figure 4.31, but external expansion has been incorporated in the integration of the nozzle with the afterbody of an aircraft. An example of this is the McDonnell-Douglas F-4, in which the nozzle exhaust expands against the aft underbody of the fuselage. This type of nozzle is envisioned for use in hypersonic aircraft as well.

4.3.2   Variable-Geometry and Ejector Nozzles

Most supersonic aircraft use afterburning for supersonic flight. Because they are also required to operate without afterburning for subsonic cruising and landing, at least a two-position variation of the nozzle throat area is required. In early engines the variation was achieved by closing a pair of eyelid-shaped segments over the end of the larger nozzle. The use of series of overlapping leaves has become standard practice.

Ejector nozzles used on many high-performance aircraft are designed so that a secondary airflow provides an aerodynamically varied expansion ratio. Two implementations of this idea are shown schematically in figure 4.32. In either case the secondary air, which may have been bled from the

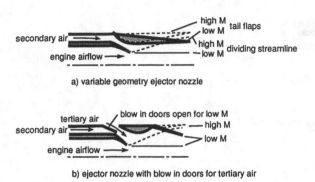

a) variable geometry ejector nozzle

b) ejector nozzle with blow in doors for tertiary air

Figure 4.32
Schematics of two types of "ejector" nozzles in which secondary airflow is used to vary the expansion ratio of the nozzle.

inlet to prevent boundary-layer separation or allowed to bypass the engine to prevent the inlet shock from "popping," flows over the outside of the afterburner, cooling it, and then passes through the outer part of the nozzle. Changing the amount of secondary air varies the nozzle area available to the engine airflow, thus varying the effective expansion ratio of the nozzle. To prevent separation in the divergent portion of the nozzle, it may be contracted for operation at low M_0, as indicated in figure 4.32.

At takeoff and when M_0 is low, the nozzle may require more air than is available from the inlet. Blow-in doors just ahead of the nozzle throat may be used to provide a tertiary airflow, which helps to fill the nozzle as shown in figure 4.32b. These doors operate in response to the pressure difference across them, so that they essentially prevent the pressure in the nozzle from falling below the ambient static pressure. This reduces the drag caused by overexpansion. At higher Mach numbers, when rising ram pressure raises the pressure of the secondary air above the ambient pressure, the blow-in doors close, and the nozzle becomes a two-stream ejector like that in figure 4.32a.

To a first approximation, the ejector nozzle can be understood by assuming that the secondary and primary airflows are isentropic and that their static pressures are equal at each axial station of the nozzle, so that they flow together through a common nozzle structure. Thus, if we denote by subscript 1 the primary flow and by subscript 2 the secondary flow, we have

$$p_1 = \frac{p_{t1}}{\{1 + [(\gamma - 1)/2]M_1{}^2\}^{\gamma/(\gamma-1)}}$$

$$= p_2$$

$$= \frac{p_{t2}}{\{1 + [(\gamma - 1)/2]M_2{}^2\}^{\gamma/(\gamma-1)}},$$

or

$$\frac{1 + [(\gamma - 1)/2]M_2{}^2}{1 + [(\gamma - 1)/2]M_1{}^2} = \left(\frac{p_{t2}}{p_{t1}}\right)^{(\gamma-1)/\gamma}. \tag{4.25}$$

We could choose either M_1 or M_2 as a parametric coordinate, but since the objective is to smoothly expand the primary flow from subsonic to supersonic conditions, we choose M_1 as the independent variable, which increases with distance along the nozzle.

The area of the primary-flow streamtube is given as a function of M_1 by equation 4.9:

$$\frac{A_1}{A_1^*} = \frac{1}{M_1}\left(\frac{1 + [(\gamma - 1)/2]M_1{}^2}{(\gamma + 1)/2}\right)^{(\gamma+1)/2(\gamma-1)}. \tag{4.26}$$

A_1^* may be interpreted as the primary-flow throat area. The flow area of the secondary streamtube is determined from its mass flow and stagnation conditions relative to the primary streamtube,

$$\frac{A_2}{A_1^*} = \frac{A_1}{A_1^*}\frac{\dot{m}_2}{\dot{m}_1}\frac{p_{t1}}{p_{t2}}\sqrt{\frac{T_{t2}}{T_{t1}}}\frac{M_1}{M_2}\left[\frac{1 + \dfrac{(\gamma - 1)}{2}M_2{}^2}{1 + \dfrac{(\gamma - 1)}{2}M_1{}^2}\right]^{\frac{\gamma+1}{2(\gamma-1)}}.$$

Using equations 4.25 and 4.26 gives

$$\frac{A_2}{A_1^*} = \frac{\dfrac{\dot{m}_2}{\dot{m}_1}\sqrt{\dfrac{T_{t2}}{T_{t1}}}\left(\dfrac{p_{t1}}{p_{t2}}\right)^{\frac{\gamma-1}{2\gamma}}\left[\dfrac{1 + \dfrac{(\gamma - 1)}{2}M_2{}^1}{\dfrac{(\gamma + 1)}{2}}\right]^{\frac{\gamma+1}{2(\gamma-1)}}}{\sqrt{\dfrac{2}{\gamma - 1}\left[\left(1 + \dfrac{\gamma - 1}{2}M_1{}^2\right)\left(\dfrac{p_{t2}}{p_{t1}}\right)^{\frac{\gamma-1}{\gamma}} - 1\right]}}. \tag{4.27}$$

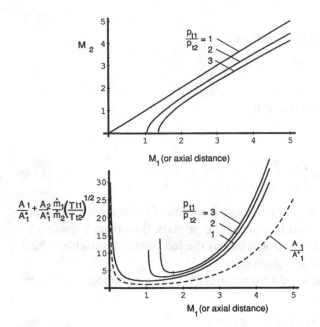

Figure 4.33
Ratios of flow areas for primary and secondary airflows to primary throat area for an ejector nozzle with an engine pressure ratio of 3.

The total flow area, $A/A_1^* = (A_1/A_1^*) + (A_2/A_1^*)$, is then the actual flow area of the nozzle. In figure 4.33 the ratio of flow area to throat flow area for the primary flow is shown as the dashed line. Above this, the total flow area is shown as a function of the primary flow Mach number for three values of the ratio of primary to secondary stream stagnation pressures, for the particular value of $(\dot{m}_1/\dot{m}_2)\sqrt{T_{t1}/T_{t2}} = 1$. Above the plot of area ratios, the secondary stream Mach number M_2 is shown for the same conditions.

Assuming for the sake of simplicity that $(\dot{m}_1/\dot{m}_2)\sqrt{T_{t1}/T_{t2}} = 1$, as in the figure, we may think of the area ratio plot as a plot of nozzle area versus axial distance. In the case where $p_{t1}/p_{t2} = 1$, each stream then occupies half the flow area, from stagnation conditions to any exit Mach number we may choose.

As the pressure ratio p_{t1}/p_{t2} is increased, we note first of all that the streams cannot coexist in the nozzle at equal static pressures unless the primary stream has been expanded to a pressure equal to or below the secondary stream stagnation pressure. For $p_{t1}/p_{t2} = 2$ this occurs at $M_1 = $

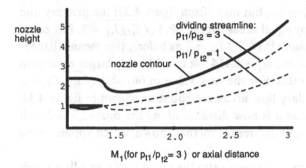

Figure 4.34
Division of flow area in a nozzle at the design point of $p_{t1}/p_{t2} = 3$, and at a lower pressure ratio of $p_{t1}/p_{t2} = 1$, for fixed nozzle geometry.

1.047, where the upper curve has a vertical asymptote. For $p_{t1}/p_{t2} = 3$ it is at $M_1 = 1.359$. To the right of these values, the streams can flow together through a common nozzle, sharing the available area in the proportion shown in the figure.

If the product $(\dot{m}_2/\dot{m}_1)\sqrt{T_{t2}/T_{t1}}$ is other than unity, the area occupied by the secondary flow scales accordingly. Thus, if $(\dot{m}_2/\dot{m}_1)\sqrt{T_{t2}/T_{t1}}$ is 2, the area of the secondary flow will be twice that shown on the figure—that is, the distance from the dashed curve to the solid one doubles.

From these results we can understand the behavior of a nozzle of fixed geometry as the pressure ratio p_{t1}/p_{t2} varies. Suppose that, as shown in figure 4.34, the nozzle area of the core flow is equal to that of the secondary flow, where the streams come together. Let $T_{t1}/T_{t2} = 2$, and suppose that $p_{t1}/p_{t2} = 3$. Our argument applies only if $M_1 > 1.359$, so let's say we have designed the primary nozzle so $M_1 = 1.4$ at its exit. From figure 4.33 we find that for $(\dot{m}_2/\dot{m}_1)\sqrt{T_{t2}/T_{t1}} = 1$ the ratio of the secondary stream area to that of the primary stream would have to be 5.62 at this point. The *actual* mass flow ratio is therefore

$$\frac{\dot{m}_2}{\dot{m}_1} = \frac{1}{(5.62)\sqrt{T_{t2}/T_{t1}}} = 0.252.$$

Using the new value of $(\dot{m}_2/\dot{m}_1)\sqrt{T_{t2}/T_{t1}} = 0.1780$, we can then compute the flow areas of the primary and secondary flows; the division between them is shown in figure 4.34 as the curve labeled $p_{t1}/p_{t2} = 3$.

Now suppose this *same nozzle* is operated with a lower pressure ratio, $p_{t1}/p_{t2} = 1.0$. With the same primary nozzle geometry, the core flow Mach number will still be $M_1 = 1.4$, but now (from figure 4.33) the primary and core flows would occupy equal areas for $(\dot{m}_2/\dot{m}_1)\sqrt{T_{t2}/T_{t1}} = 1$. For equal actual areas, and supposing that $T_{t2}/T_{t1} = 2$ as before, this means that for this lower pressure ratio, $\dot{m}_2/\dot{m}_1$ is 1.414—a considerably larger ratio than at the higher pressure ratio. For the new value of $(\dot{m}_2/\dot{m}_1)\sqrt{T_{t2}/T_{t1}} = 1$, the primary and secondary flow areas are equal, as shown in figure 4.34. For $p_{t1}/p_{t2} = 1$, the abscissa is now distance along the nozzle. The Mach number would, however, be different from that shown, which applies to the high-pressure case.

We see that for the lower pressure ratio the secondary flow fills a much larger portion of the nozzle than at the higher pressure ratio, giving the primary flow a lower area ratio appropriate to the lower core stagnation pressure.

4.3.3 Mixer Nozzles

In the discussion of the performance of turbofan engines, it was pointed out that for maximizing the specific impulse the optimum choice of the combination of fan pressure ratio and bypass ratio is the choice that yields equal jet velocities in the core and fan discharge streams. This is strictly true only for the ideal cycle, where power can be transferred without loss between the two streams. In this limit, the argument is quite simple. Let P be the total power shared by the core and fan streams; i.e., in the notation of figure 1.6,

$$P = P_c + P_f = \dot{m}(1+f)\left(c_p(T_6 - T_0) + \frac{u_6^2}{2} - \frac{u_0^2}{2}\right)$$

$$+ \alpha\dot{m}\left(c_p(T_8 - T_0) + \frac{u_8^2}{2} - \frac{u_0^2}{2}\right).$$

Stations 6 and 8 are in the jets, where the pressure is at the ambient level, so the values of T_6 and T_8 are independent of whether the gas streams arrive at their expanded condition through a jet expansion or through expansion in a turbine. In the ideal cycle, the expansion is isentropic in either case, so the fixed end pressure fixes also the end temperature. Of course T_6 and T_8 are different, but are set by the fan exit and core turbine exit conditions. (In fact, for the ideal cycle $T_8 = T_0$.) Thus, varying the bypass ratio shifts kinetic power from the core jet to the fan jet, while the total kinetic power,

$$P_{\text{kin}} = \dot{m}(1+f)\left(\frac{u_6^2}{2} - \frac{u_0^2}{2}\right) + \dot{m}\alpha\left(\frac{u_8^2}{2} - \frac{u_0^2}{2}\right),$$

is constant. The thrust is

$$F = \dot{m}(1+f)(u_6 - u_0) + \dot{m}\alpha(u_8 - u_0).$$

We want to find, within the constraints of the first equation, the choice of u_6 and u_8 that maximizes F. The bypass ratio α is to be considered fixed in this variation, since we know that for the ideal cycle the thrust increases continuously with α, as the propulsion efficiency increases. To find the optimum, we thus put

$$dP_{\text{kin}} = 0 = [\dot{m}(1+f)u_6]\,du_6 + [\dot{m}\alpha u_8]\,du_8$$

and

$$dF = 0 = [\dot{m}(1+f)]\,du_6 + [\dot{m}\alpha]\,du_8.$$

These relations must be satisfied for arbitrary variations of du_8 and du_6, so the determinant of their coefficients must be zero. This yields $u_6 = u_8$ as the condition that maximizes F for a given α.

When the fan and the core discharge through separate nozzles, as shown in figure 1.6, this condition can be realized, for a given α, by choice of the fan pressure ratio, or for a given fan pressure ratio by choice of α. This is the condition represented by equation 2.23, and so reflected in the results of figure 2.8. As was noted there, it results in rather high bypass ratios for the fan pressure ratios (about 1.60) that can be realized with single-stage fans. Thus, actual engines, if implemented with separate core and fan nozzles, tend to have higher core jet velocity than fan exit velocity, i.e., $u_6 > u_8$.

The question then arises whether it is advantageous to merge the two streams by mixing them and discharging them through a single nozzle. Such an arrangement is termed a *mixer nozzle*. It is important to note that such mixing is *not* equivalent to the transfer of energy from the core stream to the fan stream implied in the above optimization, because the mixing is an irreversible process, resulting in an entropy increase. Nevertheless, it may be that the improvement due to merging the streams may offset any loss due to mixing.

There is another reason for considering such mixing in the afterburning turbofan shown schematically in figure 2.9. Here the fan and core streams merge in a common afterburner and flow through a single nozzle. In nonafterburning operation, if the two streams do not mix before flowing through the nozzle, they emerge with different velocities, so here again there is the question whether it is advantageous to mix them before they expand through the nozzle. As we shall see, it is indeed advantageous.

To show this, we assume that the two streams mix completely, at constant total flow area, as shown in figure 2.9. For consistency with the above argument and with figure 2.9, the core stream will be denoted by subscript 5, the fan by 7, and the mixed stream by 9.

Normally, the known quantities will be p_{t5}, T_{t5}, α, p_{t7}, T_{t7}, and either M_5 or M_7. At the inlet to the mixing section the static pressures of the two streams will be equal, so specifying either M_5 or M_7 determines the other. We will take M_7 to be known; then

$$p_5 = p_7 = \frac{p_{t7}}{\left(1 + \dfrac{\gamma - 1}{2} M_7{}^2\right)^{\gamma/(\gamma-1)}} = \frac{p_{t5}}{\left(1 + \dfrac{\gamma - 1}{2} M_5{}^2\right)^{\gamma/(\gamma-1)}} \tag{4.28}$$

determines

$$M_5 = \sqrt{\frac{2}{\gamma - 1}\left[\left(1 + \frac{\gamma - 1}{2} M_7{}^2\right)\left(\frac{p_{t5}}{p_{t7}}\right)^{(\gamma-1)/\gamma} - 1\right]}. \tag{4.29}$$

The area ratio A_7/A_5 is given by (see equation 4.7)

$$\frac{A_7}{A_5} = \alpha \sqrt{\frac{T_{t7}}{T_{t5}}} \frac{M_5 \left(\dfrac{p_{t7}}{p_{t5}}\right)^{-(\gamma-1)/2\gamma}}{\sqrt{\dfrac{2}{\gamma - 1}\left[\left(1 + \dfrac{\gamma - 1}{2} M_5{}^2\right)\left(\dfrac{p_{t7}}{p_{t5}}\right)^{(\gamma-1)/\gamma} - 1\right]}}. \tag{4.30}$$

From conservation of mass (see equation 4.9),

$$(\rho u)_5^* M_5 \left[\frac{\dfrac{(\gamma + 1)}{2}}{1 + \dfrac{\gamma - 1}{2} M_5{}^2}\right]^{\gamma+1/2(\gamma-1)} (1 + \alpha)$$

$$= (\rho u)_9^* \left(1 + \frac{A_7}{A_5}\right) M_9 \left[\frac{\dfrac{\gamma + 1}{2}}{1 + \dfrac{\gamma - 1}{2} M_9{}^2}\right]^{\gamma+1/2(\gamma-1)}, \tag{4.31}$$

where, recalling equation 4.10,

$$(\rho u)^* = \sqrt{\gamma} \left(\frac{2}{\gamma + 1}\right)^{\gamma + 1/2(\gamma - 1)} \frac{p_t}{\sqrt{RT_t}}.$$

Conservation of energy gives

$$T_{t9} = \frac{1}{1 + \alpha} T_{t5} + \frac{\alpha}{1 + \alpha} T_{t7}. \tag{4.32}$$

Finally, conservation of momentum may be expressed as

$$p_5(1 + \gamma M_5{}^2) + p_5(1 + \gamma M_7{}^2)\frac{A_7}{A_5} = p_9(1 + \gamma M_9{}^2)\left(1 + \frac{A_7}{A_5}\right)$$

or

$$\frac{p_{t9}}{p_{t5}} = \frac{(1 + \gamma M_5{}^2) + (1 + \gamma M_7{}^2)(A_7/A_5)}{(1 + \gamma M_9{}^2)(1 + A_7/A_5)} \left[\frac{1 + \dfrac{\gamma - 1}{2} M_9{}^2}{1 + \dfrac{\gamma - 1}{2} M_5{}^2}\right]^{\gamma/(\gamma - 1)}. \tag{4.33}$$

Equation 4.31 provides a second explicit expression for p_{t9}/p_{t5} in the form

$$\frac{p_{t9}}{p_{t5}} = \frac{\sqrt{(1 + \alpha)\left(1 + \alpha \dfrac{T_{t7}}{T_{t5}}\right)}}{1 + A_7/A_5} \frac{M_5}{M_9} \left[\frac{1 + \dfrac{\gamma - 1}{2} M_9{}^2}{1 + \dfrac{\gamma - 1}{2} M_5{}^2}\right]^{\gamma/2(\gamma - 1)}. \tag{4.34}$$

By equating these we obtain a single expression for M_9 in terms of the prescribed parameters, which are α, T_{t7}/T_{t5}, p_{t7}/p_{t5}, and M_7.

It is readily seen from equations 4.29, 4.30, 4.33, and 4.34 that if $p_{t7}/p_{t5} = T_{t7}/T_{t5} = 1$, then $M_7 = M_5$, $A_7/A_5 = \alpha$, and $p_{t9}/p_{t5} = 1$, as should be the case. The solution in general is best obtained iteratively from equations 4.33 and 4.34.

The critical question, of course, concerns the effect of the mixer on thrust, and therefore on specific impulse. Following the procedure of section 3.8, we find

$$\frac{F_{\text{mixed}}}{\dot{m}(1 + \alpha)a_0} = M_0 \left\{\frac{1 + f + \alpha}{1 + \alpha}\left[\frac{u_{10}}{u_0} + \frac{1}{\gamma M_0{}^2}\frac{u_0}{u_{10}}\frac{T_{10}}{T_0}\left(1 - \frac{p_0}{p_{10}}\right)\right] - 1\right\}, \tag{4.35}$$

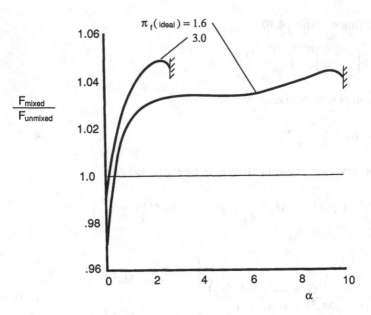

Figure 4.35
The ratio of thrust with mixing to thrust without mixing for a range of bypass ratios and for two fan pressure ratios.

where for a choked nozzle

$$\frac{T_{10}}{T_0} = \frac{2}{\gamma + 1} \theta_t \tau_t \left(\frac{T_{t9}}{T_{t5}}\right),$$

$$\frac{u_{10}}{u_0} = \frac{1}{M_0} \sqrt{\frac{T_{10}}{T_0}},$$

and

$$\frac{p_{10}}{p_0} = \left(\frac{2}{\gamma + 1}\right)^{\gamma/(\gamma-1)} \delta_0 \pi_d \pi_c \pi_b \pi_t \left(\frac{p_{t9}}{p_{t5}}\right).$$

The ratio of thrust with mixing to thrust without mixing is shown in figure 4.35 for a range of bypass ratios, and for two fan pressure ratios, for the conditions specified in figure 3.6 (except that all specific heat ratios have been taken constant at 1.4, to be consistent with the single value used in the mixing calculation), which are representative of modern turbofan engines. For the lower fan pressure ratio of 1.6, the calculation shows that the engine with mixing should produce 3–4 percent higher thrust than that without mixing in the range of bypass ratios from 2 to 10. The gain is smaller at lower bypass ratios, because mixing of the very energetic core with the less energetic fan stream at equal static pressure causes rather severe shock losses, not offset by the equalization of the exhaust velocities.

For the higher fan pressure ratio of 3 there is also a gain of more than 4 percent over the range of bypass ratios from 1 to 2.7.

Note that for $\alpha = 0$ the calculation shows a loss for both fan pressure ratios. This is because at low bypass ratios the core flow Mach number at the entrance to the mixer is greater than unity, and the mixing calculation selects the subsonic solution. Thus, at this limit there is a normal shock in the mixer, even though the bypass flow is zero.

Each curve is terminated at the bypass ratio beyond which the core flow lacks enough stagnation pressure to match the static pressure of the bypass flow, which has been assumed to be at $M_7 = 0.5$. Increasing M_7 would allow matching at somewhat higher bypass ratios, but with a rapid decrease in the thrust increment.

The computed thrust increments, which seem modest, are quite significant for commercial high-bypass engines. A 4 percent reduction in fuel consumption can be decisive in the intense competition between engine manufacturers, as it results in a significant change in direct operating costs.

To realize the gains predicted for the mixer nozzle, it is necessary to implement the mixing process without incurring viscous losses that offset the gains due to mixing, and without adding excessive size and weight. For this purpose, "lobe" or "chute" mixers, shown schematically in figure 4.36, have been developed. A careful experimental study of the flow in such a mixer is reported in reference 4.15.

As figure 4.36 suggests, large-scale mixing is brought about by the inward diversion of the duct flow between the lobes of the mixer, and the concomitant outward flow of the core gas in the lobes. This generates a segmented flow, with alternate hot and cold spokes. In addition, the adjacent inward and outward radial flows produce vortex sheets at the merger of the hot and cold streams, resulting in strong streamwise vorticity, which results in increased mixing at a smaller scale. The overall result is nearly complete mixing in a duct length downstream of the mixer about equal to the diameter of the outer duct. Because of the short flow distance for mixing, the viscous losses should be small.

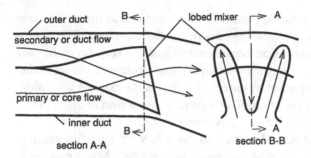

Figure 4.36
Schematic of "lobe" or "chute" mixers.

An interesting observation (from reference 4.15) concerning the flow in the mixers is that its structure is virtually independent of the stagnation temperature ratio between the inner and outer flows. At a certain level this may be understood from the fact that in an isentropic flow the flow Mach number is entirely determined by the ratio of static pressure to stagnation pressure, so as the hot and cold streams flow together at equal static pressures their Mach numbers should be independent of their respective temperatures. As heat is exchanged between the streams by mixing, in first approximation the result is to modify the density of each stream, and hence its share of the total flow area, but not the Mach number. This argument has been generalized to an approximate-substitution principle for viscous heat-conducting flows in reference 4.16.

An experimental study of ejectors using mixing lobes of this type was reported in reference 4.17. Here the emphasis was on pumping of the secondary flow by the core flow, and the resulting thrust augmentation. The authors found that the augmentation was more effective if the area of the mixing duct expanded downstream of the lobe mixer.

4.4 Combustors and Afterburners

The combustor makes possible the reaction of air with fuel at compressor outlet conditions. The fuel is usually a liquid hydrocarbon similar to kerosene. The overall mixture of fuel and air varies over a wide range with changing engine speed, and the compressor outlet air varies in pressure and temperature with altitude and flight Mach number as well as with engine speed. The high air mass flow densities in modern compressors dictate similarly high flow densities in the combustor, and these in turn imply volumetric rates of heat release much larger than those realized in most combustion systems.

When an engine is operating at its maximum power, the exit temperature from the combustor is well above the level at which the oxidation-resistant superalloys can operate for long periods. For this reason, the structure of the combustor must be cooled by a portion of the compressor discharge air, which will then mix with the combustion products to become a part of the turbine flow.

Stable combustion of hydrocarbons such as jet fuel occurs under a rather narrow range of fuel-air mixtures near the chemically correct (stoichiometric) value, so the combustor must provide a region in which such conditions exist, even though the overall fuel/air ratio may be much less than stoichiometric.

The completeness of combustion directly influences the heat release, and hence the fuel consumption of the engine, so the combustor efficiency must be very near unity for normal operating conditions. It is defined as the ratio of actual enthalpy rise of the flow from inlet to exit divided by that for complete reaction of the fuel and air to chemical equilibrium at the combustor exit conditions. Values of 0.98 or more are characteristic of modern engines.

In addition, the combustor must meet rather stringent requirements for pollutant production. These are posed in terms of amounts of carbon monoxide, unburned hydrocarbons, and nitrogen oxides, the latter being a mixture of N_2O, NO, and NO_2 represented as NO_x. Smoke, which in this case is composed of small carbon particles formed in the rich (high-fuel-content) regions of the combustor, must be below visible levels for both commercial and military aircraft.

The purpose of this discussion is to convey enough about the characteristics of the combustion process so that these constraints on the combustor, and the limitations they impose on the engine, may be understood in the context of the engine as a whole. For a comprehensive discussion of gas turbine combustion, see reference 4.18.

We begin with a discussion of the fundamental processes that take place during combustion. The actual reaction of fuel with air takes place in the gas phase. The fuel vapor must be mixed on a molecular scale with the air at a temperature that leads to a high chemical reaction rate. The reaction rate depends in a complex way on the temperature, but for many situations it can be approximated by an Arrhenius form such as

Reaction rate $\propto f(T)e^{-A/RT}$,

where A is called the activation energy and is typically of the order of 60 kcal/mole. Physically, this expression stems from two facts: first, two molecules must collide with some minimum energy in order to react; second, in a gas at temperature T the number of collisions per unit time and per unit volume in which the energy of one molecule relative to another exceeds a particular value A is proportional to $e^{-A/RT}$. This rate of collision is also proportional to the number of colliding molecules per unit volume, so the reaction rate depends on pressure as well. Depending on how many molecules are involved in a reactive collision, the rate may depend on pressure squared (two molecules), cubed (three molecules), and so on. Thus, we may write

Reaction rate $\propto p^n f(T)e^{-A/RT}$.   (4.36)

For hydrocarbon-air combustion, $n \approx 2$, and for low pressures the reaction rate becomes slow. This poses a problem for aircraft engines at very high altitudes, where the reaction rate can become limiting.

Under the usual conditions in engines, the rate of combustion is limited not by the rate of reaction but by the rate at which fuel vapor and air are mixed. Usually, the fuel is injected as an atomized spray into the hot reaction zone, where it mixes with air and with hot combustion gases. The fuel droplets vaporize; the vapor is then mixed with the air by a combination of turbulent mixing and molecular diffusion. If the temperature and the pressure in the reaction zone are sufficiently high, the reaction rate will not be limiting and the fuel vapor will react as soon as it comes in contact with sufficient oxygen.

Fundamental studies of this complex process necessarily idealize it. They can be divided into studies of combustion in premixed gases and studies of diffusion flames. Some ideas drawn from these studies will be summarized here.

4.4.1 Combustion in Premixed Gases

Suppose that a uniform mixture of fuel vapor and air has been formed at some initial temperature T_0 and pressure p_0, and that ignition is attempted (for example, by an intense electric spark). Then we may ask two questions:

(1) Does the mixture ignite and continue to burn? (2) At what rate does the flame propagate?

The answer to the first question is that at usual pressures and ambient temperatures, hydrocarbon-air mixtures will react only over a rather narrow range of fuel/air ratios—from about 0.9 to 1.2 of stoichiometric at atmospheric pressure, and not at all below about 0.2 atm at standard temperature. The equivalence ratio is defined as the fuel/air mass ratio divided by the fuel/air ratio required for complete combustion. For hydrogen, the reaction

$$2H_2 + O_2 \rightarrow 2H_2O$$

requires 4 kg of H_2 per 32 kg of O_2 or 4 kg of H_2 per $32/0.23 = 139$ kg of air, so the stoichiometric fuel air ratio for H_2 is $4/139 = 0.0288$. For octane (C_8H_{18}) it is 0.0667. In chapter 2 we found that the fuel/air ratio required to give the desired levels of turbine inlet temperature is below about 0.03, corresponding to $\phi \approx 0.5$, so that it is not possible simply to premix the fuel and air thoroughly in a combustor and then react the mixture. Rather, the fuel must be mixed with part of the air and then burned, and the combustion products then diluted with the remaining air.

Hydrogen has much wider flammability limits than hydrocarbons (roughly $0.25 < \phi < 6$ at 1 atm and standard temperature), and some special, rather expensive fuels have flammability limits intermediate between hydrogen and the heavy hydrocarbons. They have been used at times for testing and to extend the altitude limits of engines for special applications.

If the mixture does ignite, we may then ask at what rate the flame propagates into the gas mixture. This question has been the subject of a great deal of theoretical and experimental study, and the speed can be predicted with reasonable accuracy if the flow is laminar. The rates of propagation for hydrocarbon-air mixtures are of the order of 30 cm/sec near $\phi = 1$, dropping off very rapidly for $\phi < 0.5$ as shown in figure 4.37 for C_3H_8.

When the gas mixture is turbulent, as is always the case in aircraft engine combustors and afterburners, the rate can be much larger. The exact mechanisms by which the faster burning takes place are not firmly established, in spite of a great deal of intensive study. Some of the effects known to increase the burning rate are the following: If we imagine the burning to be taking place by local propagation of flame fronts into unburned mixture, the overall rate of combustion increases with the total area of these flame fronts. One effect of the turbulence at length scales large relative to the

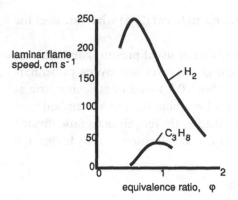

Figure 4.37
Laminar flame speed as a function of equivalence ratio for propane and hydrogen (from reference 4.19).

flame-front thickness is to distort or wrinkle the flame fronts, thus increasing their effective area and the overall burning rate. A related but distinct effect of the large-scale turbulence is the stretching of the flame fronts. As the turbulence increases the surface area of the flame front, the stretching tends to thin the flame front, thereby increasing the gradients of concentration and temperature normal to the flame front. These gradients govern the thermal and species transport, and hence the flame speed. At length scales comparable to the flame-front thickness, the turbulence may also increase the transport rates in the flame front, by augmenting the laminar diffusion and heat conduction. The result of all of these effects, and perhaps others, is an increase in the overall combustion rate as the turbulence level increases.

Both the laminar and the turbulent flame speeds are much higher for H_2 than for hydrocarbon fuels; the laminar value is shown in figure 4.37.

4.4.2   Diffusion Flames

The fuel entering the combustion region is usually only partially vaporized, and partially mixed with air, so some of the combustion takes place under conditions where the fuel (either liquid or gas) and air are separated by a boundary, at which the combustion takes place by diffusion of fuel in one direction and oxygen in the other in what is termed a *diffusion flame*.

Perhaps the most common example of a diffusion flame is a candle flame. Wax vaporizes from the wick and diffuses outward to meet air diffusing

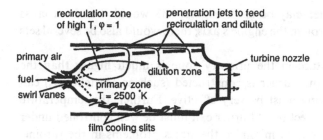

Figure 4.38
Typical gas turbine combustor, showing primary zone of near-stoichiometric combustion followed by dilution with excess air to reach T_{t4}.

inward; the result is a flame front stationary in space. Even this simple example is complicated greatly by natural convection currents set up by the heating of the gas. A fuel droplet shot into an oxidizing, hot gas behaves in a similar fashion to the candle flame. The vaporized fuel diffuses outward to meet the inward diffusing oxygen, but the process is complicated by motion of the droplet through the gas. Studies of spray combustion have shown that droplets decrease in diameter roughly according to $D^2 = D_0^2 - kt$, where D is the drop diameter, D_0 is the original diameter, and k is of the order of 0.01 cm^2/sec (see reference 4.20). If the droplets are to reside in the combustion chamber only 10^{-3} sec, say, their diameter should be less than $D_0 \approx \sqrt{kt} \approx 0.3 \times 10^{-2}$ cm to be completely vaporized.

In some modern combustors, the fuel is vaporized before injection to overcome the limitations of spray injection.

4.4.3 Primary Combustors

A typical combustor is shown schematically in figure 4.38. The upper half of the figure shows the way a recirculation zone is created by a swirl in the primary air and a high-velocity jet toward the burner axis partway downstream, and the way the combustion products are then diluted with air before entering the turbine nozzles. The lower half shows one method of cooling the burner "can": by a film of air introduced through slits. The slits, of course, are arranged to cool the entire circumference of the can, and the injection holes are arranged symmetrically on top and bottom, or around the circumference if the can is circular. Two configurations of burners have been used. In one, a series of cylindrical cans, of cross section as in figure 4.38, are arranged around the circumference of the engine. In the other, a single annular burner may be used. Figure 4.38 would then be a cross section in a plane through the engine's axis; there would also be several sets of nozzles.

The gas temperature in the primary zone is much higher than the materials of which the burner is constructed (such as Hastelloy X) can tolerate, so the cooling must be very effective. Further, the temperature profile in the radial direction at burner exit must be well controlled under all operating conditions to minimize the stress problems in the turbine. This temperature profile is sometimes deliberately made nonuniform, with lower temperatures at the blade roots (where the stress is largest) and at the tip, because the outer wall is hard to cool.

The deviation of the temperature profile at the burner exit from the desired one is characterized by a "pattern factor" which is defined as the difference between the maximum temperature (say in the circumferential direction) and the average, divided by the average. A value of about 0.20 is typical for the circumferential pattern factor.

Two other principal performance parameters for the burner are η_b, the efficiency of conversion of the fuel's chemical energy to heat, and π_b, the stagnation pressure drop. The pressure drop depends on viscous processes such as the flow through the orifices and cooling slits of the burner, and on the density changes produced by combustion. It can therefore be correlated in terms of the flow Mach number at the entrance to the burner and the stagnation temperature ratio (plus the Reynolds number for some situations). A useful rule of thumb is that the stagnation pressure drop is about one or two times the flow dynamic pressure based on the flow area of the combustor. A little ciphering will show that this is equivalent to

$$\pi_b \approx 1 - \varepsilon \left(\frac{\gamma}{2} \right) M_b^2, \tag{4.37}$$

where $1 < \varepsilon < 2$ and where M_b is the Mach number based on burner flow area. For example, if $M_b \approx 0.2$ and $\varepsilon = 2$, then $\pi_b \approx 0.94$, a typical value.

Unfortunately, the number of physical processes that influence the combustion process is so large that, despite many attempts, no effective system for correlating burner efficiency in terms of dimensionless parameters has been evolved. This is easily understood when one realizes that atomization depends on fuel surface tension, injector pressure drop, and injector shape, that the vaporization process depends on the fuel's vapor pressure and heat of vaporization, that the reaction itself depends on activation energy and chemical composition, and so on. Some general observations can be made. Because reaction rate increases with temperature and pressure, efficiency tends to decrease with decreasing pressure and inlet air temperature. It also tends to decrease with increasing combustor inlet Mach number M_b, because this decreases the residence time of the fuel-air mixture in the burner. Finally, because of the sensitivity indicated by figure 4.37, η_b tends to decrease as the fuel/air ratio is varied from the nominal design value.

Some qualitative trends can be identified by considering a family of combustors that are identical except for different pressures and mass flows if we adopt the view that the efficiency depends on the residence time in the combustor. This residence time is $\tau_{res} \approx A_b \rho_{t3} L / \dot{m}$, where L is the burner's length and A_b its flow area, and ρ_{t3} is the compressor outlet density. Rewriting this somewhat gives

$$L \propto \frac{\dot{m}/A_2}{A_b/A_2} \frac{\tau_{res}}{\pi_c^{1/\gamma}}. \tag{4.38}$$

If we keep the ratio of burner area to compressor frontal area constant and the compressor mass flow per unit frontal area constant, then L is proportional to $\pi_c^{-1/\gamma}$. If we hold π_c constant but change the size of the engine, L remains constant, so the ratio of combustor length to engine diameter decreases as the engine's size increases. For both of these reasons, combustor size decreases relative to engine size when we compare large, high-pressure-ratio engines against small, low-pressure-ratio engines. This can be seen very clearly by comparing figure 1.15 with figure 1.16 or with figure 1.17.

4.4.4 Afterburners

The higher entrance temperatures and the near-stoichiometric fuel/air ratios of afterburners enable them to operate with a simpler configuration than the primary combustor. On the other hand, their relatively high flow Mach number makes pressure losses somewhat more critical. The Mach number can be estimated from figure 4.28, which gives the ratio of nozzle throat area to compressor inlet area for a simple turbojet. If we assume that the afterburner flow area equals the compressor flow area, $A_6 = A_2$ (see figure 1.4), then $A(M_6)/A^* = A_2/A_n$, and with the values of A_n/A_2 from figure 4.28 we can find M_6 from figure 4.1 for either the afterburning

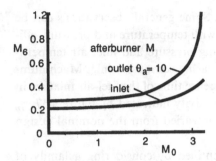

Figure 4.39
Afterburner inlet and outlet Mach numbers for the turbojet engine of figure 4.28 as functions of flight Mach number.

or the nonafterburning case. We can take the nonafterburning value as the Mach number before combustion in the afterburner, and the afterburning value as that after combustion. These values are shown in figure 4.39. Above $M_0 = 3.2$ this estimate shows the afterburner to be "thermally choked," and either A_6 would have to be increased or θ_a would have to be reduced.

A typical afterburner configuration is shown in figure 4.40. The fuel is sprayed into the turbine exit annulus, where it vaporizes and mixes. A flameholder is provided to stabilize the flame front, which stands at an angle to the flow determined by the flow velocity and the propagation velocity of the turbulent flame front, just as the angle of an oblique shock is determined. The afterburner flameholder stabilizes the flame front by producing a region of recirculating flow in which there is a large residence time. According to the model developed in reference 4.21 and sketched in figure 4.40, the recirculation zone is surrounded by a mixing zone, and the residence time of the fluid in this mixing zone determines whether a stable flame will form or whether the flame will "blow off." Because of turbulent mixing, burned and unburned material enter the mixing zone from the unburned flow. If times in the mixing zone are sufficient, a chemical reaction is initiated in the mixing-zone gas, and this reacting gas enters the flame front downstream. Thus, the criterion for stabilization is that the residence time t_{res} be greater than some chemical reaction time t_{reac}. Since t_{res} is proportional to the length L of the recirculation zone and inversely proportional to the flow velocity, the criterion for flame stabilization becomes

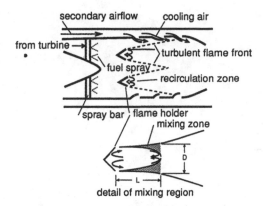

secondary airflow   cooling air

from turbine →

turbulent flame front

fuel spray

recirculation zone

spray bar   flame holder

mixing zone

D

L

detail of mixing region

Figure 4.40
Schematic of an afterburner, showing V-gutter flame holders stabilizing flame by producing a recirculation zone (lower diagram.)

$$\frac{L}{u} > \text{const} \times t_{\text{reac}}. \tag{4.39}$$

This relation is very useful, because L depends on the flameholder dimension D and the Reynolds number but not on the chemical phenomena, while t_{reac} depends on the chemistry. For typical conditions, $L/D \approx 3$. A typical variation of t_{reac} with equivalence ratio is shown in figure 4.41 for $p = 1$ atm and $T = 340°K$. The chemical time t_{reac} depends on temperature and pressure roughly according to

$$t_{\text{reac}} \propto \frac{T^{1.4}}{p}$$

for $250 < T < 400°K$ and $0.1 < p < 1$ atm. For hydrogen-air mixtures, the magnitude of t_{reac} is about one-tenth that for gasoline-air mixtures (figure 4.41). The length of the afterburner is determined by the flame spreading angle (see figure 4.40), which is almost independent of cold gas velocity and near 3° for typical conditions.

From these facts we can assemble a qualitative understanding of the compromises required in an afterburner design. We know T, p, and M at the afterburner inlet from the engine cycle. From equation 4.39, these determine a minimum flameholder dimension D once the equivalence ratio ϕ is set. The number of flameholders that can be put in is determined by

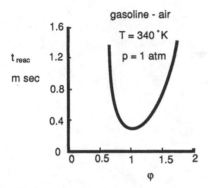

Figure 4.41
Chemical reaction time as a function of equivalence ratio ϕ for gasoline-air mixture.

the acceptable pressure drop, and the linear spreading between them then determines the afterburner length.

Because D is determined by the engine cycle independently of the engine's size, the geometry of an afterburner changes markedly with engine size; its length and its weight become larger relative to the engine as the engine's size decreases. A comparison of figures 1.15 and 1.19 makes this very clear.

4.4.5 Pollutant Formation and Control

The products of combustion from either the primary burner or the afterburner would ideally be composed of nitrogen (N_2), some unconsumed oxygen (O_2), carbon dioxide (CO_2), and water vapor (H_2O). Hydrocarbon fuels always contain a little sulfur, which appears as SO_2 in the exhaust. This sulfur content must be below about 0.1 percent in aircraft fuels to avoid corrosion of the turbine blades, so the SO_2 is a very minor constituent in aircraft exhausts. It can be a major problem in stationary gas turbines, where lower firing temperatures are used to reduce sulfur corrosion.

Except in the context of global warming, the products of ideal complete combustion, CO_2 and H_2O, are generally considered benign. At one time it was thought that cloud formation by the water in the exhaust posed a problem. It is now generally accepted that jet trails disperse soon enough in the dry stratosphere that this is not a serious concern from the viewpoint of cloud formation and its effect on the Earth's albedo. There is some possibility that water drops or ice crystals formed in the engine exhaust plume by condensation can support heterogeneous (i.e. surface) reactions among the trace constituents in the exhaust. This matter will be discussed further. With the recent concern about the effect of atmospheric CO_2 content on global mean temperature, the CO_2 exhaust of aircraft may come to be considered a pollutant; however, it is neither more nor less damaging than that emitted by other fossil-fuel-burning engines. Should it come to be perceived as a problem, the emission of CO_2 could be reduced by use of CH_4 as fuel, or eliminated by use of H_2; however, according to our current understanding, the use of either of these fuels would entail substantial economic and operational penalties.

The remainder of this section will deal with the trace constituents that result from imperfect combustion. They are primarily CO, unburned hydrocarbons (UHC), nitrogen oxides NO and NO_2 (or NO_x), and smoke. The desire to limit their emissions stems from two quite distinct concerns: air pollution near busy airports and possible damage to the Earth's ozone layer by high-flying aircraft emitting NO_x. The first of these concerns has led to studies by the US Environmental Protection Agency of the impact of aircraft emissions, and to proposals for their regulation. The history of these proposed regulations is now long and somewhat complex. Only a brief summary of the process and a statement of the regulations now in force will be given below, but both are extensively documented in the extended literature which this controversial subject has produced.

4.4.5.1 Regulation of Emissions Near Airports Section 231 of the Clean Air Act, as amended in 1970 by Public Law 91-604, directs the Administrator of the EPA to "establish standards applicable to emissions of any air pollutant from any class or classes of aircraft or aircraft engines which in his judgment cause or contribute to air pollution which endangers the public health or welfare." Regulations ensuring compliance with these standards are required to be issued by the Secretary of Transportation in accordance with section 232 of the Clean Air Act.

In July of 1973 the Environmental Protection Agency published (reference 4.22) emission standards for CO, UHC, NO_x, and smoke. These were stated in terms of the maximum allowable output of each pollutant (per unit of thrust per hour) in a standard approach-landing-taxi-taxi-takeoff cycle, which involves idle, low-power, and full-power operation. The intent was to regulate emissions below an altitude of 3000 feet, which is a typical altitude of the thermal inversion in the atmosphere over major cities. The standards are given here in table 4.2, along with the actual emissions from

Table 4.2
EPA aircraft emissions standards (1979).

Pollutant	1979 EPA standard	JT8-D	JT9D-7	CF6-50
CO	4.3	19	10.4	10.8
UHC	0.8	2.7	4.8	4.3
NO_x	3	8.0	6.5	7.7
Smoke	19–20	visible	4	13

the JT9D-7 and the CF6-50 (two large high-bypass engines in production at the time of promulgation of the standards) and those of the JT8-D (which powered all Boeing 727s and all early 737s and early McDonnell-Douglas DC-9s. The standards were to be enforced beginning on January 1, 1979, except for the regulations on smoke, which were to be phased in for various types of engines between 1974 and 1978.

The units for CO, UHC, and NO_x are grams of pollutant per kilogram thrust per hour of operation in a standard approach-taxi-taxi-takeoff cycle. The smoke number is a measure of visibility of the exhaust established by the Society of Automotive Engineers (ARP 1179, 1970), and the value of 19–20 corresponds to the threshold of visibility at full power for the 50,000-lb-thrust-class engines (CF6-50 and JT9D-7), taken from a curve of smoke number versus thrust given in reference 4.22.

The details of measurement of the gaseous pollutants and of smoke are described in reference 4.22. Briefly, they are all to be obtained by sampling of the exhaust constituents during testing of the engine at sea-level static conditions. The gaseous pollutants are determined by standard gas-analysis techniques. The smoke number (SN) is determined by drawing a sample of exhaust gas through a filter paper at a set flow rate for a set time, then measuring the change in the optical reflectance of the paper due to the deposited soot. The SN is defined as $100(1 - R_s/R_w)$, where R_s is the reflectance of the sample and R_w is the reflectance of the clean paper.

It is clear from table 4.2 that, although significant progress had been made in reduction of CO between the JT8-D and the later high-bypass engines, the actual emissions were still well above the EPA standards. The NO_x production was almost as high in the later engines as in the JT8-D, in spite of efforts to control it. As we shall see, this was due to the higher combustor inlet temperatures of the later engines. The exhaust of the early

Table 4.3
ICAO emissions standards.

CO	118
Hydrocarbons	19.6
NO_x	$40 + 2(\pi_{00})$
Smoke	$83.6(F_{00})^{-0.274}$

JT8-D was clearly visible on takeoff, but this smoke problem was largely eliminated in the later engines.

These standards generated considerable controversy in the technical and lay communities. The aircraft engine manufacturers argued that the CO and NO_x standards were not attainable without unacceptable cost and performance penalties with then-available technologies. The National Aeronautics and Space Administration launched an Experimental Clean Combustor Program to develop combustor technologies that might make the standards practically attainable. (Some results of this program will be described below.) In 1976 the EPA issued revised standards which lowered the CO to 3.0 and the HC to 0.4 (reference 4.23). But in December 1982 a new revision was issued in which the restrictions on CO and NO_x were deleted entirely, leaving only the regulations on smoke and hydrocarbons (reference 4.24).

In July 1973 the Federal Aviation Administration, acting for the Secretary of Transportation, had issued Special Federal Aviation Regulation 27 (SFAR 27) to initiate enforcement of the EPA standards. But this initial rule enforced only those standards that were to take force on or before February 1, 1974—that is, the smoke regulations. There were numerous modifications after this, but the 1979 EPA standards on CO and NO_x were never incorporated into the certification process.

In 1981 the International Civil Aviation Organization issued the standards listed in table 4.3 (see reference 4.25). In this table π_{00} is the rated pressure ratio of the engine, and F_{00} is the rated thrust in kilonewtons. The units here unfortunately are different from those in the EPA standards, being grams of pollutants per kilogram of fuel used by the engine, but the values for hydrocarbons and smoke are equivalent to the EPA standards for these emissions. Note that the standard for NO_x makes allowance for the increase of this pollutant with increasing compressor outlet temperature.

Table 4.4
Typical emissions regulations from Federal Aviation Regulations, Part 34 (August 1990).

b) Exhaust emissions of smoke from each new aircraft gas turbine engine of class TF and of rated output of 129 kilonewtons (29,000 pounds) thrust or greater, manufactured on or before January 1, 1976, shall not exceed
  $SN = 83.6(rO)^{-.274}$ (rO is in kilonewtons)
d) Gaseous exhaust emissions from each new commercial aircraft gas turbine engine that is manufactured on or after January 1, 1984, shall not exceed:
  (1) Classes TF, T3, T8 engines with rated output equal to or greater than 26.7 kilonewtons (6000 pounds)
  Hydrocarbons: 19.6 grams/kilonewton rO
  (2) Class TSS
  Hydrocarbons: $140(0.92)^{rPR}$ grams/kilonewton rO.

In August 1990, the FAA issued a new Part 34 of the Federal Aviation Regulations which establishes exhaust emission standards for civil aircraft engines, and supersedes all prior regulations (reference 4.26). It recognizes the following types of engines:

TP   Turboprop engines
TF   Turbofan or turbojet engines except T3, T8, and TSS
T3   JT3-D
T8   JT8-D
TSS Engines for supersonic aircraft

The FAA standards are only for smoke and unburned hydrocarbons, and cover both old and new engines. For the complete set, see reference 4.26. Some examples, listed by the outline headings of reference 4.26, are given in table 4.4.

The last item, referring to engines for supersonic aircraft, is a good example of the complexity of regulation in this difficult area. Here rPR is the "rated Pressure Ratio," i.e., the compressor pressure ratio at rated power. Its inclusion in the rule reflects the fact that historically the combustion efficiency has improved with increasing compressor pressure ratio. Later we shall see why this is so.

4.4.5.2   Upper-Atmosphere Emissions

While the US Supersonic Transport was under development, in the 1970s, the potential damage to the Earth's ozone layer became one of two highly publicized rallying causes for opponents of the program. The other was airport noise, which is discussed in chapter 9. A good deal of effort was put into the development of atmospheric models in an attempt to quantify the effects of NO_x emissions on the ozone layer, and this work has continued to the present. Although the conclusions of these studies were and still are subject to considerable uncertainty because of the great difficulty of modeling both the fluid-mechanical and the chemical behavior of the atmosphere, it is now generally acknowledged that NO_x emissions from future SSTs will have to be very much below the levels attainable in the 1970s in order for the deployment of a large number of SSTs to be publicly acceptable. Supersonic transports are seen as a very special threat to the environment because they are projected to fly in the stratosphere, in the range of altitudes where the ozone concentration is very high, and where the atmosphere is stable (i.e., not subject to much vertical mixing), so that NO_x emitted by the SSTs may be expected to have a long residence time in which to catalyze the destruction of ozone. It is for this reason that a relatively small fleet of SSTs is considered a serious threat to the ozone layer.

It may seem that the very large number of subsonic transports flying in the lower reaches of the stratosphere might also pose some risk, even though the ozone concentration is much lower there. But the cruise altitude of these aircraft, 35,000–40,000 feet, happens to correspond to the altitude at which the net effect of injected NO_x is about zero. There is production of O_3 by interaction of NO_x with methane diffusing up from the troposphere, and destruction of O_3 by the mechanism already discussed.

Recently it has been recognized that the integrity of the Earth's ozone layer is threatened by the release of man-made chemicals, primarily chlorofluorocarbons such as perfluorotrichloroethylene (refrigerant 12) and carbon tetrachloride. With strong experimental evidence that the ozone concentration has been significantly reduced in the polar regions, there is as of this writing an international consensus that the release of such chemicals into the atmosphere must be curtailed. It seems very likely that the standards for emission of NO_x in the stratosphere by aircraft will become more rigid than those in place at present. It may be that limitations will eventually be imposed on emissions near airports. The technical approaches to both these problems are therefore outlined in the following subsections.

4.4.5.3 Technology for Reducing Emissions Near Airports The pollutant emissions from subsonic aircraft in the landing-taxi-taxi-takeoff cycle are traceable to the wide range of operating conditions this cycle implies for the engines, from idle at relatively low engine speeds (where the compression ratio is small and the airflow and fuel flow are much less than at full power) to the full or takeoff power condition (where the fuel flow is at maximum and the compressor discharge pressure and temperature are high). Each of these operating conditions poses its own problems from the viewpoint of emissions. In general, emissions of unburned hydrocarbons and CO are worst at low power, whereas the NO_x and smoke are most troublesome at full power.

It proved extremely difficult to meet the EPA standards listed in table 4.2 with single-stage combustors of fixed geometry, such as have been standard for aircraft engines. Two-stage and parallel combustion systems therefore have been devised to meet these requirements. Some of the concepts will be described after a discussion of the physical phenomena that lead to the problems. For a comprehensive discussion of the technical status as of 1978 see reference 4.27.

At part-throttle and idle conditions the residence time in the burner is smaller than at full throttle. This can be understood from equation 4.38 by noting that $\pi_c^{1/\gamma}$ increases more rapidly with increasing compressor speed than does $\dot{m}/A_2$ (again, see chapter 5). The reduced residence time leads to incomplete combustion and to the exhaust of CO and unburned hydrocarbons. Small flow rates of fuel through the fuel injection nozzles tend to lead to poor atomization, further decreasing the combustion efficiency.

At full-throttle conditions, there is a tendency toward the formation of soot (smoke) if the primary combustion zone is fuel-rich ($\phi > 1$). This problem has been largely eliminated by "leaning out" the primary zone and by the use of higher combustor pressures. Most modern gas turbines do not smoke visibly.

The production of nitrogen oxides (NO_2, NO) is due to the reaction of nitrogen with oxygen in the high-temperature zones of the combustor. This occurs by the set of reactions

$$N_2 + O \leftrightarrow NO + N,$$

$$N + O_2 \leftrightarrow NO + O,$$

$$N + OH \leftrightarrow NO + H,$$

where the concentrations of O and OH controlling the first and third reactions are determined by thermal dissociation of O_2 and H_2O. Taken together, the three reactions convert N_2 to NO at a rate which is largely

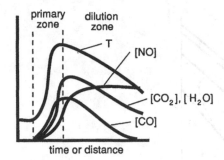

primary   dilution
zone   zone

T

[NO]

$[CO_2]$, $[H_2O]$

[CO]

time or distance

Figure 4.42
Schematic time histories of temperature and species concentrations in an air-fuel mixture
passing through a combustor.

controlled by the rate of the first reaction. This reaction is endothermic (an
energy input is required), so its rate is given by an expression like equation
4.36 with a very large value of A. This means that it proceeds only at very
high temperatures and that the change of reaction rate with temperature is
very rapid. The rate at primary-zone temperatures is much lower than that
of the reactions that limit the rate of combustion of the fuel. The process of
NO formation can thus be described schematically as in figure 4.42, which
traces a typical sample of air-fuel mixture as it progresses through the
combustor. The temperature rises rapidly as the fuel burns to produce CO,
CO_2, and H_2O in a relatively short time in the primary zone of the burner.
If there is some excess oxygen, the CO is oxidized to CO_2 as the gas is held
at this temperature. Formation of NO also begins, but at a slower rate. If
the gases were held at the peak temperature for sufficient time, the NO
concentration would build up to an equilibrium value for that temperature. Actually, the gases are cooled by mixing with excess air in the secondary zone of the burner, and the reduced temperature limits NO production
to a value below that for equilibrium at the primary-zone temperature.
Thus, the *rate* of NO formation and the *residence time* in the primary zone
control the amount of NO formed.

The rate of NO formation may be represented as

$$\frac{d[NO]}{dt} = 2k[N_2][O],$$

where [ ] denotes the concentration in particles per unit volume of the
indicated chemical species and k is a reaction-rate coefficient. According to

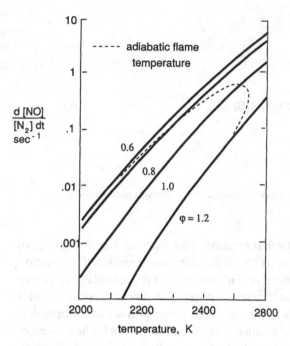

Figure 4.43
Rate of change of NO mass fraction as a function of temperature and equivalence ratio, for C_nH_{2n} initially at 700°K and 1 atm.

the scheme outlined above, [O] is determined by thermal dissociation of O_2, $O_2 \leftrightarrow 2O$, so that if the equilibrium constant for this reaction is

$$K_p(T) = \frac{p_O{}^2}{p_{O_2}}$$

we find

$$\frac{1}{[N_2]} \frac{d[NO]}{dt} = \frac{2k}{RT} \sqrt{p_{O_2} K_p}.$$

This rate of formation is plotted as a function of temperature and equivalence ratio ϕ in figure 4.43 (see reference 4.28) for C_nH_{2n}, a typical hydrocarbon composition, for an initial temperature of 700°K and pressure of 15 atm. The dashed line represents the temperature that would be attained in adiabatic combustion for each value of ϕ.

To interpret this figure, consider for example a combustor with a primary zone residence time of 1 millisecond and $\phi = 1$. It would produce [NO]/[N$_2$] $\approx 0.6 \times 10^{-3}$, or about 600 parts per million of NO. As figure 4.42 shows, this might double or triple in the dilution zone, so we would expect about $600 \times 2.5 = 1500$ parts per million of NO at the combustor exit.

It is more usual to quote the emissions in terms of grams of pollutant per kilogram of fuel burned. The stoichiometric fuel/air ratio being about 0.066, the above value in these terms would be

$$\frac{g\ NO}{kg\ fuel} \approx \left(1.5 \times 10^{-3} \frac{mole\ NO}{mole\ N_2}\right)\left(\frac{30}{28}\right)\left(\frac{0.77}{0.066}\right)(10^3) = 18.75.$$

A typical value for a large high-bypass-ratio engine (e.g. the JT9D-7) at full takeoff power is about 30 g NO/ kg fuel.

As was discussed in subsection 4.4.5.1, the airport emissions standards of the Environmental Protection Agency were stated in terms of the mass of pollutant produced in a typical takeoff-and-landing cycle per unit of thrust. The actual unit is grams of pollutant per kilogram of thrust per hour of operation in such a cycle. To estimate this value, one must estimate the NO$_x$ fraction as above for all power settings and take into account also the specific fuel consumption and time at each condition. The procedure is detailed in reference 4.22. For the JT9-D, a full-throttle value of 31.5 g NO$_x$/kg fuel corresponds to an EPA value of 4.9.

Within the conceptual framework of the conventional combustor, the way to reduce NO$_x$ is to reduce the residence time at high temperatures. This must be done without decreasing the combustion efficiency appreciably or increasing CO. If the rates of combustion of CO to CO$_2$ and of formation of NO were both kinetically limited, a decrease of NO would definitely imply an increase of CO. But if the oxidation of CO is limited by the mixing rate, then increasing the mixing rate should decrease both CO and NO. Early (1976) attempts to meet the 1979 EPA standards centered on burner designs with faster mixing rates, which were achieved by reducing the size of the burner. In one design (see reference 4.29) a large number of small swirl cans were introduced, effectively replacing the larger primary zone with a series of smaller ones. In another design (reference 4.30) the annular burner was replaced by a double-annular burner, reducing the size by a factor of 2. For either case, if the velocities are the same as in the conventional burner, the residence time is reduced by the ratio of the sizes of the primary zones.

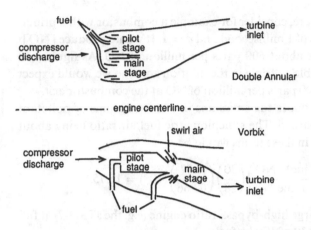

Figure 4.44
Low-emission combustor configurations from NASA's Clean Combustor Program. Top:
General Electric Double Annular Burner. Bottom: Pratt & Whitney Vorbix Combustor.

Later work carried out by the principal US manufacturers of large high-
bypass engines under contract to NASA in the Experimental Clean
Combustor Program led to the design of staged combustion systems. Two
experimental combustors which were tested are shown schematically in
figure 4.44. At the top is the Double Annular Burner concept tested by
General Electric in its CF6-50 engine. In this combustor, the outer pilot
stage operates at idle and low-power settings, and continues to act as a
pilot for the inner main stage, which uses a relatively lean mixture since its
stability is supported by the pilot stage. The pilot stage is optimized for low
CO and UHC emissions at low power settings, while the main stage is
optimized to maintain low NO_x at high power settings. Pratt & Whitney
tested a similar double annular concept, but also tested the Vorbix
Combustor shown at the bottom of figure 4.44. It used a series-type fuel
staging approach, with the pilot stage (optimized for low-power operation)
acting also as a stabilizing pilot for the main stage, which incorporated
high-intensity swirlers just downstream of the main-stage fuel-injection
point, to promote very rapid mixing and thus allow a low residence time,
which reduced NO_x formation.

In table 4.5 the results obtained for CO, NO_x and UHC with these
combustors are compared with the EPA's proposed 1979 standards. It
seems clear that the standards set to control emissions near airports could
be met by these combustion systems. However, the cost in combustor com-

Table 4.5
Comparison of NASA Experimental Clean Combustor Results and 1979 EPA Aircraft Emissions Standards.

Pollutant	1979 EPA standard	JT9D-7 (P&W) (with Vorbix Combustor)	CF6-50 (GE) (with Double Annular Burner)
CO	4.3	3.2	3.0
UHC	0.8	0.2	0.3
NO$_x$	3	2.7	4.25

plexity was judged to be large, and the concepts were considered unable to meet potential requirements for NO$_x$ reduction driven by the upper-atmosphere problem.

Furthermore, the demand for ever better fuel efficiency is driving engine designs in the direction of higher and higher compression ratios, a trend which is expected to exacerbate the NO$_x$ problem. In chapters 2 and 3 we saw that improvements in gas turbine efficiency and power output result from increased turbine inlet temperature coupled with higher compressor pressure ratios. Other things being equal, the associated increases in combustor inlet temperature result in greater NO production. A correlation of NO$_x$ in g NO$_2$ per kg fuel versus T_{t3} shows a monotonic increase from values near 2 at 400°K to 40 at 800°K (reference 4.31) for a number of actual engines. If this trend is inevitable, it may either seriously restrict the use of high pressure ratios in future engines or imply increased NO$_x$ emissions as higher pressure ratios are employed to improve fuel efficiency. However, that trend is not inevitable; rather, it appears to be a result of designing for nearly constant residence time in the burners of the engines. From the kinetic argument and from figure 4.43, we deduce that the NO$_x$ concentration in the exhaust should behave as

$$NO_x \propto \sqrt{p_3}\, e^{-2400/T_3} t_p,$$

where T_3 is in °K and t_p is the residence time in the primary zone. The residence time t_p is determined by the requirement for good combustion efficiency, as was noted in subsection 4.4.3. Correlations of combustor efficiency η_b for fixed-geometry combustors have indicated that η_b correlates as a function of $p_3^{1.75} \exp(T_3/b)/\dot{m}$, where $\dot{m}$ is mass flow through the burner and where b varies from 300°K for a fuel/air ratio of 0.016 to 150°K for 0.010. Taking $t_p \propto p_3/\dot{m}$ and assuming $p_3 \propto T_3^{\gamma/(\gamma-1)}$, we find that for constant combustion efficiency

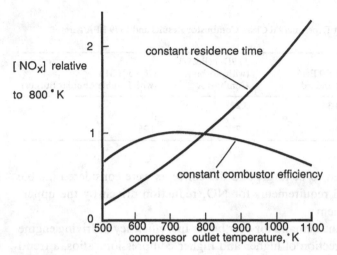

Figure 4.45
Comparison of variations of NO_x production by combustors as a function of compressor outlet temperature for constant residence time and for constant combustion efficiency.

$$NO_x \propto \frac{\exp[-(2400/T_3 + T_3/b)]}{T_3^{\gamma/4(\gamma-1)}}, \quad \eta_b = \text{const}$$

which is to be compared with the expression for constant residence time,

$$NO_x \propto T_3^{(\gamma-1)/2\gamma} \exp(-2400/T_3), \quad t_p = \text{const}.$$

These two variations are compared in figure 4.45 (see reference 4.32), from which we see that, for combustors scaled for constant η_b, NO_x actually decreases at the higher values of T_3.

4.4.5.4 Effects on the Stratosphere It should be clear from the above discussion that the most demanding requirement for NO_x reduction arises from its possible impact on the ozone (O_3) layer of the stratosphere. This layer, formed by absorption of ultraviolet radiation from the sun, shields the earth from almost all ultraviolet (wavelength < 3200 Å) radiation. The processes supporting the O_3 layer (reference 4.33) are

$$O_3 + h\nu \rightarrow O_2 + O$$

and

$$O + O_2 \rightarrow O_3,$$

where hν represents the ultraviolet photon, while reactions such as

$$O + O \rightarrow O_2$$

and

$$O + O_3 \rightarrow 2O_2$$

tend to limit the O_3 concentration. Both processes are slow, and the O_3 concentration is appreciable only in the stratosphere, where the increase of temperature with altitude makes the atmosphere stable to vertical disturbances (in contrast to the troposphere, which is thoroughly mixed vertically). In large part it is the thermal energy input due to ultraviolet absorption, which in turn maintains the high temperature in the stratosphere.

The balance between these processes can be upset by the introduction of NO, which acts to reduce the O_3 concentration by the reactions

$$NO + O_3 \rightarrow NO_2 + O_2,$$

$$NO_2 + h\nu \rightarrow NO + O,$$

and

$$NO_2 + O \rightarrow NO + O_2,$$

in which no NO is destroyed, but O_3 is combined with O to form O_2.

Early estimates of the gravity of this problem based on a fleet of 800 advanced SSTs varied from a 30 percent reduction of the total O_3 cover to nil. The high estimate was based on NO concentrations in the exhaust of current engines, which are on the order of 30 g NO_x per kg of fuel consumed.

The variation of ozone concentration with altitude is plotted in figure 4.46, where approximate flight altitudes are given for several cruise Mach numbers. It is clear that vehicles cruising in the range of Mach number from about 2.4 to 4.0 are likely to be the most threatening to the ozone, and unfortunately this is the range of most interest for commercial transports.

Under its High Speed Civil Transport Program, NASA contracted with Boeing and McDonnell-Douglas to study the potential for future high-speed civil transports (HSCTs) during the period from October 1986 to August 1988. The goal that was set for this program was that a fleet of such aircraft should have no significant impact on the ozone layer. The contrac-

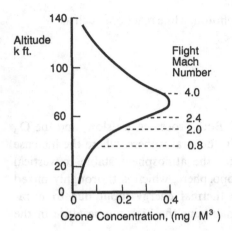

Figure 4.46
Variation of ozone concentration with altitude.

tors were charged with assessing the potential impact of their aircraft on the atmosphere. Low-NO_x combustor designs were proposed by Pratt & Whitney and by General Electric. These and their potential reductions of NO_x are described in reference 4.34, and in less detail in reference 4.35.

The engine designs being considered for future SSTs by both Pratt & Whitney and General Electric have higher compressor pressure ratios than the GE-4 engine, which was being developed for the US SST in the 1970s. The Pratt & Whitney turbine bypass engine is projected to have a pressure ratio of 15.5, versus 12 for the GE-4, and the GE variable cycle engine is projected to have a pressure ratio of 22. The new designs also have considerably higher turbine inlet temperatures; therefore, their NO_x emissions, absent improvements in combustor technology, would be higher than those of the GE-4. The baseline NO_x for the Pratt & Whitney turbine bypass engine at Mach 2.4 cruise was estimated at 32.1 g NO_x/kg fuel. This is about the same as for the JT9D-7 at full takeoff power. According to the present understanding, a large fleet of aircraft with such engines would have an unacceptably large impact on the ozone.

Several possibilities were examined for reducing the NO_x emissions. One was to derate the engines—that is, to use larger engines and operate them at reduced turbine inlet and/or compressor discharge temperature. By this means it was estimated that the NO_x emissions might be reduced as much as 45 percent, but with severe performance penalties. In any case, the 45 percent reduction would not be sufficient to meet the guideline of insignificant impact on the ozone. Much larger reductions, on the order of 95 percent or more, are needed to meet this requirement unequivocally, and reductions of about 80 percent are needed to bring the emissions from aircraft down to the 3–8 g/kg fuel demonstrated in tests of low-emissions combustors for ground-based gas turbines. The latter is the target established for the emissions-reduction portion of NASA's HSCT Program.

Such NO reductions are achievable, in principle, by means of a combustion process in which the temperature of the gases never appreciably exceeds the final turbine inlet temperature. Three possibilities have been suggested. One proposal is to take advantage of the high combustor inlet air temperature in an SST's engine (perhaps as high as $1000°K$), which broadens the flammability limits and allows combustion of a uniformly premixed, lean mixture (reference 4.36). NO_x concentrations as low as 1 g NO_x/kg fuel have been measured in the laboratory, and calculations suggest that values as low as 0.3 may be attainable. It will be difficult to utilize this process in an engine, however, because the combustor must be capable of operating also at low inlet air temperatures and over a range of fuel/air ratios.

The second possibility is to run the first stage of combustion quite rich, to depress the temperature while ensuring stable combustion, and then rapidly mix the resulting products with sufficient excess air so that the combustion can be completed without the temperature's rising appreciably above the desired combustor outlet temperature. The last scheme involves using solid catalysts to react the lean fuel-air mixture (reference 4.37). Again laboratory tests have been encouraging, but the problems of weight, size, and durability are formidable.

In the context of the HSCT Program, the engine manufacturers have identified combustor concepts which give promise of reducing the NO_x emissions to the target range (3–8 g NO_x/kg fuel in cruise). These are described briefly in reference 4.34. Some of the concepts use premixing and prevaporization. One of these employs flight-adjustable valves to regulate the airflow so as to ensure the appropriate mixtures in the primary and mixing sections of the combustor. Another avoids variable geometry, but the risks of duct flashback and auto-ignition were considered serious problems and possible barriers to successful development. Another concept is the "rich burn, quick quench" concept, with control valves to vary the airflow for the quick quench.

It is not clear at this time what course will be taken in controlling the emissions from next-generation SSTs, if such are developed. Although combustor technologies seem to be available to lower emissions to the range of 3–8 g NO_x/kg fuel, their incorporation will impose penalties in weight and complexity. Furthermore, even this target level probably is not low enough to guarantee a minimal effect on the ozone if a large fleet of aircraft is operated.

References

4.1 H. W. Liepmann and A. Roshko, *Elements of Gasdynamics*. Wiley, 1957.

4.2 R. H. Sabersky, A. J. Acosta, and E. G. Hauptmann, *Fluid Flow, A First Course in Fluid Mechanics*. Macmillan, 1971.

4.3 W. E. Moeckel and J. F. Connors, Charts for the Determination of Supersonic Air Flow Against Inclined Planes and Axially Symmetric Cones. NACA TN 1373, 1947.

4.4 H. Schlichting, *Boundary Layer Theory*. McGraw-Hill, 1960.

4.5 P. W. Runstadler, Jr., and R. C. Dean, Jr., "Straight Channel Diffuser Performance at High Inlet Mach Numbers." *ASME Journal of Basic Engineering*, September 1969: 397–412.

4.6 F. X. Dolan and P. W. Runstadler, Jr., Pressure Recovery Performance of Conical Diffusers at High Subsonic Mach Numbers. NASA CR-2299, 1973.

4.7 G. Sovran and E. D. Klomp, "Experimentally Determined Optimum Geometries for Rectilinear Diffusers with Rectangular, Conical or Annular Cross-Section." In Proceedings of the Symposium on the Fluid Mechanics of Internal Flow, General Motors Research Laboratories, Warren, Michigan, 1965.

4.8 E. E. Zukoski, personal communication, 1990.

4.9 V. G. Filipenko, Experimental Investigation of Flow Distortion Effects on the Performance of Radial Discrete-Passage Diffusers. Ph.D. thesis, MIT, 1991.

4.10 R. Decher et al., "System Aspects of Engine Installation." In The Aerothermodynamics of Aircraft Gas Turbine Engines, AFAPL-TR-78-52, Air Force Aero Propulsion Laboratory, 1978.

4.11 YF-12 Experiments Symposium, NASA CP 2054, Dryden Flight Research Center, Edwards, California, 1978.

4.12 H. Trucco, Study of Variable Cycle Engines Equipped with Supersonic Fans. ATL TR 201, NASA CR-13477, 1975.

4.13 L. C. Franciscus, Supersonic Through Flow Fan Engines for Supersonic Cruise Aircraft. NASA TM-78889, 1978.

4.14 T. S. Tavares, A Supersonic Fan Equipped Variable Cycle Engine for a Mach 2.7 Supersonic Transport. M.S. thesis, MIT, 1986.

4.15 R. W. Paterson, Turbofan Forced Mixer-Nozzle Internal Flowfield I—A Benchmark Experimental Study. NASA CR 3492, 1982.

4.16 E. M. Greitzer, R. W. Paterson, and C. S. Tan, "An Approximate Substitution Principle for Viscous Heat Conducting Flows." *Proceedings of the Royal Society, London*, A 401 (1985): 163–193.

4.17 W. M. Presz, Jr., B. L. Morin, and R. G. Gousy, "Forced Mixer Lobes in Ejector Designs." *Journal of Propulsion and Power* 4, no. 4 (1988): 350–355.

4.18 A. H. Lefebvre, *Gas Turbine Combustion*. Hemisphere, 1983.

4.19 W. Jost, *Explosion and Combustion Processes in Gases*. McGraw-Hill, 1946.

4.20 S. S. Penner, *Chemistry Problems in Jet Propulsion*. Pergamon, 1957.

4.21 E. E. Zukoski and F. E. Marble, "Experiments Concerning the Mechanism of Flame Blowoff from Bluff Bodies." In *Proceedings of the Gas Dynamics Symposium on Aerothermochemistry*. Northwest University Press, 1955. Also: E. E. Zukoski and F. H. Wright, "Flame Spreading from Bluff-Body Flameholders," in *Eighth Symposium on Combustion*. Butterworth, 1960.

4.22 *Federal Register* 38, no. 136 (1973): 19088–19103.

4.23 *Federal Register* 41, no. 159 (1976): 34722–34725.

4.24 *Federal Register* 47, no. 251 (1982): 58462–58474.

4.25 International Civil Aviation Organization (ICAO) annex 16, volume II, 1981.

4.26 *Federal Register* 55, no. 155 (1990): 32856–32866.

4.27 R. E. Jones, "Gas Turbine Engine Emissions-Problems, Progress and Future." *Progress in Energy and Combustion Science* 4 (1978): 73–113.

4.28 J. B. Heywood and T. Mikus, "Parameters Controlling Nitric Oxide Emissions from Gas Turbine Combustors." AGARD Propulsion and Energetics Panel, 41st Meeting on Atmospheric Pollution by Aircraft Engines, London.

4.29 R. W. Niedzwiecki and R. E. Jones, Pollution Measurements of a Swirl-Can Combustor. NASA TM Y-68160, 1972.

4.30 D. F. Schultz and D. J. Perkins, Effects of Radial and Circumferential Inlet Velocity Profile Distortions on Performance of a Short-Length, Double-Annular, Ram-Induction Combustor. NASA TN D-6706, 1972.

4.31 F. W. Lipfert, Correlation of Gas Turbine Emissions Data. ASME Paper 72-6T-60, 1972.

4.32 J. L. Kerrebrock, "The Effect of Compression Ratio on NO_x Production by Gas Turbines." *Journal of Aircraft*, August/September 1975.

4.33 C. E. Kolb, "The Depletion of Stratospheric Ozone." *Technology Review*, October/November 1975.

4.34 High Speed Civil Transport Study, Boeing Commercial Airplane Development, Seattle, Washington. NASA Contractor Report 4233, 1989.

4.35 Study of High-Speed Civil Transport, Douglas Aircraft Company, New Commercial Programs, Long Beach, California. NASA Contractor Report 4235, 1989.

4.36 A. Ferri, "Reduction of NO Formation by Premixing." In AGARD Conference Proceedings No. 125 on Atmospheric Pollution by Aircraft Engines, 1973.

4.37 W. S. Blazowski and D. E. Walsh, "Catalytic Combustion: An Important Consideration for Future Applications." *Combustion Science and Technology* 10 (1975).

Problems

4.1 Develop the result given as equation 3.6 by considering flow in opposite directions along the two sides of a flat heat-exchanger plate, where the flow on both sides is in constant-area channels. From subsection 4.1.4, the rate of change of pressure along the flow direction x is $dp/dx = 2\rho u^2 c_f/d$ and the rate of change of stagnation temperature is $dT_t/dx = 4\,St(T_t - T_w)/d$, where T_w is the wall temperature and d is the hydraulic diameter of the flow passage.

4.2   A simple convergent-divergent internal-compression diffuser operates at flight Mach numbers $0 < M_0 < 2.5$. What should its throat/capture area ratio be? What is its maximum pressure recovery at $M_0 = 2.5$? At what value of M_0 would its pressure recovery be unity?

4.3   The mixed-compression supersonic inlet shown in figure 4.19 is designed for $M_0 = 4$, where it produces $\pi_d = 0.59$ (including shock losses).

a. First calculate the losses for this point to check the above π_d.
b. Now suppose the inlet flies at $M_0 = 3$, with the same geometry. Determine the shock positions for best pressure recovery and calculate π_d for this Mach number.
c. Repeat part b for $M_0 = 2$.

4.4   Suppose that an aircraft using a simple internal-compression inlet (figure 4.15) designed for $M_0 = 3$ is flying at $M_0 = 2$ with the inlet started and that the shock is optimally positioned when the inlet suddenly unstarts, popping the shock. The engine is a turbojet with $\pi_c = 12$ and $\theta_t = 7.5$. By what ratio does the thrust change? To estimate this ratio, assume that the engine is ideal except for the inlet pressure loss, and that the Mach number M_2 entering the engine remains fixed.

4.5   For a flight Mach number of 3, "design" a mixed-compression, two-dimensional inlet such as that in figure 4.19. Draw a cross-sectional view of the inlet, showing shock positions and flow directions for the geometry that yields the best pressure recovery. Now estimate the ratio of the height of the boundary-layer suction slot to the capture streamtube height, required at the point where the second shock impinges on the wedge. Assume that the boundary layer is turbulent, the altitude is 10 km, and the capture height is 1 m. The momentum thickness for a turbulent boundary layer varies as $\theta^* = 0.023(Re_x)^{-1/6}\,x$, where x is the distance along the flat plate and Re_x is the Reynolds number based on the length x.

4.6   Starting with equation 4.21, compute the variation with M_0 of A_7/A_n for an ideal turbojet with $\theta_t = 7.5$ and $\pi_c = 6$. Compare your results with those of figure 4.28 and explain the differences.

4.7   Carry out the preliminary design of an afterburner for a turbojet that has a corrected mass flow of 50 kg/sec. At the afterburner design point ($h = 10$ km, $M_0 = 0.5$), the turbine exit conditions are $T_{t5} = 1000°K$, $p_{t5} = 1$ atm, and $M_5 = 0.22$. Assume that the flame holders have a drag coefficient of unity (based on the mean flow velocity at the minimum flow area), and that the maximum acceptable pressure drop is one dynamic head (based on the flow velocity ahead of the flame holders). To determine the flame holder dimension, take $L/D \approx 3$ and $\phi = 1$, and use the data of figure 4.37.

4.8   Combustors must provide for stable operation over a range of engine speeds, from idle to full power. Estimate the ratio of the combustor inlet velocity u_3 at rated rpm to that at half of rated rpm for an engine with compressor pressure ratio $\pi_c = 8$ at rated rpm. Assume that the compression is ideal, that the compressor temperature ratio $\tau_c = 1 + c_1 N^2$, where c_1 is a constant, and that $u_2 = c_2 N$, where N is the rotative speed.

4.9   The afterburner on a turbojet engine is designed to operate satisfactorily at $M_0 = 0.5$ and an altitude of 20 km. Suppose now that the requirement is modified so that it must operate only up to 10 km. Could a lighter (shorter) afterburner be used? By about what factor could it be shortened?

5 Compressors and Fans

The compressor controls the pressure ratio (and hence the thermal efficiency) of a gas turbine engine, and the mass flow as well, so it has a dominant influence on the engine's characteristics. For this reason and because it has been one of the most difficult engine components to develop, it has received great emphasis in both research and development. These efforts, which go back to the 1940s, continue unabated. Since fuel costs are a major portion of direct operating costs for commercial aircraft, there is a continuing strong motivation to raise the compression ratio of engines and to improve compressor efficiency. In military engines for multi-mission aircraft, such as air superiority fighters and attack aircraft, the mass flow per unit of frontal area and the weight of the compressor are important factors. The stability of the compression system is an additional major consideration for these military engines, which have to operate under a very wide range of conditions.

To address these complex and difficult requirements, the aircraft engine industry has long conducted a broad and deep research and development program, with some involvement of academics and support from government, primarily the NASA and the Department of Defense. Whereas in the 1970s the design and development of a new compressor involved a good deal of trial and error, the art has now advanced to the point that a new compressor can be designed with some confidence that it will perform well as designed, provided that its design does not differ too much from that of compressors with which the manufacturer has had prior experience. Modern computational fluid-dynamic and computational solid-mechanical tools, used in conjunction with extensive empirical databases, have made possible reasonably accurate prediction of both the fluid-mechanical behavior of the compressor and its structural integrity. These tools and databases are incorporated in design systems which are closely guarded as proprietary information by the major aircraft engine manufacturers.

It is not possible, therefore, to present a comprehensive summary of modern compressor design practice here. Nor would it be consistent with the aim of this textbook, which is primarily to provide an understanding of the physical behavior of aircraft engines. In contrast to the design systems, the present understanding of the physical phenomena which govern compressor behavior is rather completely documented in the literature, and it is this understanding which we attempt to convey here. Even this limited objective can be only partially realized within the limits of a textbook of this type, for the literature is extensive, complex, and in a sense quite arcane. Fortunately, a comprehensive treatise on compressor aerodynamics has recently been prepared by Cumpsty (reference 5.1), to which the reader is referred for exhaustive discussions of most of the critical aerodynamic phenomena of importance in compressors. To the author's knowledge there is no comparable treatment of the structural aspects of compressors.

The understanding of compressor phenomenology exists at two fairly distinguishable levels. First there is the understanding of the overall flow through the compressor, the way it is produced by the blading, and the general behavior of the blades in this overall flow. In a certain approximation known as the *through-flow-blade-element* approximation, this aspect of compressor behavior was well understood by the late 1960s. An appreciation of this view of compressor aerodynamics is still essential to an overall understanding of compressors, so it will be given here first essentially in its early form, then in a more modern form which makes extensive use of computation, but still within the same conceptual framework. At this level, most of the very complex fluid-mechanical phenomena are represented by greatly simplified models, in which (for example) highly unsteady flows are modeled by their time averages, and losses due to viscous phenomena are accounted for by correlations in terms of overall flow variables rather than in terms of the local properties of the flows. Such approaches still form the principle basis for preliminary design.

A second level of understanding began to develop in the 1970s with the emergence of computational and experimental techniques that have made it possible to explore in detail the complex flows through compressors (and also turbines). Among these capabilities are computational fluid dynamics, laser-doppler velocimeters, high-frequency-response pressure transducers, blow-down compressor and turbine test facilities, and the enormous increase in computational power that allows the handling of large amounts of information, both theoretical and experimental. A great deal of progress has been made in understanding the phenomena that limit and control the performance of compressors; only a portion of this understanding has been incorporated into design systems. Certainly it is far beyond the scope of this text to present all this information. However, some key areas of progress will be described, in the hope that this brief treatment will motivate some readers to explore them further and will provide some guidance to this exploration.

Two types of compressors have been widely used in gas turbines: *axial-flow* compressors, in which the air flows mainly parallel to the rotational axis of the engine, and *radial-flow* or *centrifugal* compressors, in which the air is turned from the axial to the radial direction in the compressor rotor. Axial-flow compressors are predominant in large aircraft engines because of their high mass flow capacity and their potential for high efficiency, and are also the most highly developed, so they will be the main subject of discussion here. Centrifugal compressors are used in smaller aircraft engines, in many industrial applications, and for automotive applications such as turbochargers. They have received increased attention recently. Their characteristics will be discussed, though not as extensively. Fans in the context of aircraft engines are relatively low-pressure ratio axial compressors, the distinction being that they compress air which does not pass through the thermodynamic cycle of the engine. It is useful to distinguish them from compressors, because the requirements they must meet are somewhat different from those for core compressors.

5.1 Energy Exchange, Rotor to Fluid

The feature that distinguishes compressors and turbines from the fixed aerodynamic components of the engine and the aircraft is the *energy exchange* between rotor and fluid, which makes possible the compression and expansion required for an efficient thermal cycle. Because this process of dynamic energy exchange used in all turbomachines is not discussed in the usual courses and texts in fluid mechanics, it will be discussed first from a fundamental viewpoint.

We begin with the first law of thermodynamics in the form

$$\delta e = \delta w - p\,\delta(1/\rho),$$

where e is internal energy, w is mechanical work, and ρ is the density. (Note that the energy and work are specific quantities, i.e. per unit mass.) In the inviscid limit considered here, no forces (other than the normal forces represented by p) act on the fluid, so $\delta w = 0$. If we consider the differentials of e and $1/\rho$ to be taken following an element of the fluid as it moves, the first law becomes

$$\frac{De}{Dt} = \frac{-p\,D(1/\rho)}{Dt}, \tag{5.1}$$

where

$$\frac{D}{Dt} = \frac{\partial}{\partial t} + \mathbf{u} \cdot \text{grad}.$$

The second fundamental law is the momentum equation for the fluid. Because the force acting on the fluid per unit volume is the gradient of the pressure, Newton's second law takes the form

$$\rho \left(\frac{D\mathbf{u}}{Dt} \right) = -\text{grad}\, p. \tag{5.2}$$

(Representing the fluid acceleration by $D\mathbf{u}/Dt$ is valid only if the velocity $\mathbf{u}$ is referred to an inertial coordinate system.)

Forming the scalar product of equation 5.2 with $\mathbf{u}$ gives the following expression for the rate at which the kinetic energy changes as a result of pressure forces acting on the fluid:

$$\frac{\rho \, D(u^2/2)}{Dt} = -\mathbf{u} \cdot \text{grad}\, p. \tag{5.3}$$

Combining equations 5.1 and 5.3 gives a single equation for the fluid's total energy. First introduce the enthalpy $h = e + p/\rho$ and note that

$$\frac{De}{Dt} = \frac{Dh}{Dt} - \frac{1}{\rho} \left(\frac{Dp}{Dt} - p\frac{D(1/\rho)}{Dt} \right),$$

so that equation 5.1 becomes

$$\rho \left(\frac{Dh}{Dt} \right) = \frac{Dp}{Dt} = \frac{\partial p}{\partial t} + \mathbf{u} \cdot \text{grad}\, p. \tag{5.4}$$

Adding equations 5.3 and 5.4 then yields the final relation

$$\frac{\rho \, D(h + u^2/2)}{Dt} = \frac{\partial p}{\partial t}. \tag{5.5}$$

In the approximation we have used so far,

$$h + \frac{u^2}{2} = c_p T + \frac{u^2}{2} = c_p T_t.$$

Thus, equation 5.5 shows that *in this inviscid, non-heat-conducting limit, the stagnation temperature of the fluid, and hence its stagnation pressure, can be changed only by an unsteady compression or expansion.* No steady flow process, for which $\partial/\partial t = 0$ by definition, can affect the addition to or the removal of energy from the fluid. Further, equation 5.5 shows that the energy of the fluid can be increased only by increasing the pressure. Conversely, to decrease the energy, as in a turbine, the pressure must decrease. Note that the energy of the fluid, which includes its kinetic energy, is referred to the coordinate system in which the time dependence is determined. Thus, a flow process that is steady relative to the rotor can change the energy of the fluid relative to stationary coordinates.

With these fundamental points in mind, let us consider the energy transfer between the moving blades and the fluid. For this discussion, consider an axial-flow compressor with such a large ratio of hub diameter to tip diameter that the blades may be approximated by a linear cascade, that is, an infinite row of blades moving in a straight line, as at the top of figure 5.1. If we were to follow an element of fluid as it passed through the moving rotor, we would note changes in both its velocity and its pressure. The character of the changes is more easily seen by transforming to a coordinate system stationary in the rotor, where the flow seems steady, but from an angle as shown by the dashed vectors. If the blades are shaped to turn the flow toward the axis in this coordinate system (in the direction of blade motion), they form diverging passages which (for subsonic flow) result in rising pressure as the fluid passes through the cascade. When this pressure increase is carried back to the stationary coordinate system, as we see from equation 5.5, energy is added to the flow.

The numbers 2, 3, 4, 5, 6, and 7 have been assigned to points in the engine as indicated in figures 1.4, 1.6, and 1.7. To avoid confusion with these, the successive positions along the flow direction in the compressor and later in the turbine will be designated by letter subscripts: a, b, c, d. Thus, if M_a is the Mach number ahead of the guide vanes, which are the first blades in the compressor, M_a is related to M_2 simply by the flow contraction caused by the rounded "spinner" in figure 1.4, for example.

5.1.1 The Euler Equation

Given the pressure distribution in the blade row, we could find the temperature and pressure ratios of the compressor by integrating equation 5.5 in time. There is, however, a computationally simpler way to look at the energy-transfer process, although in a sense it obscures the physical nature of the process. As is indicated in figure 5.1, we draw control surfaces upstream and downstream of the cascade and suppose that the flow across

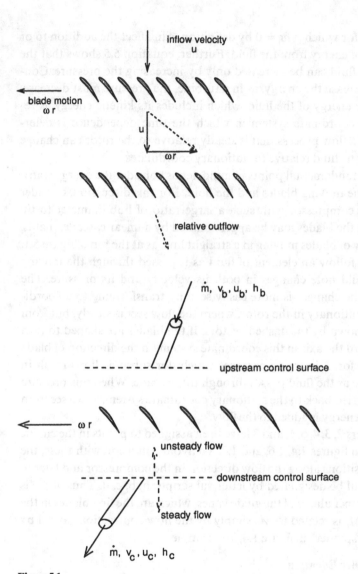

Figure 5.1
(upper) Cascade representation of a rotor blade row with absolute velocities (top) and relative velocities (bottom). (lower) A rotor cascade, showing the inlet and outlet streamtubes used in the formulation of the Euler equation and the conceptual idealization to steady flow in the rotor.

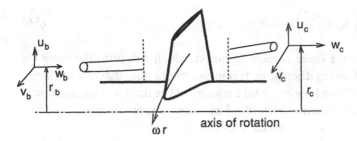

Figure 5.2
The blade row, control surfaces, and streamtube for development of the Euler equation in
rotor geometry.

these surfaces is steady, as will be true to an adequate approximation if the
surfaces are far enough upstream and downstream of an isolated row of
blades. We then identify a streamtube entering the cascade through the
upstream surface and emerging through the downstream one, and we apply the laws of conservation of (1) total fluid energy and (2) momentum in
the direction of blade motion to this streamtube. If $\dot{m}$ is the mass flow in the
streamtube, the conservation of energy requires

$$\dot{m}\left(h_c + \frac{u_c^2}{2} - h_b - \frac{u_b^2}{2}\right) = P,$$

where P is the power delivered to the fluid in this streamtube by the blades.
Conservation of momentum requires that the difference between the momentum flowing into the control volume in the upstream tube and that
flowing out in the downstream tube plus the force F acting in the rotor due
to streamtube sum to zero, or $\dot{m}(v_c - v_b) = F$. If ωr is the velocity of the
blades, then $P = F\omega r$, and we find

$$\left(h_c + \frac{u_c^2}{2}\right) - \left(h_b + \frac{u_b^2}{2}\right) = \omega r(v_c - v_b).$$

Generalizing this argument to the axisymmetric case, let the upstream tube
have mean radius r_b and tangential velocity v_b, and let the downstream
tube have mean radius r_c and tangential velocity v_c as in figure 5.2. The
energy balance is unchanged, but the momentum balance is replaced by an
angular momentum balance $\dot{m}(r_c v_c - r_b v_b) = T$, where T is the torque due
to the streamtube; now $P = \omega T$, so that

$$\left(h_c + \frac{u_c^2}{2}\right) - \left(h_b + \frac{u_b^2}{2}\right) = \omega(r_c v_c - r_b v_b). \tag{5.6}$$

This is the famous Euler turbine equation, which will form the basis for much of the following discussion. It is important to note that it is an *energy* equation, describing how the total enthalpy of the fluid in a streamtube is changed by the interaction of the fluid with the compressor rotor.

Putting $c_p T_t = h + u^2/2$, we have for a perfect gas

$$c_p(T_{tc} - T_{tb}) = \omega(r_c v_c - r_b v_b). \tag{5.7}$$

Another interesting special case is the incompressible fluid. It is still true when ρ is constant that $h = e + p/\rho$, but since e is not influenced by changes in pressure it is constant across the blade row, and the change in total enthalpy is due only to the change in pressure. So the Euler equation is

$$\frac{p_{tc} - p_{tb}}{\rho} = \omega(r_c v_c - r_b v_b), \quad \rho = \text{constant}. \tag{5.8}$$

In the following subsections we will assume that the flow is steady in coordinates moving with the blades, determine the flow patterns in these coordinates, and then deduce the pressure rise from equation 5.7 or equation 5.8. This classical procedure is satisfactory so long as the assumption of steady flow in rotor coordinates is justifiable.

5.1.2 Stage Temperature Ratio

There are normally three types of blade row in an axial compressor: the inlet guide vanes, the rotor blades, and the stator blades. (Some compressors have no inlet guide vanes.) The inlet guide vanes usually give the flow a swirl in the direction of rotor motion to reduce the flow velocity relative to the rotor blades, minimizing shock losses, and to equalize the static pressure rises in the rotor and the stator. The rotor adds energy to the flow and in the process imparts angular momentum to it. The stator removes this angular momentum and diffuses the flow to raise the pressure. The rotor-stator combination constitutes a *stage*.

It is convenient to represent the changes in velocity that occur through the blading by a *velocity diagram*. The development of such a diagram is shown in figure 5.3 for an inlet guide vane row, plus one stage. The velocity

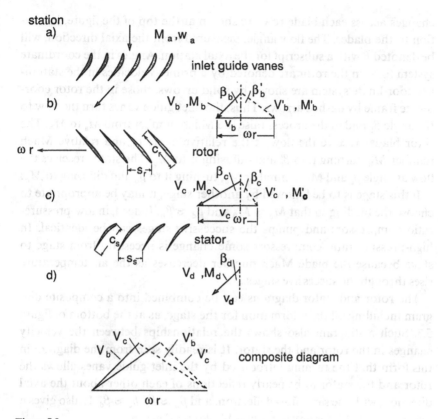

Figure 5.3
Development of a stage velocity diagram. The top shows the changes across inlet guide vanes, rotor, and stator separately; the bottom shows the composite rotor-stator diagram.

changes across each blade row are shown at the top of the figure, in relation to the blades. The flow angle, measured from the axial direction, will be denoted β with a subscript for the axial station. Angles in the coordinate system fixed in the rotor are denoted by a prime. Velocities in the stationary coordinate system are shown by solid arrows, those in the rotor coordinate frame by dashed arrows. Thus, the inlet guide vanes turn the flow to the angle β_b and in the process raise the Mach number from M_a to M_b. The rotor blades receive the flow at the relative angle β_b' and relative Mach number M_b', turning it to β_c' and diffusing it to M_c'. The stator receives the flow at angle β_c and Mach number M_c, turning it to β_d and diffusing to M_d.

If this stage is to be followed by another stage, it may be appropriate to choose the blading so that $M_d \approx M_b$ and $\beta_d \approx \beta_b$. Indeed, in low-pressure-ratio compressors and pumps the successive stages can be identical. In high-pressure-ratio compressors some change is necessary from stage to stage because the blade Mach number decreases as the air temperature rises through the successive stages.

The rotor and stator diagrams can be combined into a composite diagram including all the information for the stage, as at the bottom of figure 5.3. Such a diagram also shows the relationships between the velocity changes in the rotor and the stator. It is readily seen from the diagram in this form that the turning introduced by the inlet guide vanes allows the rotor and the stator to be nearly reflections of each other about the axial direction, with the same flow deflection, and $\beta_b' \approx \beta_c$, $\beta_c' \approx \beta_d$. It also gives a lower value of M_b' for any particular blade speed ωr.

It is characteristic of a closely spaced cascade of blades that the flow angle leaving the blade row (β_c' for the rotor and β_d for the stator) is nearly equal to the angle of the trailing edges of the blades. The difference is called the *deviation*. If the spaces between the blades form long, slender passages, the flow angle at the exit will be controlled by the geometry of the blades at the exit and will be independent of the inlet angle. Denote the chord of the blades by c and the spacing by s; then the extent to which this situation is approached depends on the ratio $\sigma = c/s$, called the *solidity*. In practice, aircraft turbomachine cascades usually have high solidities (unity or more), so it is useful to think of the blades as controlling the leaving angles β_b, β_c', and β_d in first approximation.

Expressing the velocities in the Euler equation in terms of these angles and the axial velocities, which relate to mass flow, we find

$$v_b = w_b \tan\beta_b',$$

$$v_c = w_c \tan\beta_c = \omega r_c - w_c \tan\beta_c',$$

and

$$\frac{T_{tc}}{T_{tb}} - 1 = \frac{(\omega r_c)^2}{c_p T_{tb}}\left[1 - \frac{w_c}{\omega r_c}\left(\tan\beta_c' + \frac{w_b r_b}{w_c r_c}\tan\beta_b\right)\right]. \qquad (5.9)$$

In this form, the Euler equation shows the dependence of the temperature ratio of the stage on the important dimensionless groups. The group $(\omega r_c)^2/c_p T_{tb}$ is proportional to the square of the blades' tangential Mach number, based on the upstream speed of sound. We write

$$\frac{(\omega r_c)^2}{c_p T_{tb}} = \frac{(\omega r_c)^2}{\gamma R T_b}\frac{\gamma - 1}{1 + \frac{1}{2}(\gamma - 1)M_b{}^2} = \frac{(\gamma - 1)M_T{}^2}{1 + \frac{1}{2}(\gamma - 1)M_b{}^2}$$

to emphasize this dependence, noting that M_T is based on the rotor velocity at the downstream radius of the streamtube but on the upstream speed of sound. The group $w_c/\omega r_c$ is related to the mass flow rate, or to the axial flow Mach number, since

$$\frac{w_c}{\omega r_c} = \frac{w_c}{w_b}\frac{w_b}{\omega r_c} = \frac{w_c}{w_b}\frac{M_b\cos\beta_b}{M_T}.$$

Thus, if w_c/w_b, β_b, β_c', β_d, and r_c/r_b are essentially determined by the geometry of the blading, the temperature rise of the stage depends chiefly on M_T and M_b/M_T, and it is helpful to put equation 5.9 in the form

$$\frac{T_{tc}}{T_{tb}} - 1 = \frac{(\gamma - 1)M_T{}^2}{1 + \frac{1}{2}(\gamma - 1)M_b{}^2}\left[1 - \frac{w_c}{w_b}\cos\beta_b\frac{M_b}{M_T}\left(\tan\beta_c' + \frac{w_b r_b}{w_c r_c}\tan\beta_b\right)\right].$$
$$(5.10)$$

Two important characteristics of compressors can be deduced from inspection of the Euler equation in this form:

(1) For a given geometry of the flow, the stagnation temperature rise of the stage varies as $M_T{}^2$. Much of the advancement in compressor performance over the last two decades has been made possible by improvements in materials and in supersonic aerodynamic design that have made feasible the exploitation of this effect.

(2) For given blade-exit flow angles β_c' and β_b, the stage temperature rise decreases as M_b increases (i.e., as the flow through the stage increases). The

effect increases with the magnitudes of $M_b \tan \beta'_c$ and $M_b \tan \beta_b$ in the last term of equation 5.10.

The second effect is reduced if the rotor and the stator turn the flow nearly to the axial direction, so that both $\tan \beta'_c$ and $\tan \beta_b$ are small. There are two important consequences to such a design strategy: For a given M_T and M_b it tends to maximize the temperature rise of the stage, which is generally desirable. On the other hand, it tends to reduce its stability. This can be understood by thinking of the behavior of the compressor operating against some output resistance. If the resistance is increased, tending to reduce the flow, the compressor responds by producing an increased pressure, thus countering the reduction in flow, because the factor of equation 5.10 containing $M_b \tan \beta'_c$ and $M_b \tan \beta_b$ decreases as M_b decreases. This is a stable situation, the more so the larger the turning from the axis at exit from the blade rows. If the pressure rise increased with an increase in flow, the flow would tend to increase until limited by some other effect, or conversely to decrease until limited by some effect such as stalling of the compressor.

This stability argument may be applied to the compressor as a whole, or to the local behavior in response to a fluctuation in the flow through some part of the annulus. In the first case, it shows the behavior of the compressor coupled to a discharge system. In the latter, it indicates how the compressor will behave in the presence of inlet distortion, for example. Both of these aspects are discussed at greater length later in this chapter.

As the blade Mach number is raised above the sonic value, shock losses on the blades increase the entropy rise in the stage, tending to reduce the efficiency. The advantage of high M_T is such that many modern compressors operate with $M'_b > 1$, in spite of these losses. The magnitude of M_b controls the mass flow of the compressor, per unit of area, so it is desirable to increase M_b to as near unity as other limitations will allow. The practical range for aircraft engine compressors is from 0.5 to 0.8, the highest values being attained in compressors without inlet guide vanes and with values of M'_b as high as 1.7.

From equation 5.10 it is clear that reducing β'_c is advantageous in that it results in a higher temperature ratio, but it also implies that the rotor must turn the flow through a larger angle. This implies increased losses in the rotor unless the solidity σ_r is correspondingly increased, so the choice of β'_c, along with tip Mach number and the axial flow Mach number, must take into account the losses in the rotor blade row as well as the stability considerations given above. Similarly, the choice of $\beta_b \approx \beta_d$ depends on the stator losses and the influence on M'_b as well as the desired turning. Thus, a number of design choices must be made in selecting the design-point velocity triangles for a compressor.

To approach this complex situation quantitatively, we must relate the losses in the blade row to the flow angles and Mach numbers. This discussion therefore will be resumed after the compressor through-flow is discussed, and after the characteristics of both subsonic and supersonic cascades are examined.

5.2 Compressor Geometry and the Flow Pattern

To determine the temperature rise of an existing compressor stage, one must relate the flow Mach numbers and angles contained in equation 5.10 to the geometry, the rotative speed, and any other relevant characteristics of the machine. The problem of turbomachine fluid mechanics posed in this way is called the *direct problem*. It can also be posed in the opposite sense: Given the desired flow angles and Mach numbers, determine the required geometry. This is called the *inverse problem*, and it is the problem faced in design. In this discussion the inverse problem will be considered in the main, because it is the most convenient vehicle for exhibiting the limitations and design compromises that lead to the prominent characteristics of turbomachinery. The direct problem will be discussed in a qualitative way to bring out the main features of off-design behavior.

The inverse problem (also called the *design problem*) is conventionally divided into two parts. One imagines first an axisymmetric (no variation in the tangential direction) "throughflow" with axial, tangential, and radial components of velocity; all of these can change discontinuously (or, in more refined analyses, gradually) at axial locations corresponding to the blade rows. In the limit of sudden changes, the blade rows are considered thin "actuator discs" which change the momentum and energy of the fluid, but the variations from blade to blade in the tangential direction are neglected. The throughflow is, to a certain approximation, determined by the same changes in velocity and angle that enter the Euler equation; thus the throughflow problem can be addressed without reference to the details of blade shape once the velocity triangles are set.

The second part of the problem is to determine a set of blade shapes that will produce the assumed velocity structure. (Of course there is no guarantee that an appropriate set of blade shapes exists, for just any prescribed throughflow.) The design of the blades is usually approached by idealizing the blade row as an infinite rectilinear cascade, as we did in subsection 5.1.2, the flow changes across the cascade being inferred in part from empirical cascade data and in part from two-dimensional computations. In this step, it is important to account for the effects on the flow through the blade row of radial shifts in the location of the stream tubes and their radial height, as well as the changes in the plane represented by the cascade approximation. This point will be explained further in the discussion of blading.

5.2.1   The Axisymmetric Throughflow—Radial Variations

As the flow passes through the compressor blade row, its tangential velocity is changed; its pressure and (to a lesser extent) the other velocity components are also changed. The change in tangential velocity will generally cause an unbalanced centrifugal force, which leads to a radial acceleration of the fluid. If we focus on a streamtube, such as that shown in figure 5.2, this means that the streamtube tends to shift radially as it passes through the blade row. The radial shift of the streamtubes implies changes in the radial profile of the axial velocity as well—that is, a redistribution of the flow through the compressor annulus. It is these effects, influenced by the boundary conditions imposed by the cylindrical hub and casing, that we wish to consider. One might expect that the radial accelerations induced would die out far upstream and downstream of the blade row and that the streamtube would assume a "radial equilibrium" position, which is different downstream than upstream. This is in fact the case, and we will consider first this "radial equilibrium" limit, as it shows the physical effects most simply, although it does not apply precisely to any real compressor. The methods used to deal more accurately with the throughflow will then be described.

5.2.1.1   Radial Equilibrium Throughflow   The Euler equation 5.7 can be expanded to

$$c_p T_c + \tfrac{1}{2}(u_c^2 + v_c^2 + w_c^2) - c_p T_b - \tfrac{1}{2}(u_b^2 + v_b^2 + w_b^2) = \omega(r_c v_c - r_b v_b),$$

$$(5.11)$$

where r_b and r_c are the equilibrium radii of the streamtube upstream and downstream. The centrifugal force on each fluid element must be offset by the radial pressure gradient, so $dp/dr = \rho v^2/r$ at each station. Suppose that a polytropic exponent, n, can be defined so that p is proportional to $T^{n/(n-1)}$. If the flow along the streamtube is isentropic, $n = \gamma$; but if the entropy increases, $n < \gamma$. With this relation,

$$\frac{1}{T}\frac{dT}{dr} = \frac{n-1}{n}\frac{1}{p}\frac{dp}{dr} = \frac{n-1}{n}\frac{v^2}{rRT}$$

or

$$c_p\left(\frac{dT}{dr}\right) = \frac{\gamma}{\gamma-1}R\left(\frac{dT}{dr}\right) = \frac{\gamma}{\gamma-1}\frac{n-1}{n}\frac{v^2}{r} \equiv \mu\left(\frac{v^2}{r}\right),$$

where $\mu = 1$ for isentropic flow. In most of the following discussion μ will be put to unity, corresponding to isentropic flow, to simplify the argument somewhat, but it is not difficult to include the possibility of losses through a value of μ different than 1.

Differentiating equation 5.11 with respect to r, substituting for dT/dr, and assuming that $u^2 \ll (v^2 + w^2)$, we find

$$w_c\left(\frac{dw_c}{dr}\right) - w_b\left(\frac{dw_b}{dr}\right)$$

$$= (\mu - 1)\left(\frac{v_b^2}{r} - \frac{v_c^2}{r}\right) + \left(\frac{v_b}{r} - \omega\right)\frac{d(v_b r)}{dr} - \left(\frac{v_c}{r} - \omega\right)\frac{d(v_c r)}{dr}.$$

$$(5.12)$$

Prescribing the inlet and outlet tangential velocity distributions in r thus allows the calculation of the change in axial velocity as a function of r.

CASE 1: POTENTIAL VORTEX GUIDE VANES   Suppose first that the flows both upstream and downstream of the rotor are *potential vortices*, so that $v_b r = B_b w_a r_T$ and $v_c r = B_c w_a r_T$, where B_b and B_c are constants, w_a is a reference axial velocity, and r_T is the tip radius of the blade row. It follows immediately that $d(v_b r)/dr = 0$ and $d(v_c r)/dr = 0$, so

$$w_c\left(\frac{dw_c}{dr}\right) - w_b\left(\frac{dw_b}{dr}\right) = 0 \quad \text{or} \quad w_c^2 - w_b^2 = \text{const.} \qquad (5.13)$$

Thus, the axial velocity distribution is changed only by a constant when a potential vortex increment is added by a blade row to an already existing potential vortex. In this case, whether the blade row is rotating or not is irrelevant in the determination of w_c.

To determine the temperature rise in the blade row, the assumed forms for v_b and v_c are substituted in the Euler equation 5.7 to give

$$c_p(T_{tc} - T_{tb}) = (B_c - B_b)w_a \omega r_T. \tag{5.14}$$

The important result here is that $T_{tc} - T_{tb}$ *is independent of r*, and this is true so long as the rotor adds a free vortex increment to the tangential velocity. This situation, or a generalization of it, is usually desirable in a compressor. If the blade row produced a much higher pressure ratio at one radius than at another, the portion of the annulus with the higher pressure ratio might tend to pump fluid backward through the portion with the lower pressure ratio. If losses vary with radius, the equal-pressure-rise condition implies some variation of $T_{tc} - T_{tb}$ with r; but since we are assuming isentropic flow, uniformity of pressure ratio implies uniformity of temperature ratio.

Because of the design simplicity engendered by uniform w, some early aircraft engine compressors used vortex designs with inlet guide vanes which produced the flow $v_b r = B_b w_a r_T$, the rotor modifying it to $v_c r = B_c w_a r_T$. Indeed, such blading is still used in some stationary gas turbines where weight is not a problem and where there is no great premium on mass flow per unit of frontal area, and hence none on low hub/tip radius ratios. The rotor velocity triangles at hub, mid-span, and tip are as shown at the top of figure 5.4 for a blade row with hub/tip radius ratio of 0.5. For this example the value of ωr has been chosen so that the rotor and the stator are reflections of each other at mid-span. Note that a very large turning is required in the rotor at the hub and that the rotor blades are highly twisted. This design has the disadvantage that the tangential velocity at the rotor inlet is largest at the hub, where the blade velocity is least, and smallest at the tip, where the blade velocity is greatest, so that the swirl introduced by the inlet guide vanes reduces M_b' less at the tip, where the value of M_b' is largest, than at the root. This type of guide vane is then not very effective for reducing the rotor tip relative Mach number and the associated shock losses.

CASE 2: SOLID-BODY INLET GUIDE VANES For these reasons, inlet guide vanes that come closer to producing a solid-body rotation have been used

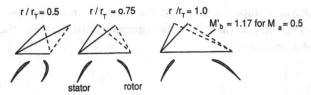

Case 1 Free-vortex inlet guide vanes; A = 0.5, B = 1.0

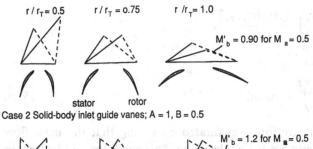

Case 2 Solid-body inlet guide vanes; A = 1, B = 0.5

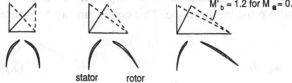

Case 3 No inlet guide vanes; A = 0, B = 0.5

Figure 5.4
Velocity triangles at hub, mid-span, and tip, for (case 1) free-vortex inlet guide vanes, (case 2) solid-body inlet guide vanes, and (case 3) no inlet guide vanes.

in many aircraft engine compressors. They also cause a more complex behavior of the axial velocity, as the following argument illustrates. For the inlet guide vanes we would have

$$v_a = 0, \quad v_b = Aw_a\left(\frac{r}{r_T}\right), \tag{5.15}$$

where station a is ahead of the guide vanes and b behind them, as in figure 5.3. From equation 5.12,

$$w_b\left(\frac{dw_b}{dr}\right) = -\left(\frac{Aw_a}{r_T}\right)\frac{d[Aw_a(r^2/r_T)]}{dr},$$

and integrating gives

$$\left(\frac{w_b}{w_a}\right)^2 = -2A^2\left(\frac{r}{r_T}\right)^2 + \text{const.} \tag{5.16}$$

The constant of integration is evaluated by noting that the mass flow through stations a and b must be the same. This condition can be written as

$$\int_{r_{Ha}}^{r_{Ta}} \rho_a w_a r\, dr = \int_{r_{Hb}}^{r_{Tb}} \rho_b w_b r\, dr. \tag{5.17}$$

To this point the flow has been treated as compressible and isentropic. To proceed further in this way, we must determine $\rho_b(r)$, and this introduces the pressure ratio p_b/p_a or the Mach number M_b. Consistency also requires that $r_{Tb} \neq r_{Ta}$ and $r_{Hb} \neq r_{Ha}$. The analysis can be carried through in this way, but for the sake of simplicity we will assume here that $\rho_b = \rho_a = $ const. It is then reasonable to assume also that $r_{Tb} = r_{Ta} = r_T$ and $r_{Hb} = r_{Ha} = r_H$. This incompressible limit, valid in fact only for low Mach numbers, gives considerable insight into the flow patterns while greatly simplifying the calculations. A further simplification can be had by noting that w_b/w_a does not differ greatly from unity for many cases of interest, so we can write

$$\frac{w_b}{w_a} = \sqrt{1 - 2A^2[(r/r_T)^2 + \text{const}]} \approx 1 - A^2[(r/r_T)^2 + \text{const}].$$

Substituting this in equation 5.17 and evaluating the constant leads to

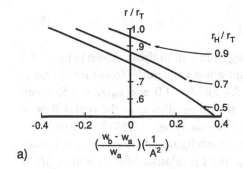

a)

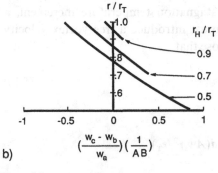

b)

Figure 5.5
Axial velocity increments due to blade rows. (a) Axial velocity increment from inlet guide
vanes producing solid-body rotation, as function of hub/tip radius ratio and radius. See
equation 5.15. (b) Axial velocity increment due to rotor (B positive) or stator (B negative)
operating in solid-body flow, but adding vortex increment. See equation 5.19.

$$\frac{w_b - w_a}{w_a} = A^2 \left(\frac{1 + (r_H/r_T)^2}{2} - \left(\frac{r}{r_T} \right)^2 \right),$$ (5.18)

so w_b varies parabolically, decreasing from hub to tip as shown in figure 5.5 for three values of $r_H r_T$. The resultant guide vane outlet (rotor inlet) velocities are shown in case 2 of figure 5.4 for $A = 1.0$ and $r_H/r_T = 0.5$, which gives about the same flow angles at the mid-radius as in the vortex flow of case 1. For the vortex flow at the mid-radius, $v_b/w_a = 0.5/0.75 = 0.667$. For the solid-body flow, $v_b/w_a = 0.75$, but from figure 5.5 we have $w_b = 1.06w_a$, so $v_b/w_b = 0.71$. For the same ωr_T, the tip relative Mach number M'_b is considerably smaller in case 2 than in case 1 (0.90 versus 1.17 for $M_a = 0.5$).

To provide a radially constant stagnation temperature increment, a rotor behind this guide vane row must introduce a free vortex velocity increment, just as for case 1, so suppose that

$$v_c = Aw_a \left(\frac{r}{r_T} \right) + Bw_a \left(\frac{r_T}{r} \right);$$ (5.19)

then, from equation 5.12,

$$w_c \left(\frac{dw_c}{dr} \right) - w_b \left(\frac{dw_b}{dr} \right) = - \left(\frac{Bw_a r_T}{r^2} \right) \frac{d(Aw_a r^2/r_T)}{dr}.$$

Integrating gives

$$\tfrac{1}{2} w_c^2 - \tfrac{1}{2} w_b^2 = -2ABw_a^2 \ln \left(\frac{r}{r_T} \right) + \text{const.}$$

To be consistent with our assumption that $(w_b - w_a)/w_a \ll 1$, we again take $(w_c - w_b)/w_a \ll 1$ and write

$$\frac{w_c^2 - w_b^2}{w_a^2} = \frac{(w_c - w_b)(w_c + w_b)}{w_a^2} \approx \frac{2(w_c - w_b)}{w_a}.$$

Thus,

$$\frac{w_c - w_b}{w_a} \approx -2AB \ln \left(\frac{r}{r_T} \right) + \text{const.}$$

Continuity of the total mass flow requires that

$$\int_{r_H}^{r_T} \left(\frac{w_c - w_b}{w_a} \right) r \, dr = 0;$$

this determines the constant, with the result that

$$\frac{w_c - w_b}{w_a} = AB\left(\frac{2\ln(r_T/r_H)}{1 - (r_H/r_T)^2} - 1 - 2\ln\left(\frac{r}{r_H}\right)\right). \tag{5.20}$$

This velocity increment is shown in figure 5.5 for the same three values of r_H/r_T chosen for $(w_b - w_a)/w_a$. For B positive, as it would be in the rotor, the trend is similar to that of $(w_b - w_a)/w_a$, so the rotor further increases the axial velocity at the hub and decreases it at the tip.

These effects can be understood in physical terms as a diffusion of the flow near the tip because the static pressure is higher there (owing to the radial pressure gradient produced by the swirl). Since the rotor adds to the swirl produced by the inlet guide vanes, it further retards the axial flow at the tip. It is reasonable to wonder, then, why this effect does not occur for case 1. Evidently the diffusing effect of the radial pressure gradient is just offset in that particular case by the increase in axial velocity that would be required by the outwardly decreasing tangential velocity, the total enthalpy being constant across the annulus.

Equation 5.20 is applicable to the stator as well as to the rotor if w_b is taken as w_c and w_c as w_d, and if B is simply assigned the appropriate negative value. In particular, if $B_s = -B_r$, so that the stator removes the swirl put in by the rotor, then $w_d - w_c = -(w_c - w_b)$ and the flow at the stator outlet is identical to that at the rotor inlet. It is then possible to put several stages, one after the other, without the axial velocity perturbations due to the blade rows becoming large.

The velocity triangles for case 2 are shown in figure 5.4 for $B = 0.50$, which (as noted) gives about the same velocity triangle at mid-radius as for case 1, with free vortex guide vanes.

CASE 3: NO INLET GUIDE VANES   As was mentioned above, many modern aircraft engine compressors and nearly all fans operate without inlet guide vanes, as this leads to high mass flow capability (and, as we shall see below, to some other characteristics which are desirable for first stages). In the present context, such stages are a special case of case 1, with zero turning in the guide vanes and with a free vortex increment introduced by the rotor. The stator may either return the flow to the axial direction or leave some of the swirl put in by the rotor. In the case of single-stage fans, it is important to remove the swirl before the flow enters the fan nozzle, so the turning in the stator is to the axial direction. There is no distortion of the axial velocity for this case, as it is case 1 with $B_b = 0$. A typical set of velocity triangles for such a rotor is shown as case 3 in figure 5.4. For $M_a = 0.5$, they imply $M'_b = 1.2$ and $M_T = 1.0$.

This approximate treatment of the throughflow permits discussion of the ways in which the choice of the guide vane flow influences the temperature rise that can be achieved with a compressor stage, or the level of shock losses for a given temperature rise. From equation 5.14, for all three cases,

$$\frac{T_{tc}}{T_{tb}} - 1 = \left(\frac{(\gamma - 1)M_a}{\sqrt{1 + \frac{1}{2}(\gamma - 1)M_a{}^2}} \right) \left(\frac{M_T}{\sqrt{1 + \frac{1}{2}(\gamma - 1)M_b{}^2}} \right) B.$$

To illustrate the differences between the three cases, we will set the values of $B = 0.5$ and $M'_b = 0.90$ equal for all three cases, the value of M'_b being selected to avoid shock losses. Then for case 1 (vortex inlet guide vanes), to maintain $M'_b = 0.90$ we scale M_a to 0.384 and M_T to 1.02, and from the above expression $(T_{tc}/T_{tb}) - 1 = 0.075$, or for isentropic flow $p_{tc}/p_{tb} = 1.29$. For case 2 (solid-body inlet guide vanes), for $M'_b = 0.90$ and $M_a = 0.5$, we have $M_T = 1.33$. From the above expression $(T_{tc}/T_{tb}) - 1 = 0.126$, and for isentropic flow this gives $p_{tc}/p_{tb} = 1.52$. For case 3 (no inlet guide vanes), for $M'_b = 0.90$, we have $M_a = 0.375$ and $M_T = 0.750$, so that $(T_{tc}/T_{tb}) - 1 = 0.0540$ or $p_{tc}/p_{tb} = 1.156$. It is apparent from these examples that if the rotor-relative inlet Mach number is limited to minimize shock losses, inlet guide vanes allow higher stage pressure ratios, and the solid-body vanes are more effective in this way than the free vortex vanes. As was noted above, however, modern transonic compressors are often designed with rotor-relative inlet Mach numbers substantially above unity, and without inlet guide vanes. Pressure ratios as high as 2.0 have been attained with acceptable efficiency in such stages.

The above radial equilibrium approach, while helpful to a general understanding of the compressor throughflow, is not satisfactory for compressor design or for analysis of experimental data. There are several reasons. First, since the velocity triangles show the radial equilibrium conditions far upstream and downstream of each blade row, they exaggerate the axial velocity changes seen by the blades. A detailed exposition of the actual variation within the context of linear throughflow theory may be found in reference 5.2, which shows that the changes actually take place over an axial distance of the order of the height of the flow annulus $r_T - r_H$, with half of the change upstream of the center of the blade row and half downstream. Second, actual compressors have converging annuli, some with a tapered casing, some with a tapered hub, and some with both. The effect of such convergence can be included approximately in a linear treatment, but the calculations become very complex. More important than either of these, however, is the need to include the effect of losses on the throughflow. That the effect of losses will be large can be anticipated by recalling that the above results followed from two physical statements; one was the condition of radial equilibrium and the other was a relation for the total enthalpy, or total pressure, in the streamtube as described by the Euler equation. The radial distribution of this streamtube total pressure is influenced by losses as well as by the energy input of the rotor, so the losses can strongly influence the radial distribution of axial velocity.

For all these reasons, practical compressor throughflow calculations are now done by numerical techniques; probably the most widely used of these is the streamline curvature method, which will be outlined in the next subsection.

5.2.1.2 Streamline Curvature Throughflow Method As in the preceding radial equilibrium description, the throughflow is modeled as an axisymmetric flow in which changes in angular momentum and in the fluid's total enthalpy are caused by the rotating and stationary blade rows. Losses are accounted for by assigning values to the entropy of the fluid in each streamtube and to its change due to losses in the blade rows. This results in a relationship between the total enthalpy and the stagnation pressure that differs from the isentropic one. Expositions of this approach are given in references 5.3 and 5.4.

The radial equilibrium condition is replaced by an expression for the acceleration of a fluid element along a direction which is usually taken to be approximately normal to the flow direction in the r–z plane. This plane is called the *meridional plane*, and the flow velocity projected on it is the *meridional velocity*. For an axial compressor the computation direction will usually be close to radial, while for a centrifugal compressor rotor it will change from nearly radial to nearly axial as the flow advances through the rotor. For the sake of simplicity the direction of computation will be assumed to be radial here, as indicated in figure 5.6 by the dashed radial lines. The streamlines similarly are indicated by the dashed lines labeled s.

Focusing on a fluid element located at the origin of the coordinate system, O, we can identify three contributions to its acceleration along the radial direction. Along the streamline direction, the acceleration is

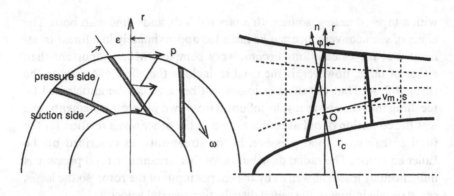

Figure 5.6
Coordinate system used in streamline curvature method (right), and schematic of radial
pressure gradients due to blade lean from radial direction (left).

$v_m(\partial v_m/\partial s)$ and its component along r is $\sin\phi v_m(\partial v_m/\partial s)$. The streamline
itself is curved, with a radius of curvature r_c, and this gives an acceleration
v_m^2/r_c along the direction n which is normal to the streamline. Its component along r is $\cos\phi(v_m^2/r_c)$. Finally there is the effect of the tangential
velocity, v_θ, which is perpendicular to the plane of figure 5.6. This implies
an inward acceleration along r of $-v_\theta^2/r$. Equating the sum of these to the
negative of the pressure gradient, we have

$$\sin\phi v_m\frac{\partial v_m}{\partial s} - \frac{v_m^2}{r_c}\cos\phi - \frac{v_\theta^2}{r} = -\frac{1}{\rho}\frac{\partial p}{\partial r} + F_b. \qquad (5.21)$$

The last term, F_b, accounts for the fact that the blades generate a pressure
field that varies periodically with the blade pitch, with discontinuities in
the pressure across the blades from the suction side to the pressure side. If
the blades are inclined to the radial direction, this discontinuity has a
radial component, with the result that the mean radial pressure gradient in
the fluid is larger than the gradient determined by the casing and hub
pressures, as indicated at the left in figure 5.6. This extra increment of radial
pressure gradient results in (or from) a radial pressure force on the blades,
due to their lean from the radial direction. In the streamline curvature
approach, the tangential blade force is represented in terms of its effect on
the angular momentum of the fluid, so it is advantageous to represent the
quantity F_b also in these terms. The tangential acceleration of the fluid
element may be written $(v_m/r)\partial(rv_\theta)/\partial m$, and so

$$F_b = \frac{v_m}{r} \frac{\partial(rv_\theta)}{\partial m} \tan\varepsilon.$$

The pressure is next related to the total enthalpy and the entropy through the thermodynamic relation

$$-\frac{1}{\rho}\frac{\partial p}{\partial r} = T\frac{\partial s}{\partial r} - \frac{\partial h}{\partial r}$$

and the definition of total enthalpy,

$$h_t = h + \tfrac{1}{2}(v_m^2 + v_\theta^2),$$

so that

$$-\frac{1}{\rho}\frac{\partial p}{\partial r} = T\frac{\partial s}{\partial r} - \frac{\partial h_t}{\partial r} + \frac{1}{2}\frac{\partial}{\partial r}(v_m^2 + v_\theta^2).$$

Finally, then, equation 5.21 becomes

$$\frac{1}{2}\frac{\partial}{\partial r}(v_m^2) = \frac{\partial h_t}{\partial r} - T\frac{\partial s}{\partial r} + v_m\frac{\partial v_m}{\partial m}\sin\phi$$

$$+ \frac{v_m^2}{r_c}\cos\phi - \frac{1}{2r^2}\frac{\partial(r^2 v_\theta^2)}{\partial r} + \frac{v_m}{r}\frac{\partial}{\partial m}(rv_\theta)\tan\varepsilon. \qquad (5.22)$$

When h_t, s, and v_θ are specified as functions of r and m, this equation can be integrated to give v_m as a function of r at each computing station.

The solution must satisfy the additional condition that the mass flow through the annulus is conserved. That condition takes the form

$$2\pi \int_{r_H}^{r_T} Bl(r)\rho v_m \cos\phi\, r\, dr = \dot{m}. \qquad (5.23)$$

where $Bl(r)$ is a factor less than unity that accounts for blockage of the annulus by the blade thickness and by viscous layers at the hub and tip. It is estimated from the blade thickness and from experience with the viscous layers.

To close this computation, h_t, s, and v_θ must be specified. This brings in the characteristics of the blading, and other considerations that affect the losses. These matters will be discussed after the phenomena associated with the blade-to-blade flow have been described.

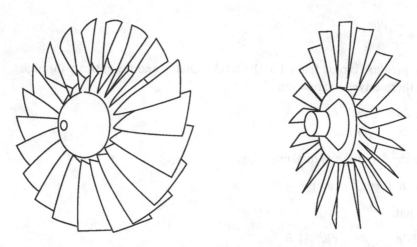

Figure 5.7
Comparison of early high-aspect-ratio and modern low-aspect-ratio blading.

5.2.2 Compressor Blading

There are a number of design parameters that characterize the blading indicated schematically in figure 5.7. As noted above, the axial Mach number controls the mass flow rate per unit of compressor annulus area, while the tangential Mach number of the rotor to a large extent determines the attainable temperature rise or pressure ratio of a stage. Tangential Mach numbers have increased substantially in recent compressor designs. It is not uncommon now for the first two rotors of a compressor to be transonic, meaning that the blade-relative Mach number is greater than unity at the tip and less than unity at the root of the rotor blades. The ratio of blade chord to spacing, termed the *solidity*, is chosen according to a balance of blade losses, deviation, and stalling behavior. Finally there is the ratio of blade span to blade chord, termed the *aspect ratio*. All other parameters being equal, a compressor rotor can be designed with a small number of low-aspect-ratio blades, as indicated at the left in figure 5.7, or with a relatively large number of high-aspect-ratio blades. There are arguments for both, but the trend over the last decade has been toward relatively low aspect ratios. Before the reasons for this are given, it may be helpful to note some of the attractions and problems of higher aspect ratios.

First, if the low- and high-aspect-ratio blade rows used the same airfoil shapes, solidity, and Mach numbers, reasoning from cascade data would suggest they should have the same aerodynamic performance. But the higher-aspect-ratio rotor would be lighter, in proportion to its chord. Since early compressors required many stages to achieve the desired compression ratios (e.g., 17 for the General Electric J-79), the weight was an important factor. Further, it was argued at times that flow in the higher-aspect-ratio blade row should better approximate the two-dimensional cascade data and analyses on which early design systems were based. On the other hand, it was also understood that these two-dimensional data and the corresponding theories did not provide reliable design methodologies; their loss predictions in particular were wide of the mark. Further on the negative side, the high-aspect-ratio blades were susceptible to flutter (a self-excited vibration which can be very destructive), and they also were very vulnerable to damage from foreign objects.

As better materials and better knowledge of transonic flows, especially in rotors, became available, they allowed higher pressure ratios per stage, and this made it possible to achieve the desired compressor pressure ratios with smaller numbers of lower-aspect-ratio rotors. Exploration of the characteristics of the lower-aspect-ratio rotors showed that they are capable of higher pressure ratios and efficiencies for the same blade speed, and that they have wider mass flow range. The reasons will be discussed in some detail below, but in brief, the higher efficiency appears to be due to modification of the shock structure, which results from the possibility of greater radial displacement of streamtubes in the low-aspect-ratio stage. The higher pressure ratio and the greater range are at least in part due to lower axial pressure gradients in the casing boundary layer, which in many cases sets the stall limit on the stage.

In addition to these aerodynamic advantages, the low-aspect-ratio rotors are less susceptible to flutter, and also less sensitive to foreign-object damage.

Whether the blading is subsonic or transonic, the design approach is to represent the blade-to-blade flow on *streamsurfaces*, which are cylindrical surfaces defined by the meridional direction and the $r\theta$ direction. For the simple case of straight cylindrical hub and casing, these surfaces are right circular cylinders. In general they are the streamsurfaces defined by the throughflow calculation. In early work the blade-to-blade flow was treated as strictly two-dimensional on these streamsurfaces, and a great deal of attention was paid to ensuring that "cascade" experiments conducted to determine the characteristics of blade sections did in fact closely approximate two-dimensional flow. Later it was realized that streamtube contraction or expansion in the radial direction (or, more generally, normal to the meridional velocity) strongly influences the behavior of the flow in the passages between the blades. For example, if the flow is subsonic, a contraction of the streamtube in the radial direction accelerates the flow, reducing the pressure rise from that which would occur in a strictly two-dimensional flow with the same blade shapes. It follows that to accurately characterize the behavior of cascades of blades, independently of the throughflow in which they will operate, it is necessary to specify the degree of streamtube contraction or expansion. This is now usually given in terms of an "axial-velocity-density ratio", which is the ratio of ρw at the blade exit to ρw at the entrance, w being the axial velocity. The need to specify this value adds an additional parameter to the already large set needed to specify a cascade geometry in two dimensions, thus compounding the difficulty of obtaining an adequate experimental data base for design. Fortunately, it is now possible to predict the general behavior of cascades numerically, using the techniques of computational fluid dynamics. The effect of the streamtube contraction can be included in these computations relatively easily, and they can be used both to predict the effect of the throughflow on the blade behavior and to correct cascade measurements for streamtube contraction.

5.2.2.1 Subsonic Blading By subsonic blading we mean blading in which the relative Mach number incident on the cascade is subsonic. When this relative Mach number is in the high subsonic range, supersonic velocities may occur on the suction (low-pressure) side of the blades, however, and such effects are important contributors to the losses. Prior to about 1980, the approach used in determining the behavior of such turbomachine blading was to define families of cascade sections, and measure their performance, described in terms of stagnation pressure loss and the deviation of the flow from the direction of the trailing edge of the blades. The details of the velocity and pressure variation on the blade surfaces, although crucial to the behavior, were not evident in these experiments, so that in a sense the development of successful blade sections relied on trial and error, and on experience (particularly under transonic conditions, where shock boundary layer interactions can give rise to high losses). With the availability of computational techniques for computing the flow in cascades, it has become possible to design cascade sections that produce favorable

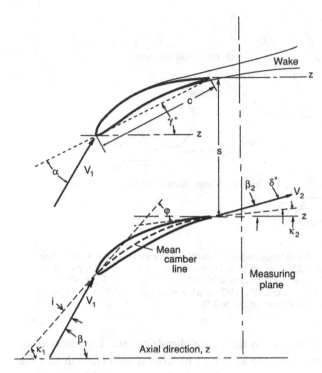

Figure 5.8
Notation for cascades.

velocity or pressure variations on the suction surfaces, minimizing these losses. One of the first successful applications of such techniques, to design "controlled diffusion" blading, is described in reference 5.5.

The notation commonly used in describing a subsonic compressor cascade is shown in figure 5.8, which has been adapted from reference 5.6. In addition to the items shown there, a complete description of the cascade requires specification of the shape of the camber line (which might, for example, be either a circular arc or a parabola, tangent to the directions k_1 and k_2) and specification of the thickness distribution, or some other description of the shape of the airfoil suction and pressure surfaces. For transonic blading a series of circular arcs is sometimes used to define the surface.

Given the geometry of the cascade, the flow is completely specified for subsonic flow if the incidence i, the upstream Mach number, and a Rey-

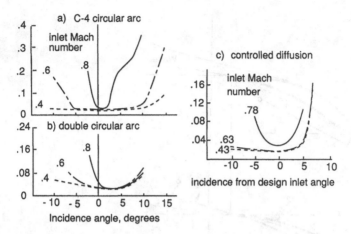

Figure 5.9
Variations of loss factor with incidence and inlet Mach number for three types of blading:
(a) C4 circular arc blade, $\phi = 25°$, maximum thickness/chord $= 0.1$, $\sigma = 1.333$, $\gamma = 37.6°$.
(b) Double-circular arc blade, $\phi = 25°$, thickness/chord $= 0.105$, $\sigma = 1.333$, $\gamma = 42.5°$
(adapted from reference 5.6). (c) "Controlled diffusion", $\phi = 43°$, $\beta_1 = 43.2°$, $\sigma = 1.429$,
thickness/chord $= 0.07$ (adapted from reference 5.5).

nolds number (usually based on the blade chord) are set. For most compressor operating conditions, the Reynolds number is high enough that the cascade performance is insensitive to its actual value, and this parameter can be suppressed. As we shall see, for fully supersonic approach velocities it is also necessary to specify the downstream pressure, relative to the upstream pressure, in order to locate shock positions.

The performance of a given cascade can then be presented in a number of ways. Perhaps the most useful is to give the total pressure loss coefficient $\bar{\omega}_1$ and the deviation angle $\delta°$. The loss coefficient is defined as

$$\bar{\omega}_1 = \frac{\Delta p_t}{p_{t1} - p_1} = \frac{p_{t1} - \bar{p}_{t2}}{p_{t1} - p_1}, \tag{5.24}$$

where $\bar{p}_{t2}$ is a mass average over the flow downstream of the cascade. The deviation is the difference between the actual downstream flow angle β_2 and the angle of the tangent to the mean camber line at the trailing edge.

Variations of loss with incidence and Mach number are given in figure 5.9 for three types of blading, two of which were described in reference 5.6 and one of which (a "controlled diffusion" type) was designed using modern computational methods (reference 5.5). For each value of M_1 there is

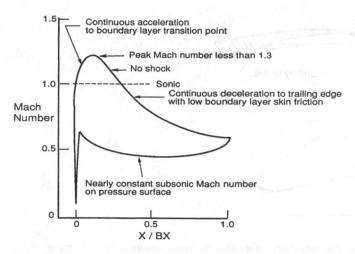

Figure 5.10
Design conditions for controlled diffusion blading (adapted from reference 5.5).

an incidence that gives minimum $\overline{\omega}_1$, and a range of incidence over which the loss is fairly low but outside which the loss rises much more rapidly. As M_1 increases, the range of incidence over which the loss is small decreases, but much less rapidly for some of the blades than for others. In particular, the C4 parabolic arc blade section shows a very narrow range of low loss incidence at high Mach numbers, although it has a very broad range at low Mach numbers. The thickness distribution of this cascade section was originally developed for low-speed isolated airfoils, and has a relatively thick forward portion which produces a supersonic flow bubble at relatively low incident Mach numbers. The double circular section is defined by two circular arcs of different radii of curvature, and therefore has a thickness distribution which is symmetrical fore and aft. It has a thinner nose than the C4 blade, and a broader range of low loss incidence at high Mach numbers. The "controlled diffusion" section is even better in this regard. The design conditions imposed on the blade in order to obtain this behavior are illustrated in figure 5.10.

A minimum-loss incidence angle is defined as the average of the values that give loss twice the minimum value. Figure 5.9 shows that the value $\overline{\omega}_{min}$ of loss at this incidence varies with M_1 and blade design, but that it varies more weakly than the $\overline{\omega}$ versus i curves. This loss occurs when the

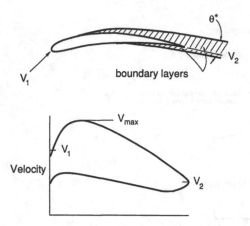

Figure 5.11
Schematic diagram of velocity distributions on suction and pressure surfaces of a blade, showing the diffusion from a maximum velocity $V_{\max}$ to the final velocity V_2 and the resultant thickening of the boundary layer on the suction surface.

flow angles minimize the adverse pressure gradients, and the consequent boundary layer thickening, on the blade surfaces.

A semiempirical analysis of cascade data has led to the observation that the minimum loss factor $\bar{\omega}_{\min}$ can be correlated in terms of a "diffusion factor" D defined as (reference 5.6)

$$D = 1 - \frac{V_2}{V_1} + \frac{|v_2 - v_1|}{2\sigma V_1}. \qquad (5.25)$$

The reasoning is as follows: An airfoil has velocity distributions on suction and pressure surfaces as indicated schematically in figure 5.11. Most of the boundary layer growth, and hence most of the wake thickness, occurs because of the diffusion of the flow along the suction surface from the peak velocity $V_{\max}$ to V_2, the velocity at the trailing edge. If θ^* is the wake momentum thickness (the width of a region of zero velocity with the same momentum defect as the wake), one would expect θ^*/c to depend on some function of $(V_{\max} - V_2)/V_1$. To relate $(V_{\max} - V_2)/V_1$ to the flow angles, we note that the lift in first approximation is

$$L \approx (\rho c/2)(V_{\max}^2 - V_2^2)$$

and that the tangential velocity change in the blade row is related to the lift by

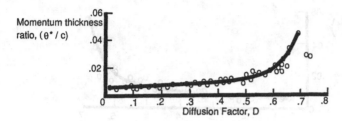

Figure 5.12
Correlation of wake momentum thickness θ^* as fraction of blade chord, with diffusion factor D at minimum loss incidence and $Re = 2 \times 10^5$. (Adapted from reference 5.6.)

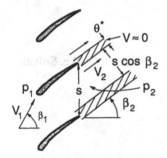

Figure 5.13
Schematic showing conversion from wake momentum thickness θ^* to the loss factor $\bar{\omega}_1$.

$$\rho V_1 (v_2 - v_1) s \sim L \sim (\rho c/2)(V_{max}^2 - V_2^2).$$

If $V_{max} - V_1 \ll V_1$ and $V_1 - V_2 \ll V_1$, then

$$\frac{v_2 - v_1}{V_1} \sim \frac{(c/s)(V_{max} - V_2)(V_{max} + V_2)}{2V_1^2} \sim \sigma \frac{V_{max} - V_2}{V_1},$$

so $(V_{max} - V_2)/V_1 \sim (v_2 - v_1)/\sigma V_1$. The simplest dependence is a linear one; hence the last term of D. The first term follows simply from the diffusion of the flow from V_1 to the lower velocity V_2 according to Bernoulli's equation. Experimental evidence that θ^*/c actually correlates with D is given in figure 5.12.

To convert from θ^*/c to the loss parameter $\bar{\omega}_1$, imagine a wake of zero velocity and width θ^*, as in figure 5.13. The (geometrical) average of downstream stagnation pressure is

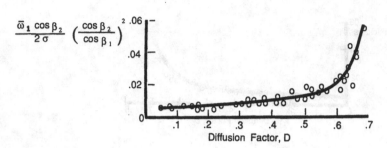

Figure 5.14
Correlation of loss parameter with diffusion factor D. (Adapted from reference 5.6.)

$$\bar{p}_{t2} = \frac{p_{t1}(s\cos\beta_2 - \theta^*) + p_2\theta^*}{s\cos\beta_2};$$

the first term represents the inviscid flow and the second the wake. Substituting in equation 5.24 gives

$$\bar{\omega}_1 = \frac{\theta^*}{s\cos\beta_2}\frac{p_{t1} - p_2}{p_{t1} - p_1},$$

and since

$$p_{t1} - p_1 = \frac{\rho V_1^2}{2} = \frac{\rho w_1^2}{2\cos^2\beta_1}$$

and

$$p_{t1} - p_2 \approx \frac{\rho V_2^2}{2} \approx \frac{\rho w_1^2}{2\cos^2\beta_2}$$

we get

$$\bar{\omega}_1 = \left(\frac{\cos\beta_1}{\cos\beta_2}\right)^2 \frac{\sigma}{\cos\beta_2}\frac{\theta^*}{c}.$$

Hence, the proper correlation is of the form

$$\bar{\omega}_1\left(\frac{\cos\beta_2}{\sigma}\right)\left(\frac{\cos\beta_2}{\cos\beta_1}\right)^2 = f(D).$$

The trend line of such a correlation is shown in figure 5.14. The loss rises rapidly above $D = 0.6$, indicating separation.

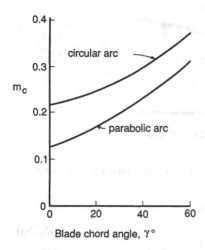

Figure 5.15
Parameter for use in Carter's rule for deviation (equation 5.26).

For design purposes we also need the deviation $\delta°$. According to Carter's rule,

$$\delta° = \left(\frac{m_c}{\sqrt{\sigma}}\right)\phi,\tag{5.26}$$

where m_c depends on the blade chord angle $\gamma°$ and is given in figure 5.15.

These data, together with figure 5.14, allow us to estimate the exit flow angle and loss for the type of subsonic blade sections for which the correlations were derived. As we shall see, extensions of the approach are very useful for transonic blading as well.

One such extension is to generalize the definition of the diffusion factor to account for the radial shifts of streamtubes that take place across blade rows, particularly those of low aspect ratio and high pressure ratio. The generalization consists in noting that the tangential force on the blade, and hence the pressure variation along the streamtube, depends on the change in angular momentum of the fluid, not merely on the change in tangential velocity. Thus, the numerator of the second term in equation 5.22 should be the difference of the outlet and inlet angular momenta, divided by some mean radius of the streamtube. The accepted generalization of equation 5.25 is then

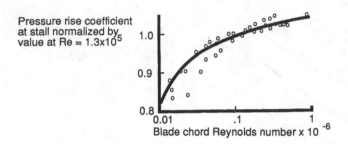

Figure 5.16
Variations of pressure rise with Reynolds number for axial stages (adapted from reference 5.7).

$$D = 1 - \frac{V_2}{V_1} + \frac{v_2 r_2 - v_1 r_1}{(r_1 + r_2)\sigma V_1}. \tag{5.25a}$$

None of these correlations, however, express any effect of Reynolds number. Yet we know that because the loss is a viscous effect, it must depend on the Reynolds number. Investigations of the effect of Reynolds number on cascade performance showed that the cascade correlations were valid only above some limiting Reynolds number, in the range of 2×10^5, there being a rapid rise in both loss and deviation below this value (reference 5.6). But data on the effect of Reynolds number on *compressor* performance, measured by the available pressure rise, indicate that the effects of low Reynolds number are considerably less severe than these cascade data would suggest. Figure 5.16 plots such data for axial compressor stages.

For a compressor blade rotating at a Mach number of unity in air at standard conditions, the Reynolds number is about 2×10^5 per cm of chord, so that blading will usually be in the "high-Reynolds-number" range in any case, except at very high altitudes and low flight Mach numbers, where the density may be less than one-tenth that at sea level. Small engines can encounter problems due to low-Reynolds-number effects under such conditions.

5.2.3 The Loss Factor and Efficiency

One essential step toward understanding the design compromises is to establish the relationship between the loss factor $\bar{\omega}_1$ and the efficiency of the compressor. To this end it is helpful first to establish the connection between $\bar{\omega}_1$ and the entropy rise. The reason for this is that the entropy, being a property of the fluid, is independent of the transformation from rotor (moving) coordinates to stator (stationary) coordinates. This is not true of the stagnation pressure, which depends on the coordinate system to which the velocity is referred. Since $\bar{\omega}_1$ refers to changes in the stagnation pressure, it must be referred to some coordinate system.

The entropy is related to the temperature and pressure of the fluid by the equation

$$s - s_0 = c_p \ln\left(\frac{T}{T_0}\right) - R \ln\left(\frac{p}{p_0}\right),$$

where the subscript 0 denotes some reference state. Since the (schematic) process connecting the stagnation state to the local thermodynamic state of the gas is isentropic, we can also write

$$s - s_0 = c_p \ln\left(\frac{T_t}{T_0}\right) - R \ln\left(\frac{p_t}{p_0}\right),$$

where

$$T_t = T(1 + \tfrac{1}{2}(\gamma - 1)M^2)$$

and

$$p_t = p(1 + \tfrac{1}{2}(\gamma - 1)M^2)^{\gamma/(\gamma-1)}.$$

If a viscous process lowers p_t from p_{t1} to p_{t2}, and if we assume $T_{t2} = T_{t1}$, the change in entropy is then

$$s_2 - s_1 = -R \ln\left(\frac{p_{t2}}{p_0}\right) + R \ln\left(\frac{p_{t1}}{p_0}\right) = R \ln\left(\frac{p_{t1}}{p_{t2}}\right).$$

The assumption of constant stagnation temperature is not generally valid in rotor coordinates, as will be explained below.

The loss factor for the rotor is defined as

$$\bar{\omega}'_b = \frac{p'_{tb} - p'_{tc}}{p'_{tb} - p_b},$$

where the primes indicate that the values are referred to rotor coordinates. Solving for p'_{tb}/p'_{tc} and substituting gives

$$s_c - s_b = -R \ln\left[1 - \bar{\omega}'_b\left(1 - \frac{p_b}{p'_{tb}}\right)\right].$$

For the stator cascade, the assumption that $T_{t2} = T_{t1}$ implies that the effects of flow unsteadiness (see section 5.1) and heat conduction in the blades are negligible. The latter assumption is easily verified; the former is more subtle. It is valid for the flow as a whole in the sense that the stator can do no work on the fluid, because it does not move. On the other hand, unsteadiness in the flow can result in energy transfer between streamtubes, both radially and tangentially, so that the stagnation temperature need not be conserved along any particular (time-averaged) streamline.

For the rotor cascade, the equivalent of equation 5.5 in rotating coordinates is

$$\frac{\rho D(h + \frac{1}{2}u^2)}{Dt} + \rho \mathbf{u} \cdot [\mathbf{\Omega} \times (\mathbf{\Omega} \times \mathbf{r})] = \frac{\partial p}{\partial t},$$

where $\mathbf{r}$ is the radius vector from the origin in the coordinate system rotating at angular velocity $\mathbf{\Omega}$ with the rotor. The term $\mathbf{\Omega} \times (\mathbf{\Omega} \times \mathbf{r})$ is the centrifugal force on the fluid particle due to $\mathbf{\Omega}$, so $\mathbf{u} \cdot [\mathbf{\Omega} \times (\mathbf{\Omega} \times \mathbf{r})]$ is the rate at which work is done on the fluid by its *radial motion* in the centrifugal force field. Thus, only if radial velocities are small is $c_p T_t = h + u^2/2$ conserved in steady flow in the rotor coordinates. The energy addition by the centrifugal field is large in radial flow compressors (indeed, this is one of their chief virtues), but within the cascade approximation often used to describe flow in axial compressors it is zero if the flow is assumed to be along surfaces of constant radius.

In the cascade approximation, then, p'_{tb} is the stagnation pressure *referred to the coordinates in which $\bar{\omega}'_b$ is determined*, so if stations b and c are ahead of and behind the rotor then p'_{tb} is the rotor inlet stagnation pressure in rotor coordinates.

Similarly, if the loss factor for the stator is $\bar{\omega}_c$ then the entropy rise across the stator is

$$s_d - s_c = -R \ln\{1 - \bar{\omega}_c[1 - (p_c/p_{tc})]\},$$

where p_{tc} is referred to the stator coordinates. The total entropy rise across the stage will be $(s_c - s_b) + (s_d - s_c)$, or

$$s_d - s_b = -R \ln\left[1 - \bar{\omega}'_b\left(1 - \frac{p_b}{p'_{tb}}\right)\right]\left[1 - \bar{\omega}_c\left(1 - \frac{p_c}{p_{tc}}\right)\right]. \tag{5.27}$$

The stage efficiency is defined as

$$\eta_s = \frac{(p_{td}/p_{tb})^{(\gamma-1)/\gamma} - 1}{T_{td}/T_{tb} - 1}$$

$$= \frac{(T_{td}/T_{tb}) \exp[-(s_d - s_b)/c_p] - 1}{T_{td}/T_{tb} - 1}.$$

Substituting for $s_d - s_b$ and taking $\overline{\omega}_b' \ll 1$ and $\overline{\omega}_c \ll 1$, we then find

$$\eta_s = 1 - \frac{[(\gamma-1)/\gamma][\overline{\omega}_b'(1 - p_b/p_{tb}) + \overline{\omega}_c(1 - p_c/p_{tc})]\tau_s}{\tau_s - 1}, \tag{5.28}$$

where $\tau_s = T_{td}/T_{tb}$.

Equation 5.28 shows some effect of the relative Mach numbers on η_s, since

$$\frac{p_{tb}'}{p_b} = [1 + \tfrac{1}{2}(\gamma - 1)M_b'^2]^{\gamma/(\gamma-1)}$$

and

$$\frac{p_{tc}}{p_c} = [1 + \tfrac{1}{2}(\gamma - 1)M_c^2]^{\gamma/(\gamma-1)}, \tag{5.29}$$

where M_b' is the Mach number relative to the rotor, and M_c is that relative to the stator. But to see the overall effect of Mach number on η_s we must take account of the variation of τ_s with M. From equation 5.10 we can write $\tau_s - 1 = cM_T^2$, where c depends on the velocity triangles and may in first approximation be assumed independent of M.

It is instructive to examine the result at the limits of small and large M^2. Expanding equation 5.29 in M^2 and substituting in equation 5.28, we find

$$\eta_s \approx 1 - \tfrac{1}{2}(\gamma - 1)\frac{M_b'^2\overline{\omega}_b' + M_c^2\overline{\omega}_c}{cM_T^2}, \quad M^2 \ll 1$$

whereas in the limit of large M

$$\eta_s \approx 1 - \tfrac{1}{2}(\gamma - 1)(\overline{\omega}_b' + \overline{\omega}_c), \quad M^2 \gg 1$$

so if we increase all the Mach numbers proportionately, keeping $\overline{\omega}_b'$ and $\overline{\omega}_c$

constant, η_s does not change initially for low Mach numbers, but at larger Mach numbers the relative importance of rotor and stator losses changes. For high Mach numbers the shock losses per se must be added to this estimate of viscous losses, and in addition the shocks may increase the viscous losses through shock-boundary layer interaction.

5.2.3.1 Stage Inefficiency Due to Cascade Losses (An Example) To illustrate the application of these ideas, let us estimate the section inefficiency of the stage represented by figure 5.4 (case 2), where the rotor-relative Mach number is less than unity at all radii. Table 5.1 lists the various quantities involved in the estimate. It has been assumed that $M_a = 0.5$, and the solidities of rotor and stator have been set to unity at the tip radius. The table shows that although both D_r and D_s are within reasonable bounds, the rotor has relatively large predicted losses at the tip and the stator has similarly large losses at the hub. Note also that the stator has $M_c > 1$ at the hub, a situation that would lead to increased losses there. As was noted in subsection 5.2.1, this stage with a pressure ratio of 2.13 at a tip relative Mach number of only 0.90 is beyond the normal limits of design. It was selected to emphasize the important effects. Lowering the pressure ratio would lower the losses at the hub and the tip and M_c at the hub, resulting in an estimate for η_s near 0.98 for the entire annulus. This is not unreasonable as an estimate of efficiency taking account only of the blade profile losses. From the discussion above it should be clear that the actual total loss levels are considerably in excess of this. Some of the reasons will be discussed below.

5.2.4 Supersonic Blading (for Transonic Compressors)

When the relative Mach number M_b' becomes greater than unity, the behavior of the compressor rotor changes qualitatively, just as the behavior of the supersonic diffuser is qualitatively different from that of the subsonic one. Except in the case of the supersonic axial flow fan, which will be discussed below, the axial Mach number $M_b' \cos\beta_b'$ is less than unity for compressor blading; thus, even though $M_b' > 1$, disturbances can propagate upstream from the rotor, and streamlines can influence each other through upstream propagation of pressure disturbances. For this reason, the throughflow problem discussed above is elliptic in the mathematical sense for the transonic rotor. But, because the blade-relative Mach number

Table 5.1
Numerical values leading to estimates of pressure ratio and efficiency for the stage of figure 5.4 (case 2), accounting only for subsonic cascade losses.

Quantity	Hub	Mid-radius	Tip
Rotor			
M'_b	0.71	0.76	0.90
V_c/V_b	1.087	0.708	0.703
$(v_c - v_b)/2V_b$	0.326	0.208	0.139
σ_r	2.0	1.5	1
D_r	0.076	0.431	0.436
$(\overline{\omega}'_b \cos\beta'_c/2\sigma_r)(\cos\beta'_c/\cos\beta'_b)^2$	0.006	0.012	0.012
$\cos\beta'_b$	0.840	0.603	0.358
$\cos\beta'_c$	0.995	0.858	0.350
$\overline{\omega}'_b$	0.017	0.021	0.072
$\overline{\omega}'_b[1 - (p_b/p'_{tb})]$	0.0048	0.0065	0.0294
Stator			
M_c	1.13	0.866	0.785
V_d/V_c	0.588	0.693	0.768
$(v_d - v_c)/2V_c$	0.221	0.192	0.159
σ_s	2.0	1.5	1
D_s	0.522	0.435	0.312
$(\overline{\omega}_c \cos\beta_d/2\sigma_s)(\cos\beta_d/\cos\beta_c)^2$	0.018	0.012	0.008
$\cos\beta_c$	0.73	0.57	0.25
$\cos\beta_d$	0.93	0.79	0.56
$\overline{\omega}_c$	0.048	0.0238	0.005
$\overline{\omega}_c[1 - (p_c/p_{tc})]$	0.026	0.0094	0.0019
η_s	0.965	0.982	0.965
τ_s	1.25	1.25	1.25
π_s	2.13	2.15	2.13

$M'_b > 1$, shocks can form on the rotor blading, causing stagnation pressure losses in addition to those due to viscous effects and strongly influencing the behavior of the flow. Fortunately, the radial interaction of the stream-tubes mitigates the transonic drag rise, apparently in much the same way that "area ruling" reduces the transonic drag rise of aircraft. It has been demonstrated, for example, that in linearized theory there is no sudden, disastrous drag rise when the blade relative Mach number at the tip passes through unity, in contrast to the situation in two-dimensional linearized airfoil theory (reference 5.8).

Early attempts to realize the potential of supersonic blading were disappointing; the efficiencies were generally well below those predicted by methods that accounted for shock losses and added to these diffusion losses. These early experiments used blading of high hub/tip radius ratio, with supersonic conditions over the entire blade height. Later blading was designed with lower hub/tip ratios; hence, it was transonic in the sense that the hub sections were subsonic even though the tips were supersonic. This terminology is at variance with the standard usage in aerodynamics, where a transonic flow is understood to be one in which the velocities are everywhere near sonic. But there is no ambiguity within the context of compressors. This blading gave much better efficiencies, apparently for the reason given above, and this type of blading has been exploited in modern compressor designs. Stages with supersonic flow over their full span have proved to have very low efficiency, and they have not been incorporated in engines to this date. The supersonic axial flow fan illustrated in figure 4.27 may change this, but it is still in the research phase.

To minimize the shock losses, sharp-nosed blading such as is shown in figure 5.17 is used for the part of the blade where $M'_b > 1$. Ordinarily the relative Mach number will be greater than unity only for the outer portions of the rotor blade, where the tangential velocity is largest. For the inner portions, $M'_b < 1$ and subsonic blading is indicated.

Many of the physical features of supersonic diffuser flow carry over at least qualitatively to the supersonic rotor. The shocks respond to changes in downstream pressure conditions and interact with the boundary layers where they impinge on the blade surfaces. There are, however, a number of unique features of the flow in the rotor. One of these arises from the transonic character of the flow alluded to above. *It is not correct to think of the flow in the supersonic portion of the blading as two-dimensional insofar as mass flow continuity is concerned.* This is because flow can occur along the

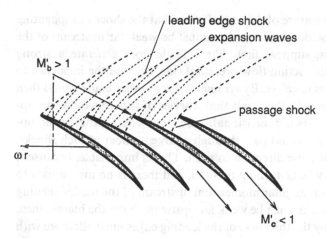

Figure 5.17
Blading for supersonic relative velocity, showing incident flow aligned with suction surface of blades and weak shocks and expansions running upstream.

blade span (perpendicular to the plane of figure 5.17) in response to pressure differences set up by different flow patterns at different radii. Thus, we must regard the flow channel depicted in figure 5.17 as one with varying height perpendicular to the plane of the drawing; this height is controlled by interaction of the channels at different radii. (This is of course true of subsonic blading also, but the effects are usually weaker there.) To specify the effect of such interaction on the conditions in one channel, the axial velocity density ratio Ω is introduced. (See, for example, reference 5.9.) It is defined as

$$\Omega = \frac{\rho_b V_b' \cos \beta_b'}{\rho_c V_c' \cos \beta_c'},$$

and it is nothing more than the ratio of mass flows per unit of annulus area at the upstream and downstream stations. If the flow were two-dimensional, Ω would be unity. If $\Omega < 1$ the channel is constricted downstream, corresponding to closing the nozzle on our schematic diffuser of figure 4.15, and if $\Omega > 1$ the channel expands downstream. These variations have the same qualitative effects on the shock in the rotor that they have on the shock in the diffuser: restricting the flow moves the shocks upstream, and lowering the backpressure causes them to move downstream.

A second unique feature of the rotor flow is that the shocks propagating forward from the blades' leading edges must be weak far upstream of the blades. To see this, suppose first that a blade does generate a strong oblique shock off its suction (low-pressure) side, because the incidence to the suction surface is negative. By symmetry, every other blade would then have to do likewise, with the result that a series of shocks would run upstream. In the ideal (two-dimensional) case they would proceed far upstream; thus, the flow would pass through a large number of such shocks, being turned in the same direction in each. This is impossible, because it would result in very large turning or swirl, and there is no mechanism to produce the resultant angular momentum upstream of the blades' leading edges; hence, the shocks must be weak far upstream. Near the blades, then, the shocks caused by the thickness of the leading edges must alternate with expansions that cancel their turning far upstream. For these purposes, "far upstream" in practice means more than a blade spacing.

The picture that emerges from this argument is shown in figure 5.17. The incident flow is compressed by the leading-edge shock, then expanded by a fan from the suction surface, up to the point on the surface that emits the last expansion wave that can pass the leading edge of the neighboring blade. The net effect of the expansion waves must be to lower the strength of the shock to nearly zero. The final turning of the flow from the initial direction is then that of the shock as negated by the expansions. If the waves were all weak, the net turning would be zero, because the expansions would just cancel the shock; in practice this is nearly the case, because the shock is fairly weak except very near the leading edge. It follows that the flow is nearly aligned with the suction side of the blade at the last expansion wave that passes the leading edge of the next blade, so that *the supersonic rotor operates with constant β_b'*.

This argument is two-dimensional and must be qualified by the observation that the three-dimensional effects of radial interaction of the streamtubes may cause variations in the strength of the shocks as they propagate upstream. Experiments with rotors tend to confirm its validity, however.

5.2.4.1 Losses in Supersonic Blading Because of the practical importance of transonic compressor stages, the magnitudes of losses in such stages and the mechanisms that produce them have been subjects of much study and discussion. One of the conclusions of this work is that it is probably not logically correct to describe the losses in terms of two-

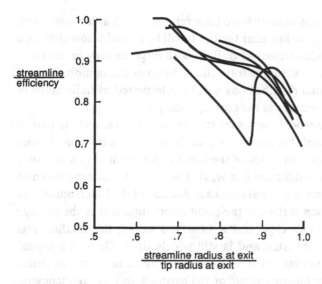

Figure 5.18
Radial distributions of efficiency in transonic compressor rotors (adapted from reference 5.10).

dimensional blade section performance, for at least three reasons. One reason is that the shock losses are determined by the angle of obliquity of the shock surface to the incident flow, and this angle is influenced by inclination of the shock surface in the meridional plane, as well as by its structure in the plane tangent to the cylindrical streamsurfaces. Another reason is that there is very strong transport of boundary-layer fluid in the radial direction in the blade wakes, so that the loss which shows up at a particular radius downstream of the rotor is not necessarily due to losses on the blades on the streamsurface through that radius. Thus, a correlation of measurements at some radial location with two-dimensional predictions on the streamsurface through that radius is suspect in principle. Finally, a significant fraction of the losses in a rotor seem to be attributable to the flows in the immediate neighborhood of the casing and the hub—that is, to effects not represented in the blade-section analysis. Nevertheless, in the absence of better techniques, the blade-section techniques are used in design, so they will be outlined here.

A review of the radial profiles of efficiency in transonic rotors was reported in reference 5.10. As figure 5.18 indicates, the efficiency measured downstream of the rotor near the hub is very high for some of these rotors (indeed, values in excess of unity have been reported) and, at the same time, that the efficiency is quite low near the tip. It will be argued below that such results are due to radial transport of fluid with large entropy in the rotor wakes; however, for now, to illustrate the techniques conventionally used for design, these radial distributions will be interpreted as indications of the loss in the blade sections at the corresponding radii.

Early attempts to model the losses in transonic rotors, such as that of reference 5.11, assumed that the losses at each radius were the sum of those due to shocks and viscous losses of the same order as those in a subsonic cascade with the same diffusion factor, D. The shock losses were taken to be the average of those of normal shocks at M'_b and at the Mach number of the flow on the suction surface at the point of impingement of the passage shock. Because it comes close to the observed values, this method was adopted for design estimates, and is still widely used. There is a logical problem, however, in that the actual shock strengths in rotors, as determined from pressure measurements at the casing with high frequency response pressure transducers, and as inferred from laser doppler velocimeter measurements of velocity, are not as strong as this model would imply. These subtleties will be elaborated in section 5.4.

As a practical design approach, suppose that we know the value of Ω or can control it so as to position the passage shock at the most favorable location; suppose further that the stagnation pressure loss can be estimated as the sum of that due to the shock and that obtained from a diffusion factor correlation generalized from that for subsonic blading, discussed in subsection 5.2.2. The exact position and strength of the shock are difficult to predict because of the effect of varying Ω. As a first approximation, it can be assumed to be a normal shock at M'_b, or the average between this value and that at the suction surface may be employed as proposed in reference 5.11. For want of better information, the deviations given by equation 5.26 can be used if $M'_c < 1$ (the usual situation, at least at the design point).

One very successful design of a high-throughflow transonic rotor using an approach similar to this is described in detail in reference 5.12. In this design the generalization of the loss-factor correlation shown in figure 5.19 was used to account in an empirical way for the increase of loss with increasing radius. The shock loss estimate of reference 5.11 was used. Otherwise the design approach was through a streamline curvature scheme equivalent to that described in subsection 5.2.1.2. The performance

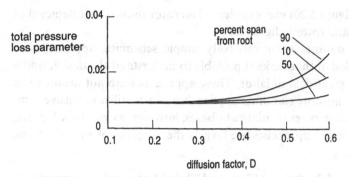

Figure 5.19
Loss parameter as function of D and percent span from root.

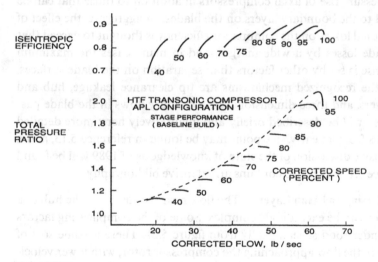

Figure 5.20
Performance of rotor with low aspect ratio and high throughflow (adapted from reference 5.12).

of this rotor (figure 5.20) was excellent. The rotor itself is that depicted as a low-aspect-ratio rotor in figure 5.7.

Section 5.3 outlines some relatively simple schematic approaches to overall stage design that make it possible to understand the design trade-offs without a great deal of labor. These approaches are not meant to be substitutes for the more elaborate approaches which utilize extensive computation and large experimental data bases; however, as was noted above, the more elaborate approaches are largely the property of aircraft engine manufacturers.

5.2.5 Non-Two-Dimensional Flows and Their Effects on Performance

There are several recognized sources of loss and several mechanisms limiting the pressure rise of axial compressors in addition to those that can be attributed to the boundary layers on the blades. In aggregate, the effect of the additional losses on the compressor efficiency is thought to exceed that of the blade losses by a wide margin, and in many cases the maximum pressure rise is set by other factors than separation on the blade surfaces. Some of the recognized mechanisms are tip clearance leakage, hub and casing layers, and three-dimensional or secondary flows in the blade passages. They will be described briefly and qualitatively here; more detailed discussions from an early viewpoint may be found in reference 5.13, and a very extensive discussion of the state of knowledge as of 1989 will be found in reference 5.1, which also contains an extensive bibliography.

5.2.5.1 Casing and Hub Layers The flow near the casing or the hub of a compressor can be exceedingly complex. Some of the complicating factors may be understood schematically from figure 5.21. There is some sort of wall layer in the flow approaching the compressor rotor, with lower velocities near the wall, so that, at the very least, the tangential velocity of the blades results in a higher incidence in the wall layer than in the core flow, and the resulting overturning of the flow near the wall generates a secondary flow (figure 5.22). In the gap between the rotor and stator, if the stator is shrouded so that there is a stationary hub at its root, the hub wall layer makes an abrupt transition from the rotating hub of the rotor to the stationary hub of the stator, with the wall shear suddenly changing direction and magnitude. The casing layer undergoes equally violent disturbances within the blade passages, being turned and diffused by the pressure gradi-

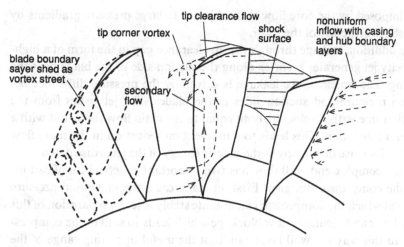

Figure 5.21
Schematic depiction of casing and hub phenomena.

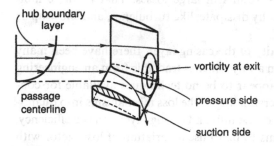

Figure 5.22
Generation of streamwise vorticity by boundary-layer turning in a blade row.

ent imposed by the core flow and subjected to large pressure gradients by the shock structure of the rotor.

In addition, leakage through the tip clearance gap in the form of a high-velocity jet generates a vortex along the suction side of the blade near the casing. The tip clearance leakage is driven by the pressure difference between pressure and suction sides of the blades. The jet issues from the suction side with a velocity about equal to the main flow there, but with a different direction. This leads to a momentum reduction in the main flow and an increase in entropy as the kinetic energy of the jet is dissipated.

This complex end wall flow has two important practical ramifications for the compressor designer. First, in many cases the maximum pressure ratio at which the compressor can operate stably is set by separation of this wall layer and resultant flow blockage which leads to stall of the compressor. In this way the wall layer can limit the useful operating range of the compressor. Second, the generation of strong velocity components which are not aligned with the main flow, and therefore cannot be diffused by the downstream blade rows, results in large losses. That is, these non-streamwise velocities eventually dissipate, like turbulence, and show up as an increase in entropy.

In the face of the complexity of the casing layer, there have been many attempts to model the flow in a way which is tractable from an engineering viewpoint. However, there appear to be no techniques available for converting such qualitative understanding of the loss mechanisms into quantitative estimates of the effect of the hub and casing layers on the efficiency. Current design practice seems to be to use a variation of loss factor with radius, such as that shown in figure 5.19. The values at the different radii are derived from experience with prior designs. Thus, the principal focus of modeling attempts has been on prediction of the effect of the wall layer on the stalling pressure rise. Most of these have idealized the wall layer as an axisymmetric layer, perturbed by the blades, which are seen to generate body forces in the layer, influencing its development in the axisymmetric pressure field generated by the blade row. As Cumpsty (reference 5.1) has pointed out, this model is so far from representing the essential physical phenomena as to be essentially useless.

Koch (reference 5.7) has had some success in modeling the effect of the wall layer on the stall of compressors by applying data from stationary diffusers, such as that discussed in subsection 4.1.3. Here the approach is to view the passage through the blades near the wall, in the coordinates of the

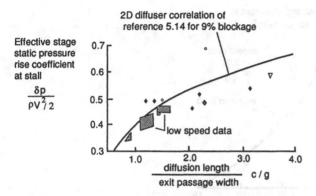

Figure 5.23
Stalling pressure coefficient of compressors versus the length-to-width ratio of the blade passages. (Adapted from reference 5.7; curve from reference 5.14.)

blade row, as analogous to that in a channel diffuser, and to apply the empirical data for stalling of such diffusers. In effect this amounts to applying the diffusion limit criteria to the core flow, ignoring the existence of the wall layer as such and the fact that in the case of the rotor the casing surface is moving relative to the blades. Nevertheless, figure 5.23 shows that it is a useful approach for a wide range of compressors. In this figure, the denominator of the pressure rise coefficient is the mean of the dynamic pressures at entrance to the rotor and stator. The diffusion length in the abscissa is essentially the chord length, and the exit passage width is the blade spacing normal to the flow direction at exit from the blade row, so that in effect the abscissa is the length-to-diameter ratio of the blade passage.

The effect of the tip clearance on this stalling pressure rise has also been examined by Koch (reference 5.7). Figure 5.24 shows that the stalling pressure rise coefficient decreases with increasing tip clearance. In this figure the values are normalized to that for a ratio of clearance to blade staggered gap of 0.055, which is the value to which the data of figure 5.23 were adjusted. This effect of tip clearance may be understood in the context of the simple channel diffuser analogy in the following way. Assuming that the tip clearance flow enters the blade passage normal to the main flow, the pressure drop required to accelerate it to the core flow speed, V, is

$$\delta p = -\frac{\dot{m}_{\text{clearance flow}}}{\dot{m}_{\text{core flow}}}\left(\frac{\rho V^2}{2}\right).$$

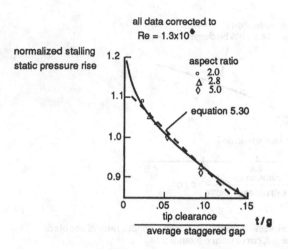

Figure 5.24
Effect of tip clearance on stalling pressure ratio (adapted from reference 5.7).

If the velocity in the clearance gap is of the same order as the free stream velocity, then we have

$$\frac{\dot{m}_{\text{clearance flow}}}{\dot{m}_{\text{core flow}}} = \frac{tc}{g^2},$$

where c is the blade chord, t is the clearance gap, and g is the staggered spacing between the blades. The core flow channel has been assigned a height equal to its width, consistent with the channel analogy. The result is then

$$\frac{\delta p \text{ at stall due to clearance flow}}{\rho V^2/2} = -\left(\frac{c}{g}\right)\left(\frac{t}{g}\right). \tag{5.30}$$

For a typical stagger angle $c/g = 2$, so this argument predicts that the stalling pressure rise coefficient should decrease by about twice the ratio of clearance to blade stagger gap. Remarkably, this is very close to the slope shown in figure 5.24, as indicated by the dashed line.

5.3 Design Choices Based on Blade Section Performance

It may seem from the above discussion of the details of flow in the blade rows that any sort of design optimization based on blade section data would be spurious at best, and at worst misleading. But in fact, methods for examining the ramifications of choice of the blade Mach number, solidity, blade loading, hub-to-tip radius ratio, etc. can be very useful, provided the results are utilized as one input to the design process, along with considerations such as have been outlined above. This section outlines an approach for judging on the basis of blade section performance how the requirements for high mass flow per unit frontal area, high pressure ratio per stage, and high efficiency interact and how they may be traded against one another.

Consider the stage of figure 5.3 with inlet guide vanes. For the sake of simplicity, assume that there are no radial displacements of streamtubes and that the axial velocity is constant throughout, so that the argument is more nearly applicable to multi-stage compressors with slowly varying hub and tip radii than to a transonic stage such as that in figure 5.7. The diffusion factors for the rotor and the stator are

$$D_r = 1 - \frac{V_c'}{V_b'} + \frac{v_c - v_b}{2\sigma_r V_b'}$$

and

$$D_s = 1 - \frac{V_d}{V_c} + \frac{v_c - v_d}{2\sigma_s V_c}.$$

From the Euler equation,

$$\frac{T_{tc}}{T_{tb}} - 1 = \frac{\omega r}{c_p T_{tb}}(v_c - v_b) = \frac{\omega r w}{c_p T_{tb}}(\tan\beta_b' - \tan\beta_c').$$

Defining a work coefficient

$$\psi = \frac{T_{tc}/T_{tb} - 1}{(\omega r_T)^2/2c_p T_{tb}} \tag{5.31}$$

and a dimensionless radius $\zeta = r/r_T$ gives

$$\psi = \zeta(\tan\beta_b' - \tan\beta_c')(2/\mu), \tag{5.32}$$

where

$$\mu = \omega r_T/w$$

and

$(v_c - v_b)/V_b' = [(\cos\beta_b'\psi)/\zeta](\mu/2).$

If we further assume that $v_c - v_d = v_c - v_b$ (that is, that the stator takes out the swirl put in by the rotor), then

$$\frac{v_c - v_d}{V_c} = \frac{\cos\beta_c\psi}{\zeta}\left(\frac{\mu}{2}\right),$$

so D_r and D_s become

$$D_r = 1 - \frac{\cos\beta_b'}{\cos\beta_c'} + \frac{\mu\psi\cos\beta_b'}{4\sigma_r\zeta}$$

and   (5.33)

$$D_s = 1 - \frac{\cos\beta_c}{\cos\beta_d} + \frac{\mu\psi\cos\beta_c}{4\sigma_s\zeta}.$$

For solid-body inlet guide vanes,

$$\frac{v_b}{w} = \frac{v_b r_T}{w}\zeta \quad \text{and} \quad \tan\beta_b' = \left(\frac{\omega r_T}{w} - \frac{v_b r_T}{w}\right)\zeta \equiv \alpha\zeta.$$

Then, from equation 5.32,

$\tan\beta_c' = \alpha\zeta - (\mu\psi/2\zeta).$

Computing $\cos\beta_b'$ and $\cos\beta_c'$ taking $\sigma = \sigma_T/\zeta$, and substituting in D_r, gives

$$D_r = 1 - \sqrt{\frac{1 + (\alpha\zeta - \mu\psi/2\zeta)^2}{1 + (\alpha\zeta)^2}} + \frac{\mu\psi}{4\sigma_{rT}\sqrt{1 + (\alpha\zeta)^2}}.$$ (5.34)

$\text{Tan}\beta_c = \mu\zeta - \tan\beta_c' = (\mu - \alpha)\zeta + (\mu\psi/2\zeta)$, and if $\beta_d = \beta_b$ then $\tan\beta_d = (\mu - \alpha)\zeta$; thus,

$$D_s = 1 - \sqrt{\frac{1 + (\mu - \alpha)^2\zeta^2}{1 + [(\mu - \alpha)\zeta + \mu\psi/2\zeta]^2}}$$

$$+ \frac{\mu\psi}{4\sigma_{sT}\sqrt{1 + [(\mu - \alpha)\zeta + \mu\psi/2\zeta]^2}}.$$ (5.35)

With the values of D_r and D_s from equations 5.34 and 5.35, the D-factor correlation gives the loss parameters

$$\lambda_r = \frac{\bar{\omega}_b \cos\beta_c'}{2\sigma_r}\left(\frac{\cos\beta_c'}{\cos\beta_b'}\right)^2,$$

$$\lambda_s = \frac{\bar{\omega}_c \cos\beta_d}{2\sigma_s}\left(\frac{\cos\beta_d}{\cos\beta_c}\right)^2,$$

which are given by figure 5.14. The loss factors are then

$$\bar{\omega}_b = \lambda_r\left(\frac{2\sigma_{rT}}{\zeta}\right)[\sqrt{1 + (\alpha\zeta - \mu\psi/2\zeta)^2}]\left(\frac{1 + (\alpha\zeta - \mu\psi/2\zeta)^2}{1 + (\alpha\zeta)^2}\right), \qquad (5.36)$$

$$\bar{\omega}_c = \lambda_s\left(\frac{2\sigma_{sT}}{\zeta}\right)\left(\frac{\sqrt{1 + (\mu - \alpha)^2\zeta^2}[1 + (\mu - \alpha)^2\zeta^2]}{1 + [(\mu - \alpha)\zeta + \mu\psi/2\zeta]^2}\right). \qquad (5.37)$$

The relative Mach numbers are also important. Since the axial velocity has been assumed constant, the axial Mach number changes only because of temperature changes; these are small enough to be neglected in an analysis like this, so the relative Mach numbers are

$$\frac{M_b'}{M_a} = \sec\beta_b' = \sqrt{1 + (\alpha\zeta)^2}, \qquad (5.38)$$

$$\frac{M_c}{M_a} = \sec\beta_c = \sqrt{1 + [(\mu - \alpha)\zeta + \mu\psi/2\zeta]^2}. \qquad (5.39)$$

If we now assume that the shock losses in the rotor for $M_b > 1$ are those for a normal shock, they can be found from figure 4.2. The fractional stagnation pressure loss, denoted say $\bar{\omega}_s'$, should be added to $\bar{\omega}_b'$ as given by equation 5.36 to give the total pressure loss of the rotor.

Finally, then, the stage efficiency can be computed from equation 5.27 using M_b' and M_c to find p_b/p_{tb}' and p_c/p_{tc}.

The general trend of diffusion factor with radius is shown in figure 5.25 for two stages, both with work coefficients $\psi = 0.5$, and with equal values of $\mu = \omega r_T/w$, one without inlet guide vanes and one with inlet guide vanes. According to equation 5.31, either would produce a temperature ratio

$$\tau_s = 1 + (\gamma - 1)M_a^2\mu\psi = 1.1$$

or a (lossless) pressure ratio

$$\pi_s \approx 1.40.$$

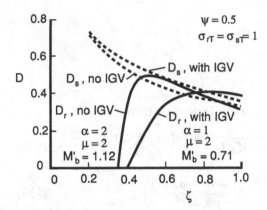

Figure 5.25
Variation of diffusion factor of rotor (D_r) and stator (D_s) with radius $(\zeta = r/r_T)$ for stages with and without guide vanes.

Perhaps most important is that the diffusion factor for the stator, D_s, increases with decreasing ζ for both designs, so if a maximum acceptable value of D_s is set, it sets a lower limit on ζ and hence on the ratio of hub to tip radius. Of course the stator solidity could be increased to reduce the loading, but this has other ramifications. A similar limit may be set by the rotor if the hub/tip ratio is large, but for small hub/tip ratio an intermediate radius is most critical for diffusion in the rotor.

These two stages, one with and one without inlet guide vanes, operating at the same blade speed $(\omega r_T = \mu w_a = 2w_a)$, produce the same pressure ratio, but the one with inlet guide vanes has a relative Mach number of only 0.71, versus 1.12 for the one without inlet guide vanes. The lower Mach number would generally result in higher efficiency, and also (as will be discussed below) offers operational flexibility if the angle of the guide vanes can be varied. Thus, the majority of core compressors use inlet guide vanes.

5.3.1 Mass Flow vs. Pressure Ratio

In engines for supersonic aircraft there is a great incentive to maximize the mass flow per unit of engine frontal area to minimize drag. This requires both that the axial Mach number M_a be as close to unity as possible and that the useful flow area $\pi(r_T{}^2 - r_H{}^2)$ be maximized. The practical limit on M_a is set by the blockage of the flow annulus by the blades. From figure 4.1 we see that if the inlet guide vanes block 16 percent of the annulus, for example, the maximum possible value of M_a is about 0.6; this is typical. With M_a set, the mass flow can be increased only by decreasing r_H/r_T; but for a given limit on D, the maximum possible value of ψ then decreases as r_H/r_T decreases, with a consequent reduction of the pressure ratio. It is important to recall, however, that this argument assumes slow variation of the hub and casing radii, so it does not apply to transonic stages such as that in figure 5.7, which achieve very high pressure ratios and also very high mass flows. Generally this type of design is most effective for the first stage. The following argument may be considered applicable to stages after the first, and in general to compressors with low blade Mach numbers.

By solving equations 5.34 and 5.35 for ψ, we can find its limiting values for specified D_r and D_s. Some typical results for a stage without inlet guide vanes, with $\alpha = \omega r_T/w_a = 2$ and with the maximum diffusion factors in rotor and stator set at 0.5, are shown in figure 5.26. From the lower curves, the stator limits ψ for r_H/r_T above 0.9 and below 0.5, while the rotor limits ψ for the intermediate hub/tip ratios. With these values of ψ_{max}, τ_s follows from equation 5.31, and, if the efficiency is assumed equal to unity, π_s is as shown in the upper part of figure 5.26. This decrease of π_s with decreasing r_H/r_T is compensated by an increase of mass flow, also shown in figure 5.26 for sea-level static conditions.

This implies that for a compressor with fixed overall pressure ratio and fixed mass flow there is a choice between compressors of small diameter with many stages and compressors of larger diameter with fewer stages. If we assume that the compressor weight is proportional to the frontal area and to the number of stages n, and that W_c is proportional to nr_T^2, then, since for a fixed overall pressure ratio $\pi_c = \pi_s^n$ or $n = \ln\pi_c/\ln\pi_s$, we have W_c proportional to $1/[1 - (r_H/r_T)^2]\ln\pi_s$. This function is plotted in figure 5.26. For this stage the lightest compressor would have $r_H/r_T \approx 0.45$, but the minimum is very broad, so that variations of r_H/r_T from 0.3 to 0.6 would lead to only small variations in compressor weight.

5.3.2 Pressure Ratio and Efficiency vs. Mach Number

Increasing the tangential Mach number of the rotor increases the temperature rise of the stage roughly as M_T^2 if the velocity triangles are kept similar (if $\omega r_T/w_a$ is kept constant), but it is more realistic to think of holding M_a constant while increasing M_T. When this is done, $\alpha = \omega r_T/w_a$ increases and M_b' increases. To see this trend consider a stage of hub/tip ratio near unity in which D_r and D_s are held constant as M_T is increased while M_a

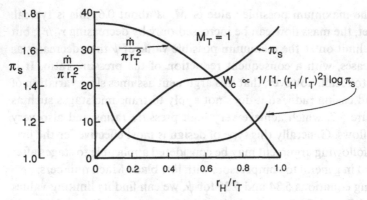

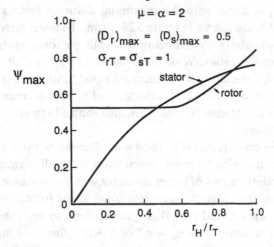

Figure 5.26
Variation of maximum work factor ψ_{max} as limited by diffusion factor, with hub/tip radius
ratio for a stage without inlet guide vanes, and the corresponding variations of pressure ratio,
mass flow per unit frontal area, and compressor weight.

is held constant. Solving equation 5.34 gives ψ as a function of M_T. With this value of ψ the solidity required for the stator to have the specified D_s is found from equation 5.35. The loss factors $\bar{\omega}_b'$ and $\bar{\omega}_c$ and the relative Mach numbers M_b' and M_c then follow, as do the efficiency and the pressure ratio.

Such results are plotted in figure 5.27 for a stage without inlet guide vanes. The pressure ratio varies from $\pi_s = 1.25$ at $M_T = 0.5$ ($\alpha = 1$) to nearly 4 at $M_T = 2$ ($\alpha = 4$). The efficiency varies from 0.976 at $M_T = 0.8$ to 0.84 at $M_T = 2$; the rapid decrease (increase in $1 - \eta_s$) is due to shock losses in the rotor that occur as M_b' rises above unity, reaching $M_b' \approx 2$ at $M_T = 2$. Although M_c is always well below unity, the stator solidity required to hold D_s to 0.5 becomes as large as 5 at the high rotor speeds, and this would pose a serious problem of flow blockage.

This calculation exaggerates the variation of η_s with M_T because the peak value of η_s would be lowered by hub and casing losses as discussed above. Nevertheless, it does give the right trend, and it demonstrates that the best blade Mach number depends on the relative importance of efficiency and stage pressure rise (or weight reduction) in a given application. For a long-range cruise application where fuel weight is more important than engine weight, the premium on compressor efficiency would drive M_T toward 1 or perhaps less, whereas a low-pressure-ratio turbo-ramjet engine might optimize at high M_T because the efficiency of the compressor is less important than the engine weight.

5.3.3 The Effect of Inlet Guide Vanes

Some of the increase in stage pressure ratio with increasing M_T can be had without the penalty due to shock losses by using inlet guide vanes to reduce the rotor relative Mach number as M_T is increased. The effect can be seen by holding α constant while increasing $\mu = \omega r_T/w_a$. This has been done in figure 5.28, where all other parameters are as in figure 5.27. The points marked by circles are identical in these two figures.

Note that $1 - \eta_s$ actually declines slightly with increasing M_T until M_c exceeds unity; then it begins to rise rapidly because of shock losses in the stator (assumed to be those for a normal shock at M_c). For $\pi_s \approx 2.4$, the predicted η_s for the stage with guide vanes is 0.96 at $M_T = 1.8$; for the one without guide vanes, $\eta_s \approx 0.92$ at $M_T = 1.45$. The decision to use inlet guide vanes or not hinges on the importance of efficiency and the advan-

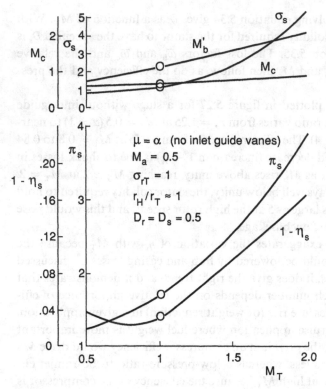

Figure 5.27
Variation of pressure ratio and stage efficiency (accounting only for cascade and shock losses) with blade tangential Mach number for a stage without inlet guide vanes and with hub/tip ratio near unity, with corresponding relative Mach numbers and stator solidity.

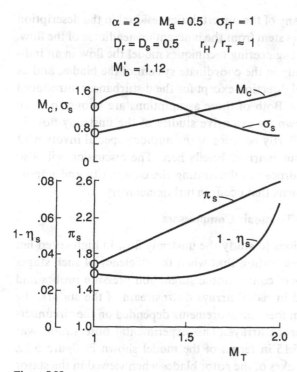

Figure 5.28
Effect of inlet guide vanes on the pressure ratio and efficiency of a compressor stage with fixed rotor relative Mach number, the rotor tangential Mach number increasing with inlet guide vane turning.

tages or disadvantages of high blade speed. The turbine can be a controlling factor in this choice.

5.4 Details of Flow in Transonic Compressors

A great deal of effort has gone into understanding the details of the flow in this type of rotor, including both experimental and computational work. The computations can now be done in two dimensions with full account for viscous effects, and the results seem to reproduce most of the structure of the flowfield as observed experimentally. This work is thoroughly reviewed in reference 5.1. Unfortunately, the conclusion is that the modeling is still not adequate for loss prediction, although the computations are very useful guides to the empirical development process.

It seems likely that many of the uncertainties residual in the description of the flow in compressors stem from the inherent unsteadiness of the flow. Most if not all practical engineering techniques model the flow in an individual blade row as steady in the coordinate system of the blades, and as uniform in the tangential direction except for the disturbances introduced by the blades themselves. Both of these assumptions are wrong to some degree, as has been shown by extensive studies of the unsteady flow in transonic compressors. Partly because of the author's special involvment in this work, it will be summarized briefly here. The discussion will also serve the pedagogical purpose of illustrating the complexity and importance of the flow phenomena that occur in turbomachinery.

5.4.1 Unsteady Flow in Transonic Compressors

One of the first motivations to study the unsteady flow in high-speed fan stages came from the observation that when the efficiency of such stages was measured by means of conventional stagnation pressure probes and thermocouples arranged in radial arrays downstream of the stators, the efficiency computed from these measurements depended on the circumferential location of the probe arrays. This experimental observation was explained in reference 5.15 in terms of the model shown in figure 5.29. Briefly, the fluid in the wakes of the rotor blades when viewed in the stator coordinates has a velocity normal to the main stream velocity, which carries it toward the pressure surface of the stators. Since this wake fluid has different stagnation temperature and pressure than the main flow when viewed in stator coordinates, its collection on the passage side near the stator pressure surface results in the fluid properties there being different than those averaged across the stator passage. Depending on their location circumferentially, pressure probes and thermocouples will then yield values different from the circumferential averages. Such effects are now accounted for by sampling at several positions behind the stator gap. This form of unsteadiness is driven by the blade-passing, and clearly is intrinsic to compressors. Another form of unsteadiness is due to the instability of the flow within the blade row. The detailed study of such flows was made possible in the 1970s by two then-new experimental techniques: the miniature silicon diaphragm pressure transducer and the laser-doppler velocimeter (LDV). The small pressure transducers allowed the construction of probes which were capable of simultaneously measuring stagnation pressure, flow Mach number, and direction, with fast enough time response to

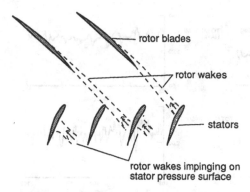

Figure 5.29
Unsteady rotor-stator interaction leading to circumferential nonuniformity, showing rotor wake fluid transported toward stator pressure surface. (Adapted from reference 5.15.)

determine these quantities behind the rotor as functions of time. Such work was first reported in references 5.16 and 5.17. Some of the results obtained for the rotor of figure 5.7 and reported in reference 5.18 are shown in figure 5.30.

The salient features of these results were the following: First, the flow downstream of the rotor is not periodic with blade passing, as it should be if the flow structure were the same on each blade; rather, it exhibits large variations from one blade passing to the next, so the flow is unsteady in the rotor coordinate system. Second, the variations in the three Mach number components (radial, axial, and tangential) and in the static and stagnation pressure were found to be much larger than would be consistent with conventional models of the rotor outflow, which assumed steady flow in the rotor coordinates. Continued work ultimately showed (reference 5.19) that these observations are consistent with the shedding of strong vortex streets, i.e. periodic arrays of vortices of alternating signs, from the blade wakes. The apparent randomness of the measurements results from the probes sampling the vortices as they are convected downstream, and the very strong tangential and radial velocity fluctuations result from the swirl in the vortices and the radial transport along their axes, respectively.

These results were confirmed by time-resolved LDV measurements (reference 5.20). Early LDV measurements of the flow in rotors had been conducted on the assumption that the flow in the rotor was steady, and the measurements were therefore done by repeated observations synchronized

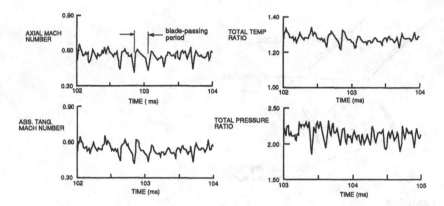

Figure 5.30
Flow variables measured downstream of the high-performance transonic compressor rotor of
figure 5.7. (Adapted from reference 5.18.)

to the rotor position, so that the velocity map obtained was in fact a time-averaged one, even though each individual velocity measurement gave the velocity at that instant.

One important consequence of this understanding is that the distribution of losses shown in figure 5.18 may be exlained by radial transport in the blade wakes. Reference 5.21 showed that the Karman vortex streets in the rotor wakes of at least some high-performance transonic compressors generate very strong flow velocities along their axes. The wake fluid from the blades tends to be concentrated in these vortical wakes, and hence is transported in the radial direction. Thus, losses which are generated on the blades at one radius may appear downstream at other radii. Although there is as yet no proof, it seems probable to the author that mechanisms such as this may in large part explain the radial distribution of losses shown in figure 5.18.

This view is by no means universal, however. The authors of reference 5.22 propose a scheme for prediction of the losses in supersonic blading that uses essentially a channel-flow argument, the channel being bounded by the streamlines in the cascade plane, and by the meridional streamsurfaces with constant spacing so that the flow is modeled as two-dimensional. The argument is that the losses in this channel are controlled by the constraints of conservation of mass flow, momentum, and energy, much as the loss in supersonic flow through a constant-area channel is constrained to be essentially that of a normal shock at the entrance Mach number. In this latter case the actual flow must satisfy the laws of conservation of mass, momentum, and energy, just as the flow through a normal shock does, because the flow area is constant and the effects of axial shear are minimal. It is not so clear that effects of radial changes in the streamlines should be negligible in the supersonic rotor, but according to reference 5.22 the scheme correlates data from a number of rotors.

It seems that in the absence of a complete three-dimensional computational capability the methods for analysis of supersonic rotor flow fields will continue to be more nearly correlative than predictive. Such methods are very useful when supported by an extensive data base, however, and will form the basis of design systems for some time to come.

Techniques for predicting the losses in supersonic blading are a subject of intensive study as of this writing, and the serious reader is advised to consult the literature for new developments.

5.5 Stage Performance: Corrected Parameters

From the Euler equation in the form of equation 5.10, it was argued that the temperature ratio across a stage should depend on the tangential Mach number of the rotor M_T, on the axial flow Mach number M_b (or M_a), and on the flow geometry as controlled by the blade angles, so we should be able to correlate, say, τ_s as a function of M_T and M_a. To obtain the pressure ratio, the efficiency must be introduced; however, if η_s is a function only of M_T, M_a, and flow geometry, then $\pi_s = \pi_s(M_a, M_T)$. In practice, stage performance can be correlated in this way as long as the Reynolds number is sufficiently large (greater than about 3×10^5 based on blade chord).

A different set of parameters is preferred, although they are equivalent to M_a and M_T. They are the corrected speed, defined as

$$\frac{N}{\sqrt{\theta}} \equiv \frac{N}{\sqrt{T_{t2}/T_r}} \text{ rpm,}$$

and the corrected weight flow, defined as

$$\frac{W\sqrt{\theta}}{\delta} \equiv \frac{W\sqrt{T_{t2}/T_r}}{p_{t2}/p_r} \text{ lb/sec,}$$

where T_r and p_r are the sea-level standard atmospheric pressure and temperature, 2116.2 lb/ft^2 and 519°R. That $N/\sqrt{\theta}$ is equivalent to M_T for a given tip diameter is obvious. By expressing W, θ, and δ in terms of the axial Mach number and stagnation conditions, we can show that

$$\frac{W_2 \sqrt{\theta_2}}{A_2 \delta_2} = \frac{c(\rho_2 u_2)}{\rho_{t2} a_{t2}} = f(M_2),$$

where for air $c = 85.3 \text{ lb/ft}^2\text{sec}$ and A_2 is to be interpreted as the flow area at the station 2 where $\rho_2 u_2$ is given.

The performance of two high-performance fan stages is shown in this format in figure 5.31 (references 5.23 and 5.24). They have no inlet guide vanes. One has a low tangential Mach number (0.96) to minimize noise. The other has supersonic tip speed and a considerably larger pressure ratio. Both have high axial Mach numbers.

Several features of these maps should be noted. As the weight flow is reduced at fixed speed, the pressure ratio rises until it reaches a limiting value, indicated by the "stall line," where the flow tends to become unsteady. At low corrected speeds the pressure ratio varies little with weight flow for both fans. This can be understood from the fact that, because these stages have no inlet guide vanes, the flow is axial both upstream of the rotor and downstream of the stator—that is, in equation 5.10 β_b is zero. If the turning in the rotor and in the stator were equal, then β_c' would also be zero, and the temperature ratio would be independent of flow. Actually $\beta_c' > 0$ except near the hub of such a fan, and because $W\sqrt{\theta}/\delta$ corresponds to M_a (the axial Mach number), increasing $W\sqrt{\theta}/\delta$ lowers τ_s, hence π_s.

But as $N/\sqrt{\theta}$ becomes larger, the constant-speed characteristics of both fans become steeper, with very little change in $W\sqrt{\theta}/\delta$ as the pressure ratio is changed. In the case of the low-tip-speed fan, this happens because the axial Mach number in the rotor passages is approaching unity as $N/\sqrt{\theta}$ exceeds the design value, so that a drop in downstream pressure does not increase the mass flow. In the case of the high-tip-speed fan, the rotor relative Mach number is greater than unity over a large part of the annulus at design speed and above, and this implies that β_b' is constant, so $W\sqrt{\theta}/\delta$ should depend only on $N/\sqrt{\theta}$.

5.6 Multi-Stage Compressors

High-pressure-ratio multi-stage compressors have been developed by placing several stages, each composed of a rotor and a stator, in series on the same shaft. The requirements that the successive stages have the same

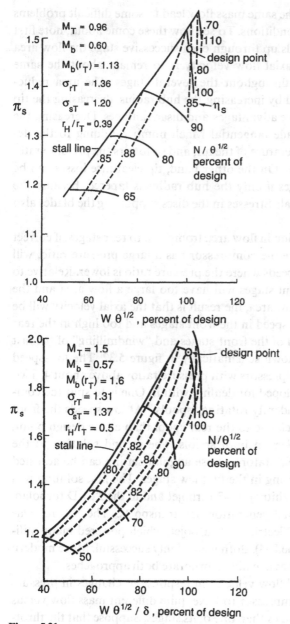

Figure 5.31
Performance maps for modern, highly loaded fan stages without inlet guide vanes: (top) with subsonic tangential Mach number (from reference 5.23); (bottom) with tangential Mach number of 1.5 (from reference 5.24).

rotative speed and pass the same mass flow lead to some difficult problems at off-design operating conditions. To see how these come about, note first that as the pressure builds up through each successive stage the flow area must be reduced if the axial flow velocity is to remain about the same fraction of blade speed throughout the several stages. The area reduction can be accomplished by increasing the hub radius, decreasing the tip radius, or both. Both have advantages and disadvantages. Decreasing the tip radius lowers the blade tangential Mach number, adding to the decrease due to rising temperature of the air, and so lowers the pressure ratio of the downstream stages. On the other hand, tip clearance losses can be excessive in the last stages if only the hub radius is tapered, because the blade heights become small. Stresses in the discs supporting the blades also increase.

In any case, the reduction in flow area from front to rear stages, if correct at the design point where the compressor has a large pressure ratio, will not be correct at lower speeds where the pressure ratio is lower. Relative to the center stages, the front stages will have too large a flow area and the rear stages too small a flow area; the result is that the axial velocity will be too low relative to blade speed in the front stages and too high in the rear stages. This leads to stall of the front stages and "windmilling" of the last ones. The situation is shown schematically in figure 5.32. The low-speed problem is serious in compressors with pressure ratios above about 4. Two methods have been developed for dealing with it. One is to split the compressor into two independently rotating "spools." At low speeds the front spool then runs slower relative to the rear spool than at the design point, and this eases the problem in both spools. The second solution to the problem is to use variable stators whose angular setting can be adjusted while the engine is operating in the first few stages. The first solution was adopted in the Pratt & Whitney JT-3 turbojet and in the JT3-D turbofan, which powered most first-generation jet transports. The second was adopted in the General Electric J-79 turbojet, which powered many military aircraft (including the F-4). Both were highly successful. More modern high-pressure-ratio engines usually incorporate both approaches.

Variations of the axial flow velocity in response to changes in pressure cause the multi-stage compressor to have quite different mass flow versus pressure ratio characteristics than one of its stages. Suppose that the throttle is opened with the compressor at design speed. The first effect is to lower the pressure and increase the axial velocity in the last stage, as shown at the

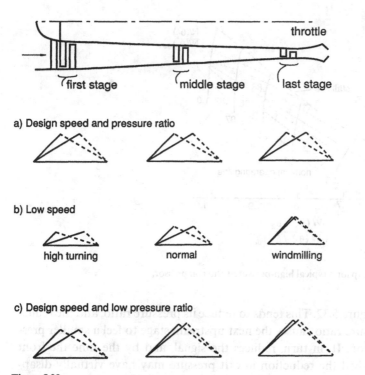

a) Design speed and pressure ratio

b) Low speed

high turning   normal   windmilling

c) Design speed and low pressure ratio

Figure 5.32
Off-design behavior of a multi-stage, high-pressure-ratio compressor, showing velocity
triangles for (a) design speed and pressure ratio, with similar angles and loading in all stages;
(b) low-speed (low-pressure-ratio) operation, with front stages highly loaded and rear stages
windmilling (negatively loaded); and (c) design speed, but low pressure ratio, with rear stages
unloaded but front stage nearly at design condition.

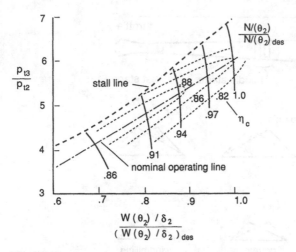

Figure 5.33
Performance map for a typical high-pressure-ratio compressor.

bottom of figure 5.32. This tends to reduce its pressure ratio, and the reduction in pressure ratio causes the next upstream stage to feel a smaller pressure reduction. It, in turn, reduces the signal, and by the time the front stage is reached the reduction in exit pressure may have virtually disappeared. As a result its flow velocity is unchanged, and the compressor mass flow is virtually unchanged by the drop in exit pressure. This leads to very steep constant speed lines for high-pressure-ratio compressors, similar to that of the single transonic stage of figure 5.20. Note, however, that the steepness of the characteristic has quite different origins in the two cases. A typical map for a modern high-pressure-ratio compressor is shown in figure 5.33.

5.7 Compressor and Compression System Stability

Repeated reference has been made to stall in the discussion of compressors, and a "stall line" has been included in the compressor and fan maps (this being defined by the limit in pressure ratio, for each compressor speed, beyond which the flow in the compressor breaks down in some way so as to make the system no longer operable). Some of the mechanisms for breakdown have been implied in the discussions of blade section stall,

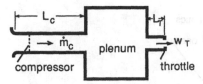

Figure 5.34
Schematic diagram of compression system, consisting of a compressor with enclosing duct, a plenum, and an exhaust duct with throttle.

which may be thought to occur for a diffusion factor somewhat above 0.6, or casing stall, which may occur when the pressure coefficient demanded of the inter-blade passage near the tip exceeds a value on the order of 0.6–0.8. But this subject is of such critical importance to the performance of compressors in aircraft engines that it merits a thorough discussion. The presentation here will draw heavily on that of reference 5.25, but with emphasis on those aspects viewed as the most important for aircraft engines, and in a somewhat simplified form emphasizing the phenomena at the expense of quantitative treatment.

The simplest system that contains all the critical elements is shown schematically in figure 5.34. It consists of the compressor, a chamber into which the compressor discharges, and a nozzle or throttle through which the flow leaves the chamber. In application to the core compressor of an engine, the chamber represents the combustion chamber, while the throttle represents the first-stage turbine nozzles. But for purposes of the initial discussion here it will be assumed that the compressor draws air from the atmosphere, and that the throttle discharges it back to ambient conditions.

As the title of this section suggests, there are two aspects to the overall stability: the stability of the flow in the compressor itself, and the stability of the system of which the compressor is a part. This distinction may be understood in simple terms by reference to figure 5.35, which plots a compressor characteristic as pressure rise coefficient versus axial velocity divided by the blade speed and, on the same coordinates, the pressure difference required to pass the same mass flow through the throttle. For low pressure ratios this characteristic is quadratic; at pressure ratios above about 2, where the throttle is choked, it becomes linear. The compressor characteristic shows schematically the pressure rise which the compressor will produce in three regions, the flow being quite different in each:

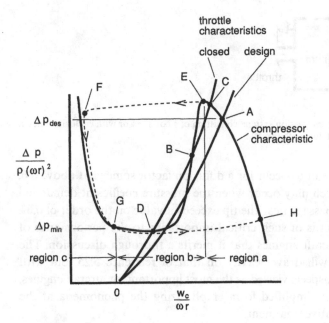

Figure 5.35
Compressor characteristic, and throttle characteristics for two different throttle areas.

(a) The normal operating region of mass flow, where the flow is reasonably uniform around the annulus, where the blades and end walls operate without massive separation and where the efficiency is high.

(b) A region of rotating stall, where the flow breaks into cells, so that some parts of the annulus have nearly normal flow while others have nearly zero flow, and the whole pattern rotates at an angular velocity which is a fraction of rotor speed. The efficiency in this region is quite low. In spite of the time variation in the rotor, the total flow through the compressor as a whole can be steady, so it is meaningful to represent this operation on a steady-state characteristic, as in figure 5.35.

(c) A region of backflow, where the blades may be massively separated and rotating stall may or may not exist.

Consider first the stability of the system as a whole, from a quasi-steady viewpoint. What follows if the mass flow is perturbed slightly from an operating point? If the system is operating at point A, and the mass flow is perturbed positively, the compressor produces a lower pressure, which results in a reduced mass flow through the nozzle, thus correcting the positive perturbation of mass flow. Hence this point is stable to such disturbances. If the throttle area were reset to attempt operation at point B, the same argument shows that the mass flow disturbance would grow rather than attenuate, and the system would tend to move to the adjacent stable point C, whereas if the mass flow disturbance were negative it would tend to move to D. Thus, at this level of argument we can say that the system is stable if at the operating point the slope of the throttle characteristic is greater than that of the compressor characteristic. But as we shall see, the situation can be considerably more complex than this simple quasi-steady argument suggests, in part because of the dynamics of the system and in part because of the coupling between the system's instability and the stability of flow in the compressor itself. Under some conditions the system can go into violent oscillations known as surge, rather than settling to a new stable point, even though the slope of the throttle curve is less than that of the compressor.

The stability of the *compressor flow* can be understood at a comparable level by imagining the flow in the annulus of the compressor to be divided into multiple parallel streams (or compressors), each of which has an identical characteristic such as that in figure 5.35, all drawing from a common plenum and discharging to another common plenum so that their inlet stagnation and outlet static pressures are equal. Suppose first that the compressor is operating at a point such as A, with a negative slope. If the mass flow in one of the parallel streams is perturbed positively, and that in another stream negatively, so that the total mass flow is unchanged, what happens to the perturbations in each stream? In the stream with increased mass flow, the compressor delivers a lower pressure rise, which for the fixed inlet and delivery pressures will result in a deceleration of the flow, correcting the initial excess mass flow and returning this stream toward its stable operating point. Similarly, in the stream with a negative mass flow perturbation, the higher pressure rise will accelerate the flow, again correcting the initial disturbance. Therefore, *at point A with negative slope the flow is stable with respect to interchange of mass flow between adjacent streams within the compressor*. The same argument shows that at point B where the slope is positive, such perturbations will grow, while at point E, where the slope is zero, the compressor is neutrally stable to exchange of mass flow between adjacent streams. Therefore, *according to this argument, as the mass flow is reduced by throttling, the flow in the compressor*

first becomes unstable at the point of zero slope. Once initiated, the instability develops into what is termed *rotating stall.*

In fact, the instability begins when the slope is still slightly negative. The
reasons are not completely understood, but one explanation is that there is
always a non-zero disturbance level in a real compressor, so that the perturbations from the operating point are not actually infinitesimal. Suppose
that at an operating point such as C the characteristic of the compressor
can be represented as

$$\psi = a + b\varphi + c\varphi^2,$$

where ψ is the dimensionless pressure rise and φ is the dimensionless flow
coefficient. Suppose that one of the multiple streams envisioned previously
suffers a negative flow perturbation, so that the perturbation of ψ is

$$\delta\psi = b\delta\varphi + c(\delta\varphi)^2.$$

If $\delta\psi$ is negative, the flow in this streamtube is unstable in that the flow will
continue to decrease. The condition for neutral stability is then that

$$b = -c\delta\varphi.$$

The curvature c is negative near the peak of the characteristic, so a negative
flow perturbation will result in instability for a negative value of b, the local
slope.

5.7.1 Rotating Stall

Once the mass flow exchange instability arises, it seems always to develop
until the mass flow in the regions of low mass flow is nearly zero, while that
in the high mass flow regions is in the normal unstalled operating range, so
the flow consists of one or more "cells" of stalled flow embedded in the
unstalled flow. One of the first studies of this phenomenon was by Emmons
et al. (reference 5.26). For single-stage fans or compressors, the cells may
initially develop in only part of the blade span, and there may be one or
more, as shown schematically in figure 5.36. With deeper throttling, a single stall cell covering the full span develops, and as the throttle is further
closed it expands to fill more of the annulus. In multi-stage compressors,
the rotating stall cell nearly always covers the full span.

The reason the stall propagates along the blade row may be seen from
the diagram at the top of figure 5.36. When the flow in a particular passage

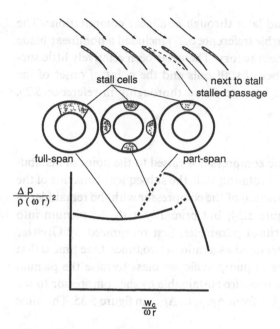

Figure 5.36
Typical rotating stall structures at different operating points of an axial compressor, and
mechanism of propagation of the stall cells (top).

stalls, it partially blocks that passage, and the flow is diverted to the neighboring passages. As indicated, this results in an increase in incidence of the
next blade in the direction of stagger and a decrease in incidence in the
adjacent blade in the opposite direction. The result is to move the stalled
region in the direction of stagger. Experience has shown that the propagation is at a fraction of the velocity tangential to the blade row ranging from
0.4 to 0.6. If the blade row is a rotor, this means that the stall is seen in
stationary coordinates to rotate in the direction of the rotor, but at a
fraction of its speed ranging from 0.4 to 0.6. A bit of cogitation should
convince the reader that simultaneous propagation of the stall in a stator
downstream of the rotor can be consistent with this general picture.

Rotating stall is a critical phenomenon in compressors, because it marks
the limit of stability as the pressure rise is increased and because its unsteady flow can excite severe vibrations in the blading. It has been studied
extensively both experimentally and theoretically. In general there has
been good success in predicting the rotational frequency of the stall, initial-

ly via linearized theories and later through nonlinear computations. The theoretical treatment of Marble (reference 5.27) included a nonlinear blade behavior in an otherwise linear theory. There has been relatively little success in predicting the number of stall cells and the point of onset of the instability. This subject has been reviewed thoroughly in reference 5.25, and in reference 5.1.

5.7.2 Surge

When the pressure rise of the compressor is raised to the point of instability, as marked by the onset of rotating stall, the subsequent behavior of the system depends on the interaction of the compressor with the remainder of the system indicated in figure 5.34, but primarily with the plenum into which it discharges. The critical parameter, first recognized by Greitzer (reference 5.28), may be understood as a ratio of two times. One time is that required for the compressor to pump sufficient mass to raise the plenum pressure from the minimum pressure sustainable by the compressor to the normal operating pressure, i.e., from Δp_{min} to Δp_{des} in figure 5.35. This time we may estimate as

$$\tau_{charge} = \frac{(\Delta p/RT)V_p}{\text{Compressor mass flow}}.$$

The other time is the flow time through the compressor, which we estimate as

$$\tau_{flow} = \frac{\rho V_c}{\text{Compressor mass flow}}.$$

The ratio of these is then

$$T = \frac{\tau_{charge}}{\tau_{flow}} = \frac{(\Delta p/\rho)}{RT} \frac{V_p}{V_c}. \tag{5.40}$$

A detailed study of the behavior of the system in figure 5.34 by Greitzer (reference 5.28), which will be outlined below, showed that the value of a parameter

$$B = \frac{\omega r}{2a} \sqrt{\frac{V_p}{V_c}} \tag{5.41}$$

largely determines whether, upon onset of instability, the system will settle into a stable rotating stall or enter a surge cycle. Since for a given compressor the pressure rise is proportional to $\rho(\omega r)^2$, it is evident that the parameter T arrived at above is proportional to B^2.

Before embarking on an outline of reference 5.28, it will be helpful to see how this ratio of times influences the development of the instability. Suppose that the compressor is operating at point C in figure 5.35, and the instability develops, initially as an incipient rotating stall. This leads to a decrease in the pressure rise which the compressor can support. Consider two extreme cases: $T \gg 1$ and $T \ll 1$.

When $T \gg 1$, the mass stored in the plenum is so large that as the flow in the compressor collapses into rotating stall, the pressure at its outlet is held close to the initial value, while the flow in the compressor is slowed and perhaps reversed, so the system rapidly transitions in a time of the order of τ_{flow} to a point such as point F on figure 5.35. The plenum discharges, over a time on the order of τ_{charge}, to the pressure which the compressor can support in rotating stall, say at point G. Then in a time like τ_{flow} the flow increases to a point on the unstalled characteristic, such as H, and the plenum is recharged in the time τ_{charge}. This cycle is likely to be self-sustaining unless some remedial action is taken, and repeated surges can be very hard on the engine. However, recovery can be effected, for example, by lowering the fuel flow, which has the same effect as opening the throttle valve in figure 5.34, so that when the system returns to the unstalled characteristic it is no longer in the unstable region. Unfortunately this is not true of the opposite limiting case.

When $T \ll 1$, the time required to discharge the plenum is so small that the pressure in the plenum can follow the delivery capability of the compressor as the rotating stall develops, and the result is that the compression system can change to a new steady state in which the compressor operates steadily in rotating stall at a point such as D on figure 5.35. In the model system of figure 5.34, this point lies on the same throttle characteristic as the original point from which the system departed. In an aircraft engine the stable rotating stall point is influenced by the combustion chamber, so it may not be as simply related to the original point.

5.7.3   Analysis and Experimental Study of Compression System Stability

In reference 5.28 the compression system diagrammed in figure 5.34 is modeled by a system of first-order differential equations, as below.

• For the compressor,

$$\frac{L_c}{A_c}\frac{d\dot{m}_c}{dt} = C\left(\frac{w_c}{\omega r}, t\right) - (p_p - p_0),\tag{5.42}$$

where C is the compressor characteristic, such as that in figure 5.35, shown explicitly as a function of both flow coefficient and time to emphasize that it is subject to relaxation toward its steady state value as the rotating stall pattern develops.
• For the throttle,

$$\frac{L_T}{A_T}\frac{d\dot{m}_T}{dt} = (p_p - p_0) - \frac{\rho w_T^2}{2},\tag{5.43}$$

where w_T is the flow velocity in the throttle.
• For the plenum,

$$\frac{\rho V_p}{\gamma p_0}\frac{dp_p}{dt} = \dot{m}_c - \dot{m}_T.\tag{5.44}$$

• For the compressor,

$$\tau\frac{dC}{dt} = C_{\text{steady state}} - C,\tag{5.45}$$

where τ is the time required for the rotating stall pattern in the compressor to relax to the steady form associated with the flow coefficient, typically several compressor rotational times.

Greitzer nondimensionalized the physical quantities, the velocities by the blade speed ωr, the mass flows by $\rho\omega r A_c$, the pressure differences by $\rho(\omega r)^2/2$, and time by the Helmholtz resonance time,

$$\frac{\sqrt{V_p L_c/A_c}}{a_0}.$$

The parameter B then emerges as the dominant one controlling the behavior of the solutions, which were obtained by numerical integration of the set of four first-order equations. It is worth noting that these equations are highly nonlinear, primarily because of the character of C. Results for two values of B are shown in figure 5.37. It was found for the compressor characteristic shown on these figures that the value of B marking the

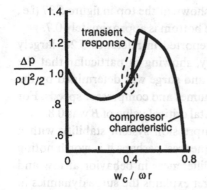

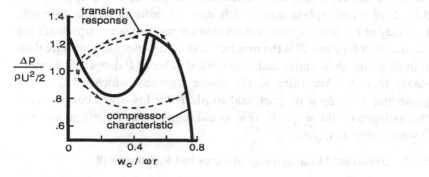

Figure 5.37
Computed rotating stall (top) and surge (bottom) behavior of a compressor for low and moderate values of *B*. (From reference 5.28.)

boundary between the type of behavior shown at the top in figure 5.37 (i.e., rotating stall) and the surge shown at the bottom is approximately 0.7.

A parallel experimental study, also reported in reference 5.28, largely confirmed the predictions of this theory, showing in particular that the boundary between steady rotating stall and surge was determined by the value of B for a wide range of plenum volumes and compressor speeds. For the compressor examined, the experimental critical value of B was 0.8.

Before closing this discussion of compression system stability with a brief discussion of its operational consequences, perhaps it is worth noting that the "physical" explanation of the differences in behavior at low and high T given above is not unique; Greitzer explains the surge dynamics in terms of a balance of pressure and inertial forces, while Cumpsty (reference 5.1) offers other explanations which may be more appealing to some. Cumpsty also offers the observation that for multi-stage compressors the parameter NB (where N is the number of stages) is more appropriate than B itself, although he states that "the critical value of B drops a little more slowly than N." According to the above argument which led to the T parameter, since Δp is proportional to $\rho U^2 N$ for low-speed compressors, the analogue would be $\sqrt{NB}$. This would suggest that the critical value of B should drop as $1/\sqrt{N}$.

5.7.4 Operational Consequences of Surge and Rotating Stall

When surge occurs in an operating engine, it causes at the very least a sudden stoppage of the airflow through the compressor. In civil transports this is evidenced by a loud bang, which is unsettling to the passengers. If repeated enough times, it can do structural damage to the engine inlet or perhaps to the fan or the fan frame. But absent such structural damage, the engine will usually recover and continue to run. In supersonic aircraft the consequences can be more serious, because the blockage due to the surge can generate a shock wave in the inlet which may produce overpressures large enough to damage the structure. The momentary very large loss of thrust due to the surge can also result in loss of control of the aircraft under some critical flight conditions. Therefore it is highly desirable to avoid surge under operational conditions.

It can be equally serious if the compressor settles into a stable rotating stall, because the engine may not recover from such a condition unless it is shut down and restarted. Not only is there a large thrust loss in the stalled operating condition, but the inability of the engine control to cope with the abnormal condition in the compressor may result in serious overheating of the combustor and the turbine. Thus, if the pilot is unaware of the occurrence of the "stagnation stall," and leaves the thrust setting at its pre-stall position, the control may increase the fuel flow in an attempt to maintain the engine speed in spite of the low pressure ratio and low efficiency of the compressor, the result being a severe overtemperature condition in the turbine. This has been a major operational problem with engines in some high-performance fighters, providing a strong incentive to understand when an engine will surge and when it will develop a "nonrecoverable" or "stagnation" stall.

Some aspects of this problem can be described properly only in the context of the interaction between the compressor and the inlet. These are deferred to chapter 8.

5.7.5 Stabilization of Compression Systems

At the time of this writing there is a very active program addressing the possibility of stabilization of compression systems by means of active control, the objective being to enable the compressor to operate stably at higher pressure ratio or lower flow rate than it can without the control.

The centrifugal compressor represents the simplest application of this idea because rotating stall plays only a minor role in its stalling behavior, which is dominated by surge. The surge behavior is rather well represented by the lumped-parameter model shown schematically in figure 5.34 and represented by equations 5.42–5.45.

One approach to the stabilization of systems including centrifugal compressors is to add to the system an element that introduces damping. Gysling et al. (reference 5.29) have demonstrated this approach, adding a variable-volume chamber at the compressor discharge, with a spring-loaded and viscously damped diaphragm. The surge mass flow was lowered by 25% with no noticeable degradation of the steady performance.

The axial-flow compressor presents a somewhat greater challenge, because of the complex behavior discussed in section 5.7. The compressor can go into a stable rotating stall state, or can surge. It seems to be true, however, that rotating stall precedes surge in most axial compressors, so that if rotating stall can be suppressed or delayed to a lower flow condition the surge should also be suppressed. Garnier, Epstein, and Greitzer (reference 5.30) advance the view (supported by data from two low-speed and one high-speed compressors) that rotating stall is itself preceded by a harmonic wave-like precursor disturbance, which rotates at a speed nearly equal to that of the fully developed rotating stall and which can be identified by its propagation speed many rotor revolutions before it develops into a large-amplitude rotating stall. This behavior is described by the theory of Moore and Greitzer (reference 5.31). Epstein (reference 5.32) proposed that the compressor might be stabilized against stall by active suppression of this precursor disturbance, and such stabilization has been demonstrated by Paduano et al. (reference 5.33) in a single-stage low-speed compressor. Day (reference 5.34) has argued that this approach may not be applicable to practical compressors, on the basis that rotating stall is sometimes initiated by spontaneous formation of small stall cells, independently of or without the appearance of the wave precursor.

This is a subject of potentially great practical importance, and one rich with opportunities for exploitation of modern sensing, computation, and control capabilities.

5.8   Centrifugal Compressors

As the name implies, a centrifugal compressor achieves part of the compression process by causing the fluid to move outward in the centrifugal force field produced by the rotation of the impeller. This part of the pressure rise differs from the pressure rise in axial-flow compressor rotors and stators; instead of arising from the exchange of kinetic energy for thermal energy in a diffusion process, it arises from the change in potential energy of the fluid in the centrifugal force field of the rotor. It is therefore less limited by the problems of boundary-layer growth and separation in adverse pressure gradients. Probably for this reason, the centrifugal compressor was the first type to attain a range of pressure ratio and efficiency useful for turbojet engines. It was used in the Von Ohain engine (1939) and the Whittle engine (1941).

An impeller (rotor) for a centrifugal compressor is sketched in figure 5.38. The air enters through the "eye" near the axis, is turned to the radial direction, and is brought to a tangential velocity near that of the rotor by the time it reaches the rotor tip. The essential feature is that all the fluid leaves the rotor at the tip radius, rather than over a range of radii as in the axial-flow compressor. If the flow Mach number were small in the impeller

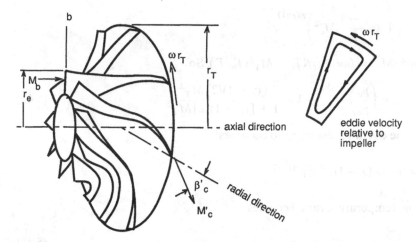

Figure 5.38
The rotor or impeller of a centrifugal compressor, showing the inlet and exit velocities, and at the right, the secondary flow generated in the impeller by the tendency for the flow to preserve zero absolute vorticity.

passages and the air had the tangential velocity of the rotor at all radii as it flowed outward, the pressure gradient in the radial direction would be $dp/dr = \rho\omega^2 r$, and if the flow were isentropic so that $\rho/\rho_b = (p/p_b)^{1/\gamma}$ we would find upon integrating that the static pressure ratio across the rotor would be

$$\left(\frac{p_c}{p_b}\right)^{(\gamma-1)/\gamma} - 1 = \frac{T_c}{T_b} - 1 = \frac{\gamma-1}{2} M_T^2, \tag{5.46}$$

where $M_T^2 = (\omega r_T)^2/\gamma R T_b$ is the square of the tip Mach number based on inlet temperature, as defined for the axial compressor. The impeller should produce this static pressure ratio even when there is very little flow through it. For $M_T = 1$ and $\gamma = 7/5$, for example, $T_c T_b = 1.2$ and $p_c/p_b \approx 1.9$.

The air leaving the impeller has kinetic energy due to its tangential velocity as well as to its small radial velocity. If this kinetic energy can be converted to thermal energy with small losses, a further pressure rise occurs just as in the stator of the axial compressor, but this process is subject to all the diffusion difficulties encountered in the axial-compressor stator and more. Suppose first that the process is isentropic; then the pressure ratio will be

$$\frac{p_d}{p_c} = \left(1 + \frac{\gamma - 1}{2} M_c^2\right)^{\gamma/(\gamma-1)},$$

where $M_c^2 = (\omega r_c)^2 / \gamma R T_c = M_T^2 (T_b/T_c)$. So

$$\frac{T_d}{T_c} - 1 = \left(\frac{p_d}{p_c}\right)^{(\gamma-1)/\gamma} - 1 = \frac{[(\gamma - 1)/2] M_T^2}{1 + [(\gamma - 1)/2] M_T^2} \tag{5.47}$$

and the overall pressure ratio becomes

$$\frac{p_d}{p_b} = [1 + (\gamma - 1) M_T^2]^{\gamma/(\gamma-1)},$$

or the temperature ratio becomes

$$\frac{T_d}{T_b} = 1 + (\gamma - 1) M_T^2. \tag{5.48}$$

Comparing equations 5.46 and 5.48, we see that half of the temperature rise of the stage occurs in the stator, and for high efficiency the static pressure ratio of the stator must equal that of the rotor. This requirement for matching the large pressure ratio of the rotor with an equally high pressure ratio in the stator has limited the efficiency of the centrifugal compressor with radial impeller vanes. It can be reduced to some extent by sweeping the rotor blade tips, as we shall see later.

The advantage of high pressure ratio is offset by the inherently low mass flow capacity per unit of frontal area that results from the radial flow geometry. Because the ratio of inlet flow area to frontal area varies as the square of the ratio of inlet tip radius to diffuser outlet radius, the mass flow capacity is a small fraction of that of an axial-flow compressor of equal diameter unless the diameter of the inlet nearly equals that of the tip; however, if the inlet is nearly equal in diameter to the tip, then the advantage of centrifugal compression is lost. Until recently, low mass flow capacity has limited the use of centrifugal compressors in aircraft engines to small shaft engines such as turboprops and helicopter engines. With increases in cycle pressure ratios, centrifugal stages have found application as the high-pressure compressors in some engines with multiple shafts. The high pressure and high density of the air in the high-pressure compressors cause their flow area to be small relative to that of the inlet stages, so the diameter of the centrifugal stage is not limiting in these applications.

5.8.1 Stage Pressure Ratio and Diffuser Mach Number

The stagnation temperature ratio of the centrifugal stage is given most conveniently by the Euler equation 5.9. If there are no pre-swirl vanes, the inlet tangential velocity is zero, so $\beta_b = 0$. In general the fluid does not leave the impeller exactly radially, so $\tan \beta'_c \neq 0$. Even for radial vanes, there is some slip; that is, $\beta'_c > 0$. If we replace the axial velocity w_c by the radial velocity u'_c to obtain the tangential velocity relative to the impeller at its outlet, equation 5.9 becomes

$$\frac{T_{tc}}{T_{tb}} - 1 = \frac{(\omega r_c)^2}{c_p T_{tb}}\left(1 - \frac{u'_c}{\omega r_c}\tan \beta'_c\right),$$

or, in terms of Mach numbers,

$$\tau_c - 1 = \frac{(\gamma - 1)M_T^2}{1 + [(\gamma - 1)/2]M_b^2}\left(1 - \frac{M'_c}{M_T}\sqrt{1 + \frac{\gamma - 1}{2}M_T^2}\tan \beta'_c\right). \qquad (5.49)$$

It has been assumed that $M'_c = M'_b$, so that T_c/T_b is given by equation 5.46.

The Mach number at the stator entrance is given by

$$M_c^2 = \frac{(\omega r_c - u'_c \tan \beta'_c)^2 + (u'_c)^2}{\gamma R T_c},$$

and this can be written as

$$M_c^2 = \frac{M_T^2}{1 + [(\gamma - 1)/2]M_T^2} + (M'_2)^2 \sec^2 \beta'_c - \frac{2M_T M'_c \tan \beta'_c}{\sqrt{1 + [(\gamma - 1)/2]M_T^2}}. \qquad (5.50)$$

This value of stator-inlet Mach number is plotted for $\beta'_c = 0$, along with the ideal pressure ratio from equation 5.49, in figure 5.39.

For π_c above about 3, M_c is greater than 1; however, it does not exceed 1.5 even at a pressure ratio of 16, so it would seem possible to design efficient centrifugal compressors with pressure ratios of this order. Nevertheless, it appears that diffusion is a serious problem for high-pressure-ratio centrifugal compressors. One way to design around this problem is to use a backward-swept impeller (i.e., $\beta'_c > 0$) and increase the tip speed to achieve the desired pressure ratio. This reduces the diffuser inlet Mach number, as may be seen from figure 5.40. In this design strategy, the limiting factor in the pressure ratio of a centrifugal compressor is the tip Mach number allowed by the materials and by the structural design of the im-

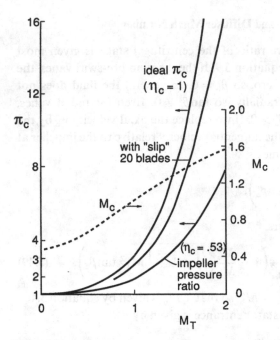

Figure 5.39
Ideal pressure ratio, stator-entrance Mach number, and impeller pressure ratio for a centrifugal compressor impeller with radial vanes as functions of tip Mach number.

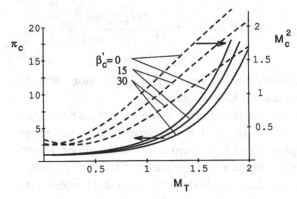

Figure 5.40
Effect of backsweep and tangential Mach number on ideal pressure ratio of centrifugal compressor.

peller. Much of the recent progress has come in these areas, and centrifugal impellers now operate at tip speeds on the order of 1700 ft/sec, with backsweep on the order of 25°.

The impeller static pressure ratio given in equation 5.36 is plotted in figure 5.39 as a reasonable lower limit to π_c. If π_c is given by this relation, then $\eta_c = 0.53$ for $\beta'_c = 0$ and $M_b = 0.5$, as assumed in figure 5.39. This might be taken as a lower limit to the efficiency of the centrifugal compressor. Of course, at shutoff the efficiency approaches zero.

5.8.2 The Impeller

The function of the impeller is conventionally divided into two parts: bringing the air to the angular velocity of the rotor and carrying it radially outward while increasing its angular momentum and static pressure. The first function is performed by the inducer (the inlet portion of the rotor), which is of nearly constant tip diameter. It acts somewhat like an axial-flow rotor without inlet guide vanes and with turning of the flow to the axial direction. If we apply the concept of diffusion factor to this part of the impeller, assuming that the flow velocity normal to the passage cross section is constant through the inducer (i.e., that the axial velocity is constant), we can write the diffusion factor for the inducer in terms of M_b, M_T, and the ratio of "eye" radius to tip radius:

$$D_{\text{ind}} = 1 - \frac{1}{\sqrt{1 + (r_e/r_T)^2(M_T/M_b)^2}} + \frac{(M_T/M_b)(r_e/r_T)}{2\sigma\sqrt{1 + (r_e/r_T)^2(M_T/M_b)^2}}. \quad (5.51)$$

This expression shows that D_{ind} increases as M_T increases for fixed r_e/r_T and vice versa. Since the mass flow capacity is based on tip (or diffuser) frontal area,

$$\frac{\dot{m}}{\pi r_T^2 (\rho u)_b^*} \approx \frac{A^*}{A} M_b \left(\frac{r_e}{r_T}\right)^2, \quad (5.52)$$

there is a conflict between mass flow capacity and pressure ratio. D_{ind} can be lowered somewhat by increasing the solidity, but only to some limiting value given by the first two terms of D_{ind}. The variations of mass flow, divided by the choked mass flow through the tip frontal area, and the required inducer solidity are plotted in figure 5.41 for $M_b = 0.5$, $M_T = 1.5$, and $D_{\text{ind}} = 0.5$. The mass flows are very low; a typical value of this parameter for an axial-flow compressor is on the order of 0.5. Large values of σ_{ind} are required for r_e/r_T greater than about 0.4. Since the curve for σ_{ind} de-

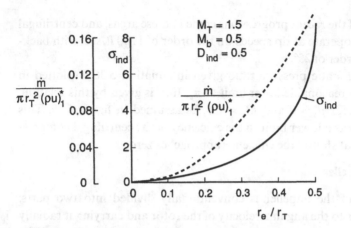

Figure 5.41
Mass flow capacity and inducer solidity required in a centrifugal impeller as functions of ratio of eye radius to tip radius.

pends on the quantity $(r_e/r_T)(M_T/M_b)$, it is clear that for a given σ_{ind} the permissible r_e/r_T decreases proportionately to any increase in M_T/M_b, so the mass flow decreases with increasing M_T.

As the air flows outward through the radial passages, its angular momentum is increased. Clearly, if the blades were very close together the flow would follow them, and for radial vanes the flow would leave radially relative to the rotor ($\beta'_c = 0$). As the spacing increases, the exit velocity inclines away from the direction of rotor motion ($\beta'_c = 0$), so the work done by the impeller decreases. This is called *slip*, and a slip factor is defined as the ratio of actual tangential velocity to $\omega r_c - u'_c \tan\beta'_c$. The slip has been estimated by calculations of incompressible flow through radial passages under the assumption that the flow remains irrotational in absolute coordinates as it passes through the rotor. For this to be so, it must have a vorticity opposite the angular velocity ω of the rotor, as sketched at the right in figure 5.38. The backward velocity relative to the tip leads to slip. Calculations by Stanitz (reference 5.35) gave slip factors of 0.90 for twenty radial blades and 0.93 for thirty blades. The effect of a slip of 0.90 on π_c is shown in figure 5.39.

5.8.3   The Diffuser

Several factors complicate the design of the diffuser. For stage pressure ratios above about 3 it must accept supersonic flow, with M_c as large as 1.4

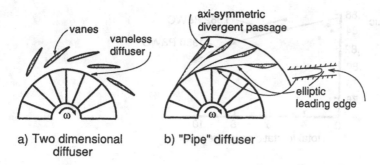

Figure 5.42
Two types of diffusers for centrifugal compressors. Left: a short, vaneless diffuser followed by
a two-dimensional vaned diffuser. Right: a "pipe" diffuser. (Adapted from reference 5.36.)

at $\pi_c \approx 10$. To maximize the mass flow per unit of overall frontal area, the
radial extent of the diffuser beyond the impeller tip should be as small as
possible. But for this last requirement, a "vaneless diffuser" (in which the
swirl velocity is decreased as the flow moves outward with constant angular momentum) could be used. A doubling of the radius would be required
to halve the velocity in such a diffuser, and such a large area is usually
unacceptable. As a compromise, a short vaneless diffuser is often combined
with a vaned two-dimensional diffuser, as sketched in figure 5.42a. Some
high-pressure-ratio compressors have used "pipe" diffusers formed by axisymmetric channels nearly tangential to the rotor tip, as sketched in figure
5.42b. These have given surprisingly good performance, better than that of
equivalent two-dimensional diffusers. Although the reasons are not fully
understood, they may be connected with the "sweep" of the contours presented to the supersonic flow. Sweep has been used for the leading edges of
hypersonic ramjet diffusers.

It is difficult to estimate what efficiencies are ultimately possible with
centrifugal compressors at high pressure ratios. A historical summary from
reference 5.37 is shown in figure 5.43, and a performance map for a typical
centrifugal compressor with design pressure ratio of 5 is shown in figure
5.44.

5.9   Supersonic-Throughflow Fan

At its design point, the supersonic-throughflow fan would operate with
supersonic axial velocity throughout the flow path (that is, the flow would

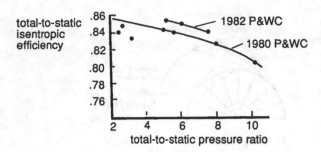

total-to-static isentropic efficiency

total-to-static pressure ratio

Figure 5.43
Historical summary of efficiency vs. stage pressure ratio for centrifugal compressors (from reference 5.36).

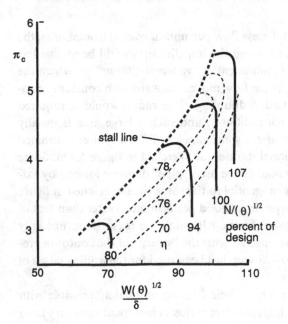

Figure 5.44
Performance map for a typical centrifugal compressor with design pressure ratio of 5.

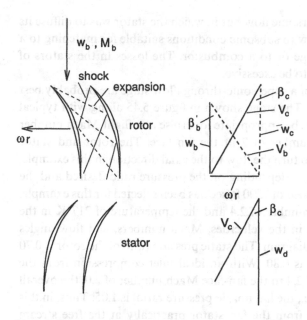

Figure 5.45
Blade shapes and velocity triangles for a supersonic-throughflow fan.

be supersonic in both absolute and rotor-relative coordinates). For a design such as that shown schematically in figure 4.27, the Mach number at the fan face might be 2 or a bit less for a flight Mach number of 2.4. As indicated in figure 5.45, the flow is turned toward the axial direction in the rotor; since the flow area expands, the pressure drops and the flow is accelerated relative to the rotor. The flow is also supersonic relative to the stator, which also turns toward the axis to remove the swirl imparted by the rotor and so also has increasing flow area and dropping pressure. Thus we have a "compressor" which adds energy to the flow while lowering the static pressure. This can result in a useful fan design because of the compression in the inlet ahead of the fan. In principle at least, it affords the advantage of lower diffusion requirements in the blade passages than for a subsonic stage of comparable work; in addition, it offers a reduction of shock losses, since the flow does not have to be diffused to subsonic conditions at any point in the stage.

In this respect the supersonic-throughflow fan differs fundamentally from early supersonic compressors, in which the rotor was conceived to operate with fully supersonic flow but in which the stator was to diffuse its supersonic incident flow to subsonic conditions suitable for matching to a following subsonic stage or to a combustor. The losses in the stators of such stages turned out to be excessive.

The characteristics of a supersonic-throughflow stage are probably best seen from an example. The stage shown in figure 5.45 along with typical velocity triangles might be appropriate for cruise at a flight Mach number of 2.4 with a Mach number of 2 at the fan face. The rotor and stator bladings are assumed to turn the flow to the axial direction in this example; this choice might change, depending on the pressure ratio desired and the blade speed. A blade speed of 1500 ft/sec has been selected for this example. With the flight Mach number of 2.4 and the temperature of 217°K in the stratosphere, it results in the velocities, Mach numbers, and flow angles shown on the velocity diagram. The static pressure ratio of the rotor is 0.70 and that of the stator is 0.80. With an ideal inlet compression from the flight Mach number of 2.4 to the fan-face Mach number of 2.0, the overall static pressure ratio (i.e., the fan nozzle pressure ratio) is 1.05. Thus, in this case the flow emerges from the fan stator practically at the free stream pressure.

The designs of both the rotor and stator for such a stage are controlled by the need to turn the flow through the desired angles with minimum shock losses and with a reasonable degree of uniformity in the outflow. The turning is by a weak leading-edge shock from the pressure surface followed by additional compression waves, and by expansion waves from the suction surface, as indicated schematically in figure 5.45. Since the wave angles are rather low at the incident Mach numbers of 2.46 (rotor) and 3.09 (stator), a solidity on the order of 3 or more is indicated. A simple calculation shows that the diffusion factor is very low, so diffusion in the normal sense is not a limit for such stages.

For the case of axial flow at the rotor and stator exits, the Euler equation 5.10 gives a temperature ratio of

$$\frac{T_{tc}}{T_{tb}} - 1 = \frac{(\gamma - 1)M_b^2 \tan^2(\beta_b')}{1 + \frac{1}{2}(\gamma - 1)M_b^2},$$

where β_b' is the flow angle into the rotor and M_b is the fan-face axial Mach number. This relation is plotted in figure 5.46. For the stage shown in figure 5.45, the temperature ratio is 1.46 and the pressure ratio is 3.76 at the design point.

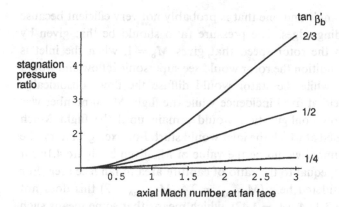

Figure 5.46
Pressure ratio of supersonic-throughflow fan as a function of fan-face axial Mach number and rotor-inlet flow angle.

For low-loss operation, β'_b must be nearly constant so the shocks are weak; it follows that the temperature ratio of the stage would vary with flight Mach number according to the variation of M_b as shown in the figure for constant β'_b. For the design in figure 5.45 the pressure ratio is 1.74 when $M_b = 1$. To maintain such a constant inflow angle, the rotative speed of the fan would have to be adjusted as the air speed at the fan face changes with changing flight speed. Alternatively, the fan speed could be maintained, with some increase in losses due to stronger shocks in the rotor.

A complete discussion of the behavior of such stages and their matching to a core engine is beyond the scope of this book, but a brief discussion (somewhat speculative) of the starting and low-speed behavior seems in order. Evidently there are two independent variables that influence the operating state of the stage: the flight Mach number and the rotative speed. The flight Mach number, by setting the stagnation pressure at the fan face, determines the pressure ratio against which the stage works, while the rotative speed and the flight speed together set the relative Mach number in the rotor. The combination of these two variables admits of a rather complicated array of operating conditions. To simplify the discussion somewhat, it will be assumed here that the rotor always operates at such a speed that its incidence angle β'_b is constant at the design value, so the rotative speed increases as the air velocity at the fan face increases.

Beginning at static conditions, as the rotor speed is increased, the stage should behave pretty much as an ordinary subsonic compressor, albeit one with low stagger angle and one that is probably not very efficient because of the sharp leading edges. The pressure ratio should be that given by figure 5.45 up to the rotor speed that gives $M_b = 1$, when the inlet is choked. At this condition the rotor would see supersonic inflow and would contain a shock, while the stator would diffuse the flow subsonically. With the rotor held at fixed incidence while the flight Mach number was increased, this flow configuration would remain until the flight Mach number was reached at which the inlet would start. For fixed geometry, the starting Mach number would be the value of M_0 given by figure 4.16 for the value of A_0/A_t equal to the ratio of capture area to fan face area. For the example considered here ($M_{0\text{design}} = 2.4$, $M_{2\text{design}} = 2$) this does not occur until $M_0 = 3.3$ ($A_0/A_t = 1.42$), which means that some means such as bleed would be required to start the inlet. If the shock is passed to the rotor face by such means, it should pass into the rotor and take a position determined by the backpressure. As the flight Mach number is increased with a corresponding increase in the ratio of fan face stagnation pressure to ambient pressure, the shock should move downstream and out of the stator, leaving the stage operating with fully supersonic flow.

References

5.1 N. A. Cumpsty, *Compressor Aerodynamics*. Longmans, 1989.

5.2 F. E. Marble, "Three Dimensional Flow in Turbomachines." In *Aerodynamics of Turbines and Compressors* (High Speed Aerodynamics and Jet Propulsion Series, vol. 10). Princeton University Press, 1964.

5.3 W. R. Hawthorne and R. A. Novak, "The Aerodynamics of Turbomachinery." *Annual Reviews of Fluid Mechanics* 1 (1969): 341.

5.4 L. H. Smith, "The Radial Equilibrium Equation of Turbomachinery," *Transactions of ASME, Series A*, 88 (1966).

5.5 D. E. Hobbs and H. D. Weingold, "Development of Controlled Diffusion Airfoils for Multistage Compressor Applications," *ASME Journal of Engineering for Power* 106 (1984): 271–278.

5.6 S. Lieblein, "Experimental Flow in Two-Dimensional Cascades." In Aerodynamic Design of Axial Flow Compressors, NASA SP-36, 1965.

5.7 C. C. Koch, "Stalling Pressure Rise Capability of Axial Flow Compressor Stages." *ASME Journal of Engineering for Power* 103 (1981): 645–656.

5.8 J. E. McCune, *Journal of Aerospace Sciences* 25 (1984): 544.

5.9 A. A. Mikolajczak, A. L. Morris, and B. V. Johnson, Comparison of Performance of Supersonic Blading in Cascade and in Compressor Rotors. ASME paper 70-GT-79.

5.10 J. L. Kerrebrock, "Flow in Transonic Compressors." *AIAA Journal* 19 (1981): 4–19.

5.11 G. R. Miller, G. W. Lewis, and M. J. Hartmann, "Shock Losses in Transonic Rotor Rows." *ASME Journal of Engineering for Power* 83, no 3 (1961): 235–242.

5.12 A. J. Wennerstrom, "Experimental Study of a High-Throughflow Transonic Axial Compressor Stage." *ASME Journal of Engineering for Gas Turbines and Power* 106 (1984): 552–560.

5.13 W. D. Rannie, "The Axial Compressor Stage." In *Aerodynamics of Turbines and Compressors* (High Speed Aerodynamics and Jet Propulsion Series, vol 10). Princeton University Press, 1964.

5.14 G. Souvran and E. D. Klomp, "Experimentally Determined Optimum Geometries for Rectilinear Diffusers with Rectangular, Conical or Annular Cross Section." In *Fluid Mechanics of Internal Flow*, ed. G. Souvran. Elsevier, 1967.

5.15 J. L. Kerrebrock and A. A. Mikolajczak, "Intra-Stator Transport of Rotor Wakes and Its Effect on Compressor Performance." *Journal of Engineering for Power*, October 1970: 359.

5.16 J. L. Kerrebrock, A. H. Epstein, D. M. Haines, and W. T. Thompkins, "The MIT Blowdown Compressor Facility." *Journal of Engineering For Power* 96, no. 4 (1974): 394–405.

5.17 W. T. Thompkins and J. L. Kerrebrock, "Exit Flow From a Transonic Compressor Rotor." In *Unsteady Phenomena in Turbomachinery* (AGARD Conference Proceedings No. 177, 1975).

5.18 W. F. Ng and A. H. Epstein, "Unsteady Losses in Transonic Compressors." *Journal of Engineering for Gas Turbines and Power* 107, no. 2 (1985). See also W. F. Ng, Time Resolved Stagnation Temperature Measurement in a Transonic Compressor Stage, Ph.D. thesis, MIT, 1983.

5.19 A. H. Epstein, J. Gertz, P. R. Owen, and M. B. Giles, "Vortex Shedding in Compressor Blade Wakes." *AIAA Journal of Propulsion and Power* 4, no. 3 (1988): 236–244.

5.20 M. Hathaway, J. Gertz, A. H. Epstein, and A. Strazisar, "Rotor Wake Characteristics of a Transonic Flow Fan." *AIAA Journal* 24, no. 11 (1986).

5.21 P. A. Kotidis and A. E. Epstein, Unsteady Radial Transport in a Transonic Compressor Stage. ASME International Gas Turbine Conference, Brussels, 1990. See also MIT Gas Turbine Laboratory Report 199, 1989.

5.22 C. Freeman and N. A. Cumpsty, A Method for Prediction of Supersonic Compressor Blade Performance. ASME Gas Turbine Conference and Exposition, Toronto, 1989. See also a summary in reference 5.1.

5.23 K. G. Harley and E. A. Burdsall, High Loading Low-Speed Fan Study II: Data and Performance—Unslotted Blades and Vanes." NASA CR 72667 (PWA-3653).

5.24 J. P. Nikkanen and J. D. Brooky, Single Stage Evaluation of Highly Loaded High Mach Number Compressor Stages V. NASA CR 120887 (PWA-4312).

5.25 E. M. Greitzer, "The Stability of Pumping Systems." *ASME Transactions, Journal of Fluids Engineering* 103 (1981): 193–242.

5.26 H. W. Emmons, C. E. Pearson, and H. P. Grant, "Compressor Surge and Stall Propagation." *ASME Transactions* 77 (1955): 455–469.

5.27 F. E. Marble, "Propagation of Stall in a Compressor Blade Row." *Journal of the Aeronautical Sciences* 22, no. 8 (1955): 541–554.

5.28 E. M. Greitzer, "Surge and Rotating Stall in Axial Flow Compressors." *Engineering for Power* 98, no. 2 (1976): 190.

5.29 D. L. Gysling, J. Dugundji, E. M. Greitzer, and A. H. Epstein, Dynamic Control of Centrifugal Compressor Surge Using Tailored Structures. ASME Paper 90-GT-122, 1990.

5.30 V. H. Garnier, A. H. Epstein, and E. M. Greitzer, Rotating Stall Anticipation and Initiation in Axial Compressors. ASME Paper 90-GT-156, 1990.

5.31 F. K. Moore and E. M. Greitzer, "A Theory of Post-Stall Transients in Axial Compression Systems, Part I and II." *ASME Journal of Engineering for Gas Turbines and Power* 108 (1986): 231–239.

5.32 A. H. Epstein, J. E. Ffowcs-Williams, and E. M. Greitzer, "Active Suppression of Aerodynamic Instabilities in Turbomachines." *Journal of Propulsion and Power* 5, no. 2 (1989): 204–211.

5.33 J. Paduano, A. H. Epstein, L. Valavani, J. P. Longley, E. M. Greitzer, and G. R. Guenette, Active Control of Rotating Stall in a Low Speed Axial Compressor. ASME Paper 91-GT 88, 1991.

5.34 I. J. Day, Active Suppression of Rotating Stall and Surge in Axial Compressors. ASME Paper 91-GT-87, 1991.

5.35 J. D. Stanitz and G. O. Ellis, Two Dimensional Compressible Flow in Centrifugal Compressors with Straight Blades. NACA TN 1932, 1949.

5.36 D. P. Kenny, A Comparison of the Predicted and Measured Performance of High Pressure Ratio Centrifugal Compressor Diffusers. ASME Paper 72-GT-54.

5.37 D. P. Kenny, The History and Future of the Centrifugal Compressor in Aviation Gas Turbines. Society of Automotive Engineers paper SAE/SP-804/602.

5.32 I. H. Hunter and N. A. Cumpsty, "Casing Wall Boundary-Layer Development Through an Isolated Compressor Rotor." *ASME Journal of Engineering for Power* 104 (1982): 805–818.

Problems

5.1 Calculate the stagnation temperature ratio for a compressor rotor with $r_H/r_T \approx 1.0$, $M_T = 1.0$, $\sigma = 1$, and $D = 0.5$, assuming constant axial velocity, no inlet guide vanes, and $M_b = 0.5$. If the polytropic efficiency is 0.9, find the stagnation pressure ratio. If the mass flow rate of this compressor rotor is 100 kg/sec, what power is required to drive it?

5.2 When a gas turbine is used to power a helicopter, two alternatives are (a) to use a power turbine geared to the rotor (a turboshaft drive) and (b) to duct the engine exhaust gas through the rotor to nozzles at the blade tips, the jets driving the rotor. Compare these two schemes from the standpoint of efficiency of utilization of the energy of the gas generator exhaust gas.

1. Write the Euler equation to apply to the tip jet rotor drive.
2. Derive an expression for the specific fuel consumption for the tip jet drive, containing the rotor tip Mach number as a parameter.
3. Compare and plot the specific fuel consumption of the tip jet drive versus the geared turboshaft drive as a function of tip Mach number. Assume that all processes are ideal. For $P/\dot{m}c_p T_0$, you should find

$$\left(\frac{P}{\dot{m}c_p T_0}\right)_{\text{tip jet}} = (\gamma - 1)M_T^2\left(\sqrt{1 + \frac{2W^*}{(\gamma - 1)M_T^2}} - 1\right),$$

where P is the power delivered to the rotor by the tip jet drive, W^* is (as defined in section 2.7) the value of $(P/\dot{m}c_p T_0)_{\text{shaft}}$ for the shaft drive system, and M_T is the rotor tip tangential Mach number. Note that this result implies

$(P/\dot{m}c_p T_0)_{\text{tip jet}} \leq (P/\dot{m}c_p T_0)_{\text{shaft}}/2.$

Why is this so?

5.3 A high-performance transonic compressor stage is to be designed to give an ideal stagnation pressure ratio of 1.50 at sea-level static conditions with a tip tangential speed of 400 m/sec and mass flow of 100 kg/sec. It is to have a hub/tip radius ratio at the rotor inlet of 0.50, inlet guide vanes that produce a flow angle $\beta_a = 30°$ at the tip, and an axial Mach number = 0.5 at the rotor inlet. The result of your "design" should be (a) velocity triangles for the hub, tip, and mid radii; (b) sketches of the cascades at the same three radii, giving the flow angles, blade spacings/chord, and rough blade shapes; and (c) a sketch of the annulus cross section showing the contours of the hub and the casing and the blade locations. You might proceed as follows:

1. From the Euler equation, find the tangential velocity change across the rotor, and hence β_c. From the guide vane outlet angle, get A.
2. Assuming incompressible flow, find the axial velocity profiles at rotor inlet and exit by the method of subsection 5.2.1 (note the utility of figure 5.5) and draw velocity triangles at hub, tip, and midspan.
3. Choose the solidity for rotor and stator so the largest "D factor" for each is about 0.5.
4. Assuming that the axial Mach number is constant at 0.5, sketch the annulus cross section.

5.4 A single-stage compressor produces a maximum pressure ratio of 1.5 with an efficiency $\eta_c = 0.85$ when operating in air at near ambient conditions (300°K and 1 atm). Its mass flow is 100 kg/sec. If the rotative speed is held constant, what maximum pressure ratio and mass flow could it be expected to produce at inlet conditions of 1000°K and 0.5 atm? Would the efficiency change?

5.5 A turbojet engine is to be optimized for vertical takeoff. The requirement is that the engine operate for 5 minutes at full thrust. The optimum engine minimizes the sum of engine weight plus fuel weight for the 5 minutes of operation.
    From the information presented in figures 5.17 and 5.18, estimate the best hub/tip radius ratio, tip tangential Mach number, and pressure ratio for the compressor for this engine. Consider the following as parameters: π_c, n (number of stages), M_T, and r_H/r_T. Begin by writing expressions for specific impulse and $F/\dot{m}a_0$ from the simplified cycle analysis, but including the effect of η_c. Then estimate the compressor weight. Assume that the engine weight is a constant times the compressor weight. Make any plausible assumptions or estimates that seem necessary.

5.6 For a compressor with overall pressure ratio $\pi_c = 25$, estimate the number of stages required as a function of tangential Mach number M_T in the first stage, and plot this relationship. Use the results given in figure 5.18 for $r_H/r_T = 1.0$. Assume that the tip diameter of all stages is the same, and that the polytropic efficiency is 0.9.

5.7 Multi-stage, high-pressure-ratio compressors often have variable-angle stator blades in the first few stages to help relieve the low-speed operating problems referred to in section 5.5. Sketch the velocity triangle for the inlet stage of such a compressor with inlet guide vanes, and indicate which way the inlet guide vanes and the first-stage stators should be rotated when the compressor goes from full speed to idle.

5.8   With modern materials and methods of stress analysis, it is possible to design centrifugal compressor impellers that have "backward-curved" blades ($\beta'_c > 0$) and yet operate at large tip speeds, giving pressure ratios on the order of 8. Efficiency can be improved by the reduction of M_c that results from the backward curvature. To understand this trend, determine the variation of M_c^2 with β'_c for fixed T_{tc}/T_{tb}. Is there a value of β_c that minimizes M_c^2?

6 Turbines

Much that has been said about compressors applies equally well to turbines, but two factors lead to major differences between turbines and compressors. First, the high gas temperature at a turbine inlet introduces material problems much more serious than those associated with a compressor, and has led to blade cooling in modern aircraft engines. (The high temperature also leads to lower tangential Mach numbers for turbine blades than for compressor blades with the same blade speed, and this eases the aerodynamic problems somewhat.) Second, the pressure falls through the turbine rather than increasing as in the compressor. This dropping pressure thins the boundary layers, reducing separation problems and rendering the aerodynamic design less critical.

The turbine efficiency is less critical to the performance of a turbojet engine than the compressor efficiency (see section 3.7), and because of this and other factors there was considerably less detailed aerodynamic development of the turbine than of the compressor in early work on aircraft engines. The situation has now changed because turbine efficiencies are critical in high-bypass turbofans, and the weight of the turbine can be a major part of the total weight of such engines. The cost of the turbine can be a major part of the total cost of any engine because of the difficult and expensive materials. So there is a great incentive to reduce the number of turbine stages while increasing efficiency.

To clarify the relationship between the turbine and compressor, consider a turbojet engine with equal compressor and turbine diameters. The compressor-turbine power balance requires that $\theta_0(\tau_c - 1) = \theta_t(1 - \tau_t)$. As we saw, for a single compressor stage, $\tau_s - 1 \propto M_T^2$; the constant of proportionality depends on the blading geometry. The same is true for a multistage compressor, as figure 5.33 shows, so we can write $\tau_c - 1 = M_{Tc}^2 f_c$ (geom)n_c, where M_{Tc} is the tangential Mach number of the compressor and n_c is the number of compressor stages. Similarly, $1 - \tau_t = M_{Tt}^2 f_t$(geom)$n_t$, where M_{Tt} is the tangential Mach number of the turbine rotor. The power balance then becomes

$$\theta_0 M_{Tc}^2 f_c(\text{geom})n_c = \theta_t M_{Tt}^2 f_t(\text{geom})n_t.$$

For equal blade speeds $\theta_0 M_{Tc}^2 = \theta_t M_{Tt}^2$, so the effects of higher temperature and lower Mach number on the power capability of the turbine relative to that of the compressor just cancel. If f_c were equal to f_t, the number of compressor and turbine stages could be equal. Actually, because of the better behavior of the boundary layers in the turbine, the blade loading can be higher than in the compressor, and $f_t > f_c$, with the result that $n_t < n_c$. In early axial-flow turbojets the ratio of compressor to turbine stages was as high as 15. It is decreasing as the design of high-work compressor stages improves; a ratio of 4 or 5 is more typical of advanced engines.

Another requirement is that the mass flow of the turbine equal that of the compressor. This sets a minimum to the flow area at the turbine outlet (station 5 of figure 1.4) relative to the compressor inlet area A_2. If we assume that $M_5 = M_2$, the area ratio is simply

$$\frac{A_5}{A_2} = \sqrt{T_{t5}/T_{t2}} \, (p_{t2}/p_{t5});$$

for the ideal turbojet cycle this can be written in terms of π_c, θ_0, and θ_t as

$$\frac{A_5}{A_2} = \frac{\sqrt{\theta_t/\theta_0}}{\pi_c}\left(1 - \frac{\pi_c^{(\gamma-1)/\gamma} - 1}{\theta_t/\theta_0}\right)^{(\gamma+1)/2(\gamma-1)}. \tag{6.1}$$

This result is plotted in figure 6.1. The turbine exit flow area is less than the compressor inlet flow except for very low π_c; but even at $\pi_c = 10$ it is about half the compressor flow area, and that it must be this large has important implications for the stress in the turbine blades and for the stage pressure ratio.

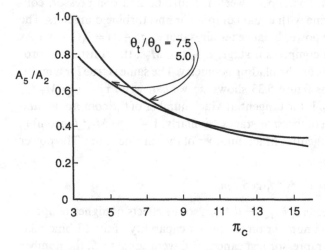

Figure 6.1
Ratio of turbine outlet area to compressor inlet area for a turbojet engine (or gas generator) with equal compressor and turbine diameters, as a function of compressor pressure ratio.

Aircraft engines use axial-flow turbines almost exclusively, as do most large stationary and automotive gas turbines. Radial-flow turbines (the analogue of the centrifugal compressor) are used in some small turboshaft engines, in turbochargers, and in auxiliary power units. Only the axial-flow turbine will be discussed here.

6.1 Turbine Stage Characteristics

A turbine stage consists of a nozzle row and a rotor. Sometimes a down-stream stator is added to enforce zero swirl at the exit. As indicated in figure 6.2, the nozzle vanes turn the flow while dropping the pressure and raising the Mach number. The rotor blades, moving in the same direction as the tangential velocity from the vanes, turn the flow back to remove the angular momentum put in by the vanes. They may simply turn the flow without a further drop in pressure, in which case the turbine is called an *impulse turbine*, or they may further drop the pressure. The ratio of the rotor pressure drop to the total stage pressure drop (or sometimes the ratio of the kinetic energy changes) is called the *degree of reaction*. Velocity diagrams for impulse and 50 percent reaction stages are shown in figure 6.2.

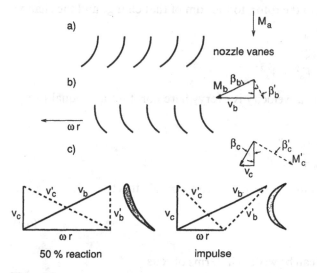

Figure 6.2
Construction of velocity triangles for a turbine stage, with typical composite diagrams for a 50% reaction stage and an impulse stage.

The Euler equation 5.9 applies to the turbine stage as well as to the compressor stage. It is more conveniently written for turbines as

$$1 - \frac{T_{tc}}{T_{tb}} = \frac{(\omega r_b)^2}{c_p T_{tb}} \left[\frac{w_b}{\omega r_b} \tan\beta_b + \left(\frac{r_c}{r_b} \right)^2 \left(\frac{w_c}{\omega r_c} \tan\beta_c' - 1 \right) \right], \tag{6.2}$$

so that both sides are positive. The last term in brackets represents v_c, the tangential velocity at the rotor outlet. Usually this is small at the design condition because any swirl energy left in the exhaust gas as it leaves the engine nozzle is lost. Thus, for the sake of simplicity we will assume $\tan\beta_c' = \omega r_c / w_c$ and write

$$1 - \tau_t = \frac{(\gamma - 1) M_T M_b}{1 + \frac{1}{2}(\gamma - 1) M_b^2} \sin\beta_b, \tag{6.3}$$

where $M_b = V_b / \sqrt{\gamma R T_b}$ is the total Mach number leaving the nozzle vanes. This form of the Euler equation shows that for maximum work per stage (small τ_t) we want large M_T (high blade speed), large M_b, and large turning in the vanes (large $\sin\beta_b$).

6.1.1 Degree of Reaction, Stage Pressure Ratio, Blade Temperature, and Mass Flow

We will define a degree of reaction R as the ratio of kinetic energy change in the rotor, relative to the rotor, to the sum of that change and the change in the vanes. That is,

$$R = \frac{(V_c')^2 - (V_b')^2}{(V_c')^2 - (V_b')^2 + V_b^2 - V_a^2}.$$

Assuming that the axial velocity is everywhere constant and equal to w_a, we get

$$R = \frac{\sec^2\beta_c' - \sec^2\beta_b'}{\sec^2\beta_c' - \sec^2\beta_b' + \sec^2\beta_b - 1}.$$

Using the condition $\tan\beta_c' = \omega r_c / w$, we can simplify this to

$$R = 1 - \frac{M_b \sin\beta_b}{2M_T} \tag{6.4}$$

so that equation 6.3 can be written in terms of R as

$$1 - \tau_t = \frac{(\gamma - 1)M_T^2}{1 + \frac{1}{2}(\gamma - 1)M_b^2} 2(1 - R). \tag{6.5}$$

For a given blade speed, the impulse turbine ($R = 0$) gives the largest temperature drop. On the other hand, because there is little pressure drop in the rotor, viscous effects tend to be worse for low degrees of reaction, so the efficiency is lower for low reaction than for about 50 percent reaction ($R = 0.5$).

Another factor controlled by the degree of reaction is the difference between the turbine inlet stagnation temperature (which we have characterized by θ_t) and the stagnation temperature felt by the rotor blades. This temperature is relevant to the problem of turbine rotor stress. With the stagnation temperature in rotor coordinates denoted as T_{tr}', it is

$$T_{tr}' = T_b[1 + \frac{1}{2}(\gamma - 1)(M_b')^2] = T_{tb}\left(\frac{1 + \frac{1}{2}(\gamma - 1)(M_b')^2}{1 + \frac{1}{2}(\gamma - 1)M_b^2}\right),$$

and since $T_{tb} = T_{ta}$,

$$\frac{T_{tr}'}{T_{ta}} = \frac{1 + \frac{1}{2}(\gamma - 1)M_b^2 \cos^2\beta_b \sec^2\beta_b'}{1 + \frac{1}{2}(\gamma - 1)M_b^2}$$

$$= 1 + \frac{\frac{1}{2}(\gamma - 1)M_T^2}{1 + \frac{1}{2}(\gamma - 1)M_b^2}(4R - 3). \tag{6.6}$$

The second equality results from a little manipulation using equation 6.4; it shows that the rotor temperature increases with increasing R. For example, taking $M_T = 0.5$ and $M_b = 1$ shows that $T_{tr}'/T_{ta} = 1 + 0.042(4R - 3)$, and the difference between an impulse turbine ($R = 0$) and a 50 percent reaction turbine ($R = 0.5$) is 0.08 or about 100°K at $T_{ta} = 1400$°K. In terms of the changes possible with improved materials, this is a large difference; it is equivalent to about 10 years' effort on alloys.

The mass flow per unit of annulus area is also related to the degree of reaction for a given blade speed. To see this we must examine the compressible flow turning in the nozzles. Suppose for simplicity that the vanes have zero thickness and that the radial height of the passages is constant, so that the flow is two-dimensional in first approximation. We can then deduce a connection between β_b and M_b. Continuity of the axial flow requires that $\rho_b V_b \cos\beta_b = \rho_a V_a$. The constancy of T_t gives

$$T_b[1 + \frac{1}{2}(\gamma - 1)M_b^2] = T_a[1 + \frac{1}{2}(\gamma - 1)M_a^2]$$

and for isentropic flow

$$\frac{\rho_b}{\rho_a} = \left(\frac{T_b}{T_a}\right)^{1/(\gamma-1)}.$$

Combining these, we find

$$\cos\beta_b = \frac{M_a}{M_b}\left(\frac{1 + \frac{1}{2}(\gamma-1)M_b^2}{1 + \frac{1}{2}(\gamma-1)M_a^2}\right)^{(\gamma+1)/2(\gamma-1)}.$$

For a given M_a, β_b first increases ($\cos\beta_b$ decreases) as M_b is increased; then β_b decreases again for M_b large (because $(\gamma + 1)/(\gamma - 1) > 1$), so there is a value of M_b for which β_b is largest and hence a value for which $M_b \sin\beta_b$ is largest for a given M_a. Since the mass flow capacity increases as M_a increases (up to 1), a compromise must be made between stage temperature ratio, which improves with increasing $M_b \sin\beta_b$, and mass flow capacity. From the expression for $\cos\beta_b$ we have

$$M_b^2 \sin^2\beta_b = M_b^2 - M_a^2\left(\frac{1 + \frac{1}{2}(\gamma-1)M_b^2}{1 + \frac{1}{2}(\gamma-1)M_a^2}\right)^{(\gamma+1)/(\gamma-1)}.$$

Differentiating with respect to M_b and putting the result to zero, we find the M_b that maximizes $M_b \sin\beta_b$ for fixed M_a, and substituting this into $M_b^2 \sin^2\beta_b$ gives

$$\begin{aligned}
(M_b^2 \sin^2\beta_b)_{\max} &= \frac{2}{\gamma-1}\left[\left(1 + \frac{\gamma-1}{2}M_a^2\left[\frac{1 + \frac{\gamma-1}{2}M_a^2}{\frac{\gamma+1}{2}M_a^2}\right]^{(\gamma-1)/2}\right) - 1\right] \\
&\quad - M_a^2\left[\frac{1 + \frac{\gamma-1}{2}M_a^2}{\frac{\gamma+1}{2}M_a^2}\right]^{(\gamma+1)/2} \\
&= \frac{2}{\gamma+1}(M_b^2 - 1) \\
&= (1-\tau_t)^2\left[\frac{1 + \frac{\gamma-1}{2}M_b^2}{(\gamma-1)M_T}\right]^2 = [2M_T(1-R)]^2. \quad (6.7)
\end{aligned}$$

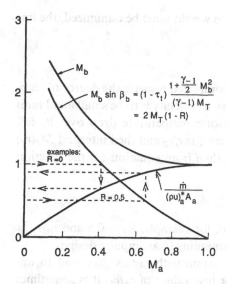

Figure 6.3
Maximum tangential Mach number and mass flow as functions of inlet axial Mach numbers
(and of reaction for fixed blade speed).

The last two equalities follow from equations 6.3 and 6.5. This relation
is plotted in figure 6.3 along with the mass flow density, which is just
$A^*/A(M_a)$. We see that as M_a increases, $M_b \sin\beta_b$ decreases, with a corresponding decrease in $1 - \tau_t$, but the mass flow density, of course, increases.
For a given M_T the impulse turbine has lower mass flow than the 50
percent reaction turbine, but greater work. Examples for $M_T = 0.5$ are
shown in the figure.

M_b is always greater than unity for this maximum-work condition, as
indicated by the second equality of 6.7 and as plotted in figure 6.3. The
optimum nozzle therefore always is convergent-divergent; within the given
assumptions it turns the flow first to the value of β giving $M = 1$ then back
to a lower value. For $M_a > 0.5$, however, the optimum M_b is not much
above 1, so a simple convergent nozzle gives nearly the maximum tangential velocity.

The optimum degree of reaction for any particular application will depend on the relative importance of efficiency, temperature drop per stage,
mass flow capacity, and blade temperature. Degrees of reaction near 50
percent are usually preferable in subsonic cruise engines, where efficiency is

a dominant requirement; where engine weight must be minimized, the impulse turbine may be better.

6.1.2 Radial Variations

In fact, the above comments are oversimplified because the degree of reaction must vary with radius if the exit swirl velocity is to be small at all radii and the turbine is to produce a uniform temperature drop over its full annulus. If we put $M_T = M_T(r_T)\zeta$ where ζ is r/r_T and then interpret $M_T(r_T)$ as the tip tangential Mach number, then from equation 6.5 the reaction must be

$$R = 1 - (1 - R_T)/\zeta^2,$$

where R_T is the reaction at the tip. If, for example, $R_T = 0.5$ and $r_H/r_T = 0.707$, the hub will have $R = 0$, corresponding to an impulse design.

Because this rapid variation of reaction with radius may lead to unacceptable efficiencies at the hub for low values of r_H/r_T, it is sometimes necessary to accept exit swirl from the rotor, either providing exit vanes to remove it or allowing it to carry through the next turbine stage just as the inlet guide vane swirl did in the compressor.

When the flow differs from a free vortex, the angular velocity changes across blade rows induce radial variations of axial velocity just as in the compressor. The particular case of constant $(1 - \tau_t)$ and zero exit swirl corresponds to a free vortex, so these effects are small for such a design, but designs with nonzero swirl leaving the rotor will in general lead to streamline shifts. Of particular practical importance is the case for β_b equal to a constant, which allows the nozzle vanes to be of constant shape over their length.

6.2 Turbine Blading

The requirements imposed on turbine blading by cooling (such as larger than otherwise desirable leading and trailing edge thicknesses) and the problems of manufacture from refractory alloys have tended to dictate against great aerodynamic refinement in turbine blading. The available systematic data for turbine cascades are for uncooled configurations; thus these data are not directly applicable to the cooled stages of a turbine. They are applicable to the uncooled stages, however, and in turboshaft engines or high-bypass turbofan engines the efficiency of these stages is very important to engine performance. In recent years, considerable success has been had with calculation of the pressure distribution and the heat-transfer distribution on turbine blades by computational fluid dynamic (CFD) methods, and this is now a standard part of the design procedure for high-performance turbines, as will be elaborated below.

Less success has been had in predicting the losses by direct computation, although knowledge of the pressure distribution on the blades does help. Still, recourse is had to empirical correlations, usually held as proprietary information by the engine companies, for the final design. Not much of this information is available in the literature. A comprehensive survey of the correlations of turbine efficiency available in the literature at the time of its compilation can be found in reference 6.1. Three main sources of loss are usually identified exclusive of those associated with cooling: profile losses, due to viscous shear on the blade sections; secondary flow losses, similar to those suffered in compressors, and discussed in subsection 5.2.5; and losses due to leakage past the blade tips. Both tip clearance and secondary flow losses are included in correlations of efficiency as a function of aspect ratio (the ratio of blade span to blade chord). Such correlations are reviewed in reference 6.1.

The leakage is more critical in turbines than in compressors because of the generally larger pressure difference across a blade row. Many turbine rotors are shrouded—that is, the blade tips carry an annular ring made up of segments, one on each blade, which carry a knife-edge seal that runs close to an abradable material such as metallic honeycomb. Such seals are used in the engines shown in figures 1.15–1.20. The efficiency is decreased about 1 percent for each percent of leakage past the rotor, but unfortunately there are no simple means available for calculating the leakage. It is estimated by taking the blade clearance gap as an orifice in parallel with the rotor. Methods for estimating secondary flow losses are the subject of active research, but none seem to be sufficiently developed to be useful in estimating turbine efficiency.

Even though blade inlet Mach numbers are generally small, shock losses arise from locally high Mach numbers in the turbine blades. As the blade loading is increased, the velocity and the Mach number over the suction surface rise, eventually reaching supersonic values. When this happens there is an embedded region of supersonic flow, which can lead to shocks, as sketched in figure 6.4. The shocks can, in turn, cause boundary-layer

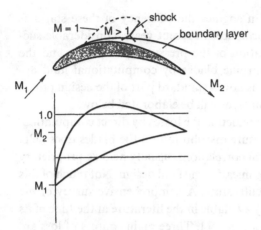

Figure 6.4
Mach number distribution and shock–boundary layer interaction on highly loaded turbine blade.

separation in the adverse pressure gradient on the aft portion of the blade; the net result is a large increase in loss. This effect is important in limiting blade loading in modern aircraft turbines.

6.2.1 Choice of Solidity

When the velocity changes across the turbine stage have been selected, the primary remaining design variable is the solidity. Increasing the solidity reduces the loading on the individual blades, lessening the possibility of separation and shock losses; however, it increases the surface area, potentially increasing the viscous losses. Thus, there is some optimum solidity. A commonly used approach to such optimization is to express the loading of a turbine blade row in terms of the Zweifel coefficient (reference 6.2), which is defined for the rotor in the notation of figure 6.2 as

$$\psi_z = \frac{\int_0^1 (p_p - p_s) \, d(x/c)}{p_{t1} - p_2},$$

where x is distance in the axial direction, c is the axial chord, p_p and p_s are the pressures on the pressure and suction surfaces respectively, p_{t1} is the stagnation pressure at the inlet to the blade row, and p_2 is the static pressure at the exit. The stagnation pressure and velocities are defined in the coordinates of the blade row. This coefficient in principle measures the actual load carried by the blade in comparison to what may be thought of as a reasonable maximum load, namely the difference between the inlet stagnation pressure and the downstream static pressure, acting over the entire axial chord of the blade. Zweifel argued that a value of about 0.8 for this coefficient produced minimum losses in the turbine blading.

In practice the coefficient is evaluated during the design process in terms of the velocity change across the blade row and the solidity. The numerator is the tangential force acting on the blade, per unit of span and chord. In terms of the inlet and outlet tangential velocities and the axial mass flux, this must be equal to $\rho_2 w_2 (v_1 - v_2) s$, where s is the blade spacing; thus, we can write the Zweifel coefficient as

$$\psi_z = \frac{\rho_2 w_2 (v_1 - v_2)}{p_{t2} - p_2} \frac{s}{c},$$

where the assumption has been made that the stagnation pressure relative to the blades is the same at 2 as at 1. A little manipulation then shows that

$$\psi_z \frac{c}{s} = \frac{\gamma M_2{}^2 \cos\beta_2 \sin\beta_2 (v_1/v_2 - 1)}{\{1 + [(\gamma - 1)/2] M_2{}^2\}^{\gamma/(\gamma-1)} - 1}. \qquad (6.8)$$

In the incompressible limit as $M_2 \to 0$, this becomes

$$\psi_z \frac{c}{s} = \left(\frac{v_1}{v_2} - 1\right) \sin(2\beta_2), \qquad (6.9)$$

which is the form in which the Zweifel coefficient is often expressed. As noted, the value of this approach is said to be that the efficiency of a turbine cascade optimizes as a function of solidity for values of Ψ_z between 0.8 and 1.0 for a wide range of turbine designs. Thus, given the desired tangential velocity ratio across the blade row and the leaving flow angle, one can compute the best solidity from this expression. Unfortunately there seems to be little if any data available in the literature which validates this approach, but it is supported by a correlation in terms of the diffusion factor, as we shall see presently. And it is widely used.

Since the profile losses in a turbine are due principally to separation on the suction side aggravated by shock–boundary layer interaction due to locally supersonic flows, a correlation of losses versus the diffusion factor suggested in reference 6.3 is appealing. For the turbine,

$$D = 1 - \frac{V_2}{V_1} + \left| \frac{v_2 - v_1}{2\sigma V_1} \right|,$$

where, just as for the compressor, the term V_2/V_1 represents the effect of mean velocity change through the blade passage, and $(v_2 - v_1)/2\sigma V_1$ represents the adverse pressure gradient on the suction side due to the blade force required to produce the flow deflection. Generally for a turbine $V_2/V_1 \geq 1$; it is unity for impulse blading and larger for 50 percent reaction (see figure 6.2). Thus, the flow acceleration reduces D and the deflection increases it. Figure 6.5 presents the turbine cascade loss data of reference 6.4 in the format of $(\overline{\omega}_2 \cos\beta_2/2\sigma)(\cos\beta_2/\cos\beta_1)^2$ versus D.

Here we choose $\overline{\omega}_2 \equiv (p_{t2} - p_{t1})/(p_{t2} - p_2)$ as the loss parameter rather than $\overline{\omega}_1 \equiv (\overline{p}_{t2} - p_{t1})/(p_{t1} - p_1)$, so that the stagnation pressure loss is measured against the largest dynamic pressure. The figure includes impulse blading and 50 percent reaction blading with a variety of deflection angles and solidities $1 < \sigma < 3$. The compressor loss correlation is included for comparison, as is the turbine correlation of reference 6.3. Im-

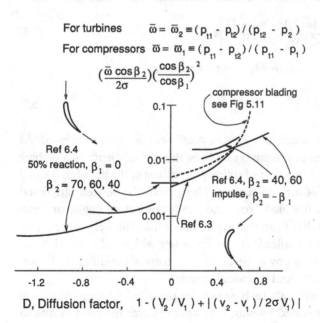

Figure 6.5
Turbine cascade loss data presented as D-factor correlation and compared with compressor cascade data.

pulse blading has D values in the general range of compressor blading and correspondingly large loss factors. Reaction blading has negative D for the range of solidities covered in the data and correspondingly small loss factors corrected for the flow angles.

The corrected loss factor plotted in figure 6.5 measures the ratio of boundary-layer momentum thickness θ^* to chord. This does become very small for negative D as our physical arguments suggest, but this does not mean that $\bar{\omega}_2$, which measures the fractional stagnation pressure drop in the cascade, is small, because the negative D is associated with turning to large β_2. For $\beta_2 = 70°$, $\cos^3 \beta_2 \approx 0.04$, so that $\bar{\omega}_2 = 50\sigma$ times the value plotted in figure 6.4. For $\sigma = 1$, which corresponds to $D = -0.06$, $\bar{\omega}_2 = 0.05$.

In fact, corresponding to the optimum Zweifel coefficient and the solidity that is consistent with it according to equation 6.9, there should be a solidity that produces the minimum loss factor according to the correlation of figure 6.5. That this is indeed the case is shown by figure 6.6, which shows the loss factor computed as a function of σ for two types of blade rows, one representing an impulse turbine and the other a 50% reaction turbine. It can be seen that there is a distinct minimum of the loss factor for each case, both at a solidity just under 2.

It is instructive to relate this D-factor correlation to the Zweifel-coefficient approach. This is readily done by solving equation 6.9 for σ and

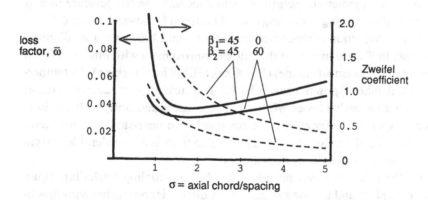

Figure 6.6
Loss factor as a function of solidity predicted by loss correlation of figure 6.5, showing optimum solidities for impulse and 50% reaction bladings. Zweifel coefficients of 1.0 and 0.5 for the impulse and 50% reaction bladings are indicated.

then substituting this in the expression for D. After a little manipulation one finds

$$D = 1 - \frac{V_2}{V_1}\left(1 - \frac{\psi_z}{4\cos\beta_2}\right). \qquad (6.10)$$

The Zweifel coefficient as given by this equation is plotted as a function of the solidity in figure 6.6, from which it can be seen that the Zweifel coefficient corresponding to the solidity that gives minimum loss by the D-factor correlation is about 1.0 for the impulse turbine and 0.5 for the 50% reaction turbine. Considering the uncertainties in this comparison, this may be regarded as good agreement.

In summary, it seems that either the D-factor correlation of figure 6.5 or the Zweifel-factor rule will give a reasonable estimate of the best solidity, especially if refined by means of experience with the actual blade sections that are to be employed.

6.3 Turbine Cooling

The desire for higher thrust per unit of air flow provides a powerful incentive to increase the turbine inlet temperature. Fuel consumption also improves if the increased temperature is accompanied by an increase in compressor pressure ratio. Over the years there has been a gradual improvement in materials, permitting small increases in temperature in new engines and uprating of existing ones. This trend is shown in figure 6.7.

The introduction of air-cooled turbines around 1960 gave a small initial increase in T_{t4} and increased the rate of improvement with time. The latest commercial transport engines (PW 4000, GE 90) have turbine inlet temperatures at takeoff power near 1650°K. The turbine inlet temperatures at cruise are somewhat lower. Military engines use significantly higher values. Some experimental engines have operated with near-stoichiometric temperatures (in the range of 2300–2500°K), so there is a considerable margin left for improvement in service engines.

Of the many schemes proposed for turbine cooling (including liquid thermosiphon and boiling systems), only direct air cooling has seen practical application. Air is bled from the compressor, carried aft, and introduced into the turbine (rotor) blades through their roots, as sketched in figure 6.8. In entering the blades, the air also cools the rim of the turbine

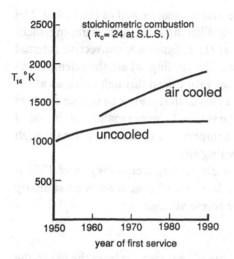

Figure 6.7
Trend of turbine inlet temperature with time.

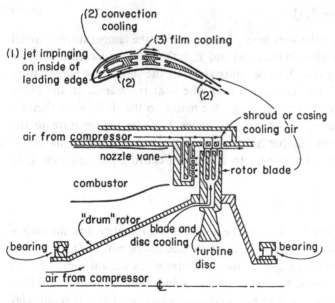

Figure 6.8
Schematic of air-cooled turbine, with cross section of cooled airfoil section at top.

disc. The first-stage nozzle vanes are also cooled, as well as the casing. The cooling air can be used in a number of different ways, such as impingement cooling of the leading edge, shown at (1) in figure 6.8; convective internal cooling of blade surface, as at (2); and film cooling, where the attempt is to sheath the blade with a film of cool air introduced through holes, as at (3). The last method differs from the first two in that it aims to reduce the heat transfer to the blade surface, while the internal cooling maintains the metal temperature below the outside gas temperature by heat transfer through the blade material to the internal cooling air.

An excellent review of the state of air cooling technology as of 1978 is presented in reference 6.5. To the author's knowledge there is no similarly comprehensive presentation of more recent vintage.

6.3.1 Internal Cooling

Consider first internal cooling. The rate of heat transfer from the gas to the blade surface is expressed in terms of a Stanton number St defined so that if q_w is the rate of heat transfer to the surface then

$$q_w = St[\rho u c_p (T_r - T_w)], \tag{6.11}$$

where T_r is the "adiabatic recovery temperature" (the temperature attained by an insulated surface in the flow) and T_w is the wall temperature. For our purposes, we may take T_r to be approximately the stagnation temperature relative to the blade (see section 6.1). The heat transferred to the blade surface must be conducted through the metal, so that if k_s is the thermal conductivity then $q_w = -k_s \, \mathrm{grad}\, T$, where T is the temperature in the metal. The Stanton number has a particularly useful physical interpretation. The total heat transferred to a blade of chord c per unit span is of order

$$c q_w = [\rho u c_p (T_r - T_w)s] \, \sigma \, St,$$

and the quantity in brackets is the total thermal energy flux through a blade channel measured relative to the blade temperature. Thus, we can say that the Stanton number (times solidity) gives the ratio of the heat transferred to the blade to the heat that would result from cooling all flow through the blade passage to the blade temperature. That it is small relative compared to unity is our good fortune.

The Stanton number is controlled by the boundary-layer behavior. If the Prandtl number is unity, or if in turbulent flows the Reynolds analogy applies, then St $\approx C_f/2$, where C_f is the friction coefficient. For flow over a flat plate,

$$2\,\text{St} = C_f = \frac{0.66}{(\text{Re}_x)^{1/2}} \quad \text{(laminar)}$$

and   (6.12)

$$2\,\text{St} = C_f = \frac{0.0592}{(\text{Re}_x)^{1/5}} \quad \text{(turbulent)}.$$

Although these relations do not apply quantitatively to the turbine blade, they indicate trends. A detailed discussion of turbine blade heat transfer will be found in reference 6.6.

We see that if the boundary layer is laminar $q_w \propto x^{-1/2}$, while if it is turbulent $q_w \propto x^{-1/5}$. Of course, q_w is not infinite at the leading edge of the blade, but it can be very large; hence the impingement cooling. For $p_{t4} = 20$ atm and $\theta_t = 6$, $\text{Re}_x = 3 \times 10^7$ per meter at $M = 1$. (A useful fact to remember is that $\text{Re}_x = 1.2 \times 10^7$ per meter for air with $T_t = 288°\text{K}$ and $p_t = 1$ atm at $M = 1$.) Transition from laminar to turbulent flow occurs for Re_x between 3×10^5 and 10^6 on a flat plate. The Reynolds number for transition on a turbine blade is difficult to predict but probably lies below 10^6, so transition will occur on most turbine blades. When it does, the heat-transfer rate increases, so that the distribution of q_w on a blade is somewhat as sketched in figure 6.9.

The problem of designing the cooling system is to schedule the internal flow so that with such a distribution of q_w the blade has as nearly uniform a

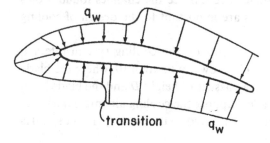

Figure 6.9
Distribution of heat transfer on a cooled turbine blade, showing sudden increase at boundary-layer transition to turbulent state.

temperature as possible. Because the thermal stresses caused by nonuniform temperature can be a limiting factor, it is important to properly
schedule the cooling air. Indeed, if thermal stresses set a limit on the temperature difference between two parts of the airfoil, then the precision with
which the cooling can be controlled determines the permissible difference
between gas temperature and blade temperature; hence, it determines the
maximum permissible turbine inlet temperature, because the variations in
blade temperature are a fraction of the gas-blade temperature difference.

The heat transfer to the cooling fluid at points such as (2) in figure 6.8
can be estimated in first approximation from results for flow in long tubes,
such as

$$2 \, \mathrm{St} = C_f = 0.023 \, \frac{1}{(\mathrm{Re}_D)^{1/5}}, \qquad (6.13)$$

where the Reynolds number is now based on the hydraulic diameter of the
passage (one-fourth the area divided by the circumference).

There are, however, strong effects on the heat transfer in internal passages, due to rotation. Reference 6.7 presents a comprehensive study of
these effects in rotating radial passages with roughening, which is intended
to increase the heat transfer. It is found for outward flow that rotation
lowers the heat transfer on the low-pressure side of the passage (that is, the
leading side), while it increases the heat transfer on the high-pressure or
trailing side. For inward flow, the heat transfer is again lower on the low-
pressure side and higher on the high-pressure side, but now the low-
pressure side is the trailing side and the high-pressure side is the leading
side. Generally, increasing the difference between surface and fluid bulk
mean temperature increases these effects. Since the changes found are as
much as factors of 4, these effects are important to the design of cooling
systems.

To see how the various requirements on the cooling system interact,
consider the schematic cooled blade shown in figure 6.10, where the blade
surface is cooled by flow through a passage of height D and the blade has a
chord c and spacing s in a cascade. Suppose the cooling air enters the blade
at temperature T_c. The total heat transferred to the blade (per unit span) is
estimated as

$$cq_w = \sigma \, \mathrm{St}[\rho u c_p (T_r - T_w)s];$$

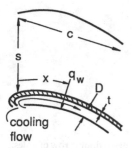

Figure 6.10
Schematic of internal cooling of a turbine blade.

this must be transferred to the cooling air, with mass flow $\dot{m}_c$ (per unit span), so

$$cq_w = \dot{m}_c c_p (T_w - T_c)$$

if the cooling air is heated all the way to blade temperature. It follows that the ratio of cooling to turbine mass flow must be at least

$$\frac{\dot{m}_c}{\rho u s} \geq \sigma \, \text{St} \left(\frac{T_r - T_w}{T_w - T_c} \right). \tag{6.14}$$

A typical value of St is 0.005. For $T_r - T_w = 200°\text{K}$ and $T_w - T_c = 400°\text{K}$, we find $\dot{m}_c / \rho u s = 0.0025$; but because there are two sides to the blade, the required cooling flow according to this estimate would be about 0.005 of the turbine mass flow. The actual cooling flow is somewhat larger than this, partly because the cooling air must still be colder than the blade when it leaves it and partly because of the difficulty of getting sufficient heat transfer surface in the cooling passages. A value of about 3 times that given by equation 6.14 is a reasonable estimate.

6.3.2 Film Cooling

The objective in film cooling is to lower the adiabatic recovery temperature of the boundary layer by mixing cool air into it and thus reducing the heat transfer to the surface. The cool air is introduced through a series of small holes inclined at an angle to the surface, as shown in figure 6.11. Their spacing is usually several hole diameters, the diameter being from 0.05 to 0.1 cm or larger. There may be one row, or a staggered double row as shown in the figure.

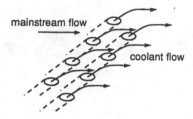

Figure 6.11
Schematic of cooling configuration for film cooling, showing a staggered double row of holes.

Ideally, such cooling eliminates the constraints imposed by heat transfer through the blade material and the resulting thermal stresses. In practice, the film cooling produces a thermal stress pattern of its own because the air flowing through the small holes cools the metal immediately surrounding them more than the average for the blade as a whole. Introduction of air through the surface of the blade also disrupts the boundary layer flow, tending to increase both shear (pressure loss) and heat transfer. The overall gain results from a balance of these undesirable effects against the reduction of adiabatic recovery temperature. The holes normally are inclined downstream from the perpendicular to the surface, to encourage the jet to lie along the surface, there being a tradeoff here between this aerodynamic benefit and the larger stress concentration and greater manufacturing difficulty as the angle increases.

Since the diameter of the cooling holes is normally greater than the thickness of the boundary layer, the flow picture is one of discrete jets that penetrate the boundary layer, are turned by the free stream, and then flow along the surface, mixing with the boundary-layer fluid. Attempts have been made to use transpiration cooling, in which the scale of the holes and their spacing would be small relative to the boundary layer thickness so that the cool air would be introduced into the low-velocity portion of the boundary layer. But such a fine distribution of pores is difficult to achieve and is highly susceptible to plugging and surface damage. No such ideal transpiration cooling has been achieved in practice.

The two conflicting effects of boundary-layer disruption and cooling are shown clearly by the experimental data of figure 6.12, where the heat flux to a film-cooled surface is plotted as a function of $m = \rho_c u_c / \rho_\infty u_\infty$ where $\rho_c u_c$ is the mass flow density from the coolant holes and $\rho_\infty u_\infty$ is that in the external flow. Three curves are shown. In all cases the wall was at 811°K

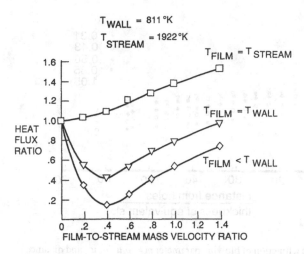

Figure 6.12
Effects of film-cooling mass flow and cooling airflow temperature on heat flux to film-cooled flat plate (from reference 6.8).

and the external flow at 1922°K. For the upper curve the "cooling" air was at stream temperature. We see that increasing the cooling flow increases the heat flux, partly because of heat transfer in the cooling holes and partly because of stirring of the boundary layer. When the cooling air is at wall temperature, heat flux is reduced for small cooling flows but increased when the stirring effect becomes dominant at high flows. For a cooling-air temperature below wall temperature, the trends are similar but the cooling is more effective.

The effectiveness of film cooling is usually described quantitatively in terms of an "adiabatic film effectiveness," defined as

$$\eta_{ad} \equiv \frac{T_r - T_{rf}}{T_r - T_c}, \tag{6.15}$$

where T_r is the adiabatic wall temperature (the temperature the wall would reach if insulated from all but the free stream) and T_{rf} is the temperature the wall will reach with the film cooling but no other cooling. As before, T_c is the cooling-air temperature. We see that η_{ad} varies between 0 (no cooling by film) and 1 (wall cooled to T_c). A heat-transfer film coefficient is then defined as

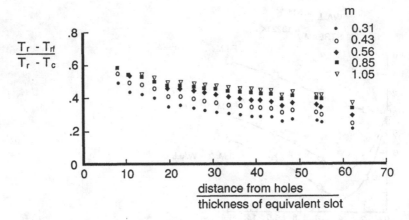

Figure 6.13
Adiabatic film effectiveness as a function of blowing parameter $m = \rho_c u_c / \rho_\infty u_\infty$ and distance downstream from holes divided by thickness of slot with same flow area as row of holes. For subsonic nozzle blade, two rows of holes 0.484 and 0.584 inches from leading edge on suction surface; total length of suction surface 2.2 inches. (From reference 6.6.)

$$h \equiv \frac{q_w}{T_{rf} - T_w}. \qquad\qquad (6.16)$$

The designer wishes to find q_w, having T_r, T_c, the cooling mass flow ratio $m = \rho_c u_c / \rho_\infty u_\infty$, and the geometry of the cooling holes. From a series of experiments a correlation of η_{ad} such as that given in figure 6.13 would be found; the parameters are distance downstream from the injection holes and $m = \rho_c u_c / \rho_\infty u_\infty$. The correlation depends somewhat on the configuration of the holes, but this dependence is reduced by nondimensionalizing the downstream distance by the width of an "equivalent slot" which has flow area equal to that of the array of holes. The heat flux q_w is then estimated from knowledge of h, which is usually taken to be that for a well-developed boundary layer with no injection.

6.3.3 Impingement Cooling

Impingement cooling is often used to cool the leading edges of turbine nozzle vanes or first-stage rotor blades, where the heat transfer rates are very high (as indicated in figure 6.9), and where film cooling is difficult because of uncertainties in the local flow due to variation of the location of the stagnation point. As is shown schematically in figure 6.14, the idea is

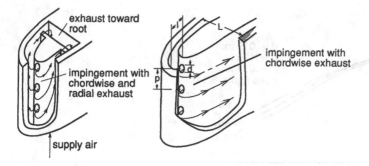

Figure 6.14
Schematics of impingement cooling applied to the leading edge of a turbine blade. Radial exhaust at left, chordwise exhaust at right. The dimensions refer to the correlation in figure 6.15 for chordwise exhaust.

simply to increase the local mass flux density across the inside of the blade surface by directing a jet against it. The flow in the passage can be quite complex, however, owing to the interaction of the jet flows with the mean flow in the passage. In some configurations the flow exhausts in the chordwise direction, so that the overall flow is more or less two-dimensional. In others the flow exhausts in the spanwise direction toward the tip of the blade, and in this case the jets exist in a crossflow which is stronger at larger radii.

Both of these configurations have been studied experimentally. The Nusselt number at the line of impingement has been correlated in reference 6.9 for the geometry at the right of figure 6.14 in the form

$$Nu_{stag} = 0.44 \ Re^{0.7} \ (d/p)^{0.8} \ exp[-0.85(l/d)(d/p)(d/L)^{0.4}],$$

where the quantities d, p, l and L are as defined in figure 6.14.

The variation in the chordwise direction from the line of impingement is then as shown in figure 6.15. Here the Nusselt number is defined as $Nu = hd/k$, where h is the film coefficient, such that the local heat flux $q = h(T_w - T_c)$. T_c is the stagnation temperature of the cooling jets and T_w is the local wall temperature. The Reynolds number $Re = \rho u d/\mu$ is based on the jet's diameter d and velocity u.

References 6.10 and 6.11 report a study of the radial exhaust configuration which was carried out in a rotating system so that the effects of buoyancy and coriolis forces on the cooling flow were modeled. Figure 6.16 plots the Nusselt number, defined as above, averaged along the chordwise

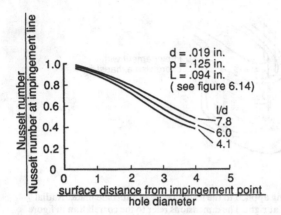

Figure 6.15
Variation with distance from the line of jet impingement of the Nusselt number for impingement cooling with chordwise exhaust, averaged in the radial direction (from reference 6.9).

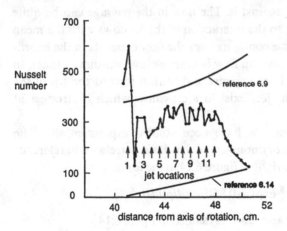

Figure 6.16
Nusselt number averaged in chordwise direction as a function of radius for radial exhaust,
showing effects of individual jets and of rotation (from reference 6.11).

direction near the leading edge of the blade, as a function of radius. Here the impact on cooling of the individual jets can be seen. The effect of rotation was found in general to be a reduction of the Nusselt number from that found in non-rotating experiments, as indicated by comparison with the results of reference 6.9. Some complex variations were also found, such as that at the smallest radius in figure 6.16. It is thought that this effect is due to deflection of the jet by the buoyancy force.

6.3.4 Design of Blade Cooling System

The design of the air cooling system for a turbine blade row might proceed as follows. First, from the aerodynamic design of the stage and from knowledge of the acceptable material temperature, the heat-transfer rate is estimated at each point on the blade and on the end walls. The possibility of absorbing this heat load by means of internal convection cooling and the resultant thermal stresses are then assessed. Where convection cooling is possible it is the preferred design because its impact on efficiency is generally smaller than that of film cooling.

However, for some difference between adiabatic recovery temperature and the permissible metal temperature, the internal cooling will no longer suffice and film cooling must be used to reduce the rate of heat transfer to the blade. The procedure is then to determine the effectiveness required to bring about this reduction, and to include arrays of holes just upstream of or at the points of peak heat flux, which provide this effectiveness. In modern blade designs, film cooling is used at the leading edge and in the regions of transition on the suction and pressure surfaces. It is normal in the development process for unexpected hot spots to emerge, and these are generally controlled by local application of film cooling.

6.3.5 Effect of Cooling on Turbine Efficiency

The previous definition of efficiency must be modified somewhat to deal with the cooled turbine. We shall redefine it as the ratio of the actual turbine work per unit of total airflow, primary plus cooling, divided by the ideal work that would be attained in expanding that total airflow through the actual pressure ratio. We assume that the cooling flow expands through the same pressure ratio as the primary flow. Thus, if ε is the ratio of cooling airflow to total airflow,

$$\eta_t = \frac{(1 - \varepsilon)(T_{t4} - T_{t5}) + \varepsilon(T_{tc} - T_{t5})}{[(1 - \varepsilon)T_{t4} + \varepsilon T_{tc}][1 - (p_{t5}/p_{t4})^{(\gamma-1)/\gamma}]}, \tag{6.17}$$

where T_{tc} is the stagnation temperature of the cooling air and T_{t5} is the mean outlet temperature for the total flow. With the temperatures and ε given, π_t can be computed from this relation if η_t is known.

The cooling flow can influence η_t in three ways:

1. The cooling air emerging from the blades can change their drag characteristics, probably increasing drag.
2. The cooling air itself suffers a pressure loss in passing through the cooling passages, so that it has lower stagnation pressure when mixed into the downstream flow, thus decreasing π_t for a given τ_t.
3. The entropy of the flow as a whole is increased by the transfer of heat from the hot primary flow to the cooling flow.

Consider the third mechanism. The increase in entropy per unit mass of total flow that results from the transfer of an amount of heat ΔQ from the primary flow at turbine inlet temperature T_{t4} to cooling-air temperature T_{tc} is

$$\Delta s = \Delta Q[(1/T_{tc}) - (1/T_{t4})].$$

Taking the estimate of subsection 6.3.1 for the heat transfer gives

$$\frac{\Delta \eta_t}{c_p} = -2\sigma\,\mathrm{St}\left(\frac{T_{t4}}{T_{tc}} - 1\right)\left(1 - \frac{T_w}{T_{t4}}\right) \tag{6.18}$$

per blade row; that is, we must double this value for a cooled stage.

As was explained in subsection 5.2.4, the entropy change in the stage is related to the stagnation pressure and temperature changes across it by the equation

$$\Delta s = (1 - \varepsilon)c_p \ln\frac{T_{t5}}{T_{t4}} + \varepsilon \ln\frac{T_{t5}}{T_{tc}} - R \ln\frac{p_{t5}}{p_{t4}};$$

thus,

$$\frac{p_{t5}}{p_{t4}} = \left\{ e^{-(\Delta s/c_p)}\left[\left(\frac{T_{t5}}{T_{t4}}\right)^{1-\varepsilon}\left(\frac{T_{t5}}{T_{tc}}\right)^{\varepsilon}\right] \right\}^{\gamma/(\gamma-1)} \tag{6.19}$$

expresses the effect of the heat transfer on the turbine pressure ratio.

To account for the effect of pressure drop in the cooling air (the second mechanism), we may add the entropy change due to this pressure drop to the change due to cooling. This entropy change can be written

$$\frac{\Delta s_f}{c_p} = \left(\frac{\gamma-1}{\gamma}\right)\varepsilon \ln\frac{p_{t4}-\Delta p_f}{p_{t4}} = -\ln\left(1-\frac{\Delta p_f}{p_{t4}}\right)^{\varepsilon(\gamma-1)/\gamma},$$

and since $\varepsilon \ll 1$, we have

$$\frac{\Delta s_f}{c_p} \approx \left(\frac{\gamma-1}{\gamma}\right)\varepsilon\frac{\Delta p_f}{p_{t4}}. \tag{6.20}$$

Substituting equation 6.19 into equation 6.17 and assuming

$$\frac{\Delta s}{c_p} = \frac{\Delta s_c + \Delta s_f}{c_p} \ll 1,$$

we find

$$\eta_t = 1 - \left(\frac{\tau_t}{1-\tau_t}\right)\left(\frac{\Delta s}{c_p}\right)$$

$$= 1 - \left(\frac{\tau_t}{1-\tau_t}\right)\left[2\sigma\,\text{St}\left(\frac{T_{t4}}{T_{tc}}-1\right)\left(1-\frac{T_w}{T_{t4}}\right) + \left(\frac{\gamma-1}{\gamma}\right)\varepsilon\left(\frac{\Delta p_f}{p_{t4}}\right)\right]. \tag{6.21}$$

The significance of this relation is best seen by means of a typical example. Take $\sigma = 1$, $\text{St} = 0.005$, $T_w/T_{t4} = 0.7$, $T_{t4}/T_{tc} = 2.5$, $\tau_t = 0.82\,(\pi_t = 0.5)$, and $\Delta p_f/p_{t4} = 0.5$. From equation 6.11 we estimate $\varepsilon \approx 0.01$, and η_t becomes

$$\eta_t \approx 1 - (0.82/0.18)\,[(0.01)(1.5)(0.3) + (0.01)(0.14)] = 1 - 0.027;$$

thus, there is a 2.7 percent degradation of efficiency due to the cooling.

The first of the three mechanisms is hardest to quantify, and we must rely mainly on experimental data. In an experiment reported in reference 6.12 the kinetic energy efficiency was measured for a nozzle cascade with injection through each of 12 rows of holes. The kinetic energy efficiency was defined as

$$\eta_{KE} = \frac{\text{Actual exit kinetic energy}}{\text{Ideal exit kinetic energy for main flow plus coolant flow}}.$$

There was about a 0.2 percent reduction in η_{KE} for each percent cooling flow on the pressure side, and about 0.5 percent on the suction side.

Thus, to the 2.7 percent loss per percent of cooling flow estimated from equation 6.21 we should add about another 0.5 percent, giving a total of some 3.2 percent loss in turbine efficiency per percent of cooling flow.

There is a shortage of systematic data in the literature against which to test an estimate such as this. However, a comparison of the efficiency of a cooled turbine and an uncooled turbine of the same design has been published (reference 6.13). The conclusion can be drawn from the data that a 2.5 percent rotor cooling flow reduced the efficiency about 6.5 percent.

6.4 Turbine Design Systems

The design of a modern air-cooled turbine is an intricate process that taxes the best capabilities of the large engineering organizations maintained by aircraft engine manufacturers. It is not possible to do it justice here; however, an outline of the general approach taken may be useful. Once the major engine cycle parameters (such as pressure ratio, bypass ratio, and turbine inlet temperature) have been set and the general layout of the engine has been decided upon, the designer might go through the following steps, though not necessarily in this order.

(1) A *meanline design*, which consists in developing the velocity triangles for each blade row, estimating the efficiency from correlations based on D factor or Zweifel coefficient, and determining the annulus height, stress levels, and cooling requirements in each blade row. It may be necessary to explore a large number of configurations, with different blade speeds and numbers of stages at this level, to arrive at a design that seems attractive. The result of this stage will be a preliminary definition of the turbine flow-path, including the blade speed and annulus height at each stage, the blade aspect ratio and solidity, and the gas properties in each stage.
(2) A *throughflow analysis*. Just as for the compressor, if the tangential velocity increments put in or taken out by the blade rows differ from free vortices, the axial velocity distribution will be non-uniform. It may be desirable to use such non-free-vortex distributions in order to avoid extremes of reaction at the hub or tip radii. Such designs (sometimes termed *controlled-vortex* designs) offer substantial improvements in overall performance, but require that the flow be analyzed by techniques similar to those described in subsection 5.2.1.2.
(3) *Blade section design*. With the more accurately defined throughflow as the base, the blade shapes are analyzed in detail. This yields shapes that minimize shock and separation losses. This step is carried out with two-

dimensional numerical analyses combined with boundary-layer calculations.

(4) *Three-dimensional inviscid analysis of blade passage flow.* Present
capabilities do not permit viscous three-dimensional analyses in the design phase, but a three-dimensional inviscid calculation can reveal flow
anomalies in the individual blade rows and facilitate shaping of the blades
to avoid problems.

(5) *Cooling and durability design.* The cooling system is now designed as
outlined in subsection 6.3.4 to produce some prescribed blade temperature
distribution. With this temperature distribution, the state of thermal stress
in the blades, their attachments, and the discs is then analyzed using a finite
element method, implemented as a version of Nastran. This analysis deals
with the transients experienced in startup and shutdown as well as under
full power and cruise conditions. When the stress calculations reveal a
problem, the cooling design is revised until a satisfactory solution is developed. It may be that at this stage no acceptable solution is available (because, for example, too high a blade speed was selected in the mean line
design phase). In this case, the designer must return to this step and increase either the number of stages or the solidity, or must find another
solution.

6.5 Turbine Similarity

The representation of empirical turbine performance in terms of corrected
parameters is entirely analogous to that of compressors (section 5.4). Tangential Mach number is represented by the corrected speed, $N/\sqrt{\theta}$, where
θ is the inlet stagnation temperature divided by the reference standard and
N is rpm for a given turbine. Axial Mach number is represented by the
corrected weight flow $W\sqrt{\theta}/\delta$. A map for a typical 50 percent reaction (at
mid-radius) single-stage turbine is shown in this format in figure 6.17.

The abscissa has been taken as $(W\sqrt{\theta}/\delta)(N/\sqrt{\theta})$ instead of as $W\sqrt{\theta}/\delta$
because the mass flow is nearly independent of speed for $\pi_t > 2.5$; thus, all
speed characteristics collapse onto a single line, and the turbine has the
mass flow characteristic of a choked nozzle. A separate plot would then be
needed to show η_t as a function of $N/\sqrt{\theta}$. The choking does not occur in
the same way as in a simple nozzle because of the energy extraction by the
rotor. It may occur at the nozzle exit or at the rotor blade exit, or it may be

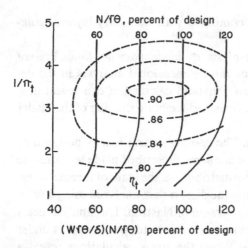

Figure 6.17
Typical turbine performance map.

a result of flow limitations at both positions. If the nozzles choke with increasing $1/\pi_t$ without choking in the blades, then we find the characteristic, shown in figure 6.17, that $W\sqrt{\theta}/\delta$ is independent of $N/\sqrt{\theta}$ for large $1/\pi_t$. If, on the other hand, the rotor outlet or exit annulus were choked, $W\sqrt{\theta}/\delta$ would depend on $N/\sqrt{\theta}$ for a given $1/\pi_t$ because changes in rotor speed would affect the stagnation pressure and temperature, and hence $(\rho u)^*$, in the downstream portions of the turbine. We see from figure 6.17 that for low $1/\pi_t$ the mass flow decreases with increasing $N/\sqrt{\theta}$ for fixed $1/\pi_t$. This is because the stagnation pressure drop across the rotor is larger at larger $N/\sqrt{\theta}$ (recall that $1 - \tau_t \propto M_T^2$), so for given upstream and downstream pressures there is less pressure drop across the nozzles at large $N/\sqrt{\theta}$, and hence less mass flow.

Because η_t does not vary as rapidly with off-design variations as in a compressor, the turbine characteristic can be approximated for preliminary design calculations by a constant η_t and a choked mass flow characteristic, so that

$$W\sqrt{\theta}/\delta = \text{const}, \eta_t = \text{const}$$

is an approximate description of the turbine characteristics for purposes of engine performance calculation. This approximation will be used in the discussion of component matching in chapter 8.

References

6.1 J. H. Horlock, *Axial Flow Turbines*. Kruger, 1973.

6.2 O. Zweifel, "The Spacing of Turbomachine Blading, Especially with Large Angular Deflection." Brown Boveri Review 32 (1945): 12.

6.3 W. L. Stewart, W. J. Whitney, and R. Y. Wong, "A Study of Boundary Layer Characteristics of Turbomachine Blade Rows and Their Relation to Overall Blade Loss." *Journal of Basic Engineering* 82-D (1960): 588.

6.4 D. G. Ainley and G. C. R. Mathieson, A Method of Performance Estimation for Axial Flow Turbines. ARC R & M No. 2974, Her Majesty's Stationery Office, 1957.

6.5 M. Suo, "Turbine Cooling." In *The Aerothermodynamics of Aircraft Gas Turbine Engines*, ed. G. C. Oates. AFAPL TR-78-52, Air Force Aero Propulsion Laboratory, Wright-Patterson Air Force Base, Ohio.

6.6 R. D. Lander, R. W. Fish, M. Suo, L. L. Grimme, and J. E. Muska. AIAA Paper 72-9, 1972.

6.7 J. H. Wagner, B. V. Johnson, R. A. Graziani, and F. C. Yeh, Heat Transfer in Rotating Serpentine Passages with Trips Normal to the Flow. ASME Paper 91-GT-265, 1991.

6.8 M. E. Crawford, H. Choe, W. M. Kays, and R. J. Moffatt, Full Coverage Film Cooling Heat Transfer Studies—A Summary of the Data for Normal-Hole Injection and 30° Slant-Hole Injection. Report HMT-19, Stanford University, 1975.

6.9 R. E. Chupp, H. E. Helms, P. W. McFadden, and T. R. Brown, "Evaluation of Internal Heat Transfer Coefficients for Impingement Cooled Turbine Airfoils." *Journal of Aircraft* 6 (1969): 203–208.

6.10 J. C. Kreatsoulas, Experimental Study of Impingement Cooling in Rotating Turbine Blades. Ph.D. thesis, MIT Department of Aeronautics and Astronautics, 1983.

6.11 A. H. Epstein, J. L. Kerrebrock, J. J. Koo, and U. Z. Preiser, "Rotational Effects on Impingement Cooling." Symposium on Transport Phenomena in Rotating Machinery, Honolulu, 1985.

6.12 Herman W. Prust, Jr., Two-Dimensional Cold-Air Cascade Study of a Film-Cooled Turbine Stator Blade. II, Experimental Results of Full Film Cooling Tests. NASA TM x-3153, 1975.

6.13 H. Nouse et al., Experimental Results of Full Scale Air-Cooled Turbine Tests. ASME Paper 75-GT-116, 1975.

6.14 W. D. Morris and T. Ayhan, "Observations on the Influence of Rotation on Heat Transfer in the Coolant Channels of Gas Turbine Rotor Blades." *Proceedings of the Institute of Mechanical Engineers* 193 (1979): 303–311.

Problems

6.1 The analogue of the centrifugal compressor is the radial inflow turbine. Supposing that such a device has radial vanes and zero exit swirl, derive an expression for its stagnation temperature ratio as a function of tip tangential Mach number. What is its effective degree of reaction?

6.2 A turbine is to be "designed" to drive a compressor that has the following characteristics at its design point:

Weight flow	100 kg/sec
Pressure ratio	8
Efficiency	0.85
Tip speed	400 m/sec
Tip diameter	0.7 m
Inlet conditions	1 atm, 300°K

The turbine inlet temperature is to be 1400°K.
  There are a number of design options. You might consider
  (a) zero exit swirl from rotor,
  (b) 50 percent reaction ($R = 0.5$) at mid-span (gives M_T at mid-span),
  (c) $M_b = 1$ at mid-span (gives M_a).
Sketch the velocity triangles at hub, mid-span, and tip, and the shape of the annulus. Also, estimate the blade root stress assuming that the blades are of constant cross-section, and calculate the horsepower output of this turbine.

6.3 Using flat-plate boundary-layer relations for heat transfer and a free-stream Mach number of 1, estimate the heat transfer to the surface of a turbine nozzle vane kept at a uniform surface temperature of 1200°K by internal cooling when the turbine inlet pressure and temperature are 25 atm and 1600°K. Plot q_w as a function of distance from the leading edge for a blade with a 4 cm chord. Take $\mu = 0.6 \times 10^{-4}$ kg/m sec. If the thermal conductivity $k = 0.4$ watt/cm °K, how thick can the blade's skin be if the temperature drop in the skin is not to exceed 100°K?

6.4 Using the data of figure 6.13, find the factor by which the heat flux to a wall is reduced by film cooling from a set of two rows of holes, at a distance of 10 equivalent slot widths from the holes, for a mass blowing parameter $m = 0.43$, when the adiabatic wall temperature $T_r = 1600$°K, the wall temperature $T_w = 1200$°K, and the coolant temperature $T_c = 1000$°K.

6.5 To design (approximately) the nozzle-vane cooling for a turbine, proceed as follows:
  1. Estimate, using flat-plate relations for boundary layers, the heat-transfer rate for an internally cooled blade, taking the metal temperature to be uniform. Assume that the transition from laminar to turbulent occurs at some Re_x on the order of 2.5×10^5.
  2. Calculate the internal convection cooling airflow required to cool the trailing edge, taking reasonable thicknesses for the blade walls and the cooling passage. Model the blade by a two-dimensional structure for this purpose, and use equation 6.13 for the internal flow.
  3. Now calculate the amount of cooling this internal airflow will give the rest of the blade surface—that is, find q for a reasonable thickness.
  4. Using the film cooling correlation of figure 6.13, find the distribution of film cooling holes and air needed to cool the "hot spots" on the blade (for example, those at the leading edge and at transition).
  5. Make a sketch of the vane cross section. Carry out this procedure for a vane with a chord of 4 cm, with $T_r = 1600$°K, $T_w = 1200$°K, $T_c = 1000$°K, and $p_{t4} = 25$ atm. Take $k = 40$ watts/m °K and $\mu \approx 0.6 \times 10^{-4}$ kg/m sec.

6.6 Equation 6.6 shows that the rotor-relative temperature increases with the degree of reaction R for fixed M_T and M_b. Suppose, however, that the temperature ratio τ_t is held constant as R is varied. Then how does T_{tr}/T_{ta} vary with R? Does your result imply that increasing R relieves the high-temperature, high-stress problem of the turbine rotor?

7 Engine Structures

The requirements for low weight in combination with operation at high tangential velocities (and, in the case of the turbine, at high temperatures) impose severe constraints on the design of turbomachinery, which are in turn reflected in structural characteristics unique to aircraft engines. Even a casual perusal of the cross-sectional drawings in figures 1.15–1.20 shows that an engine is composed of closely related components of rather complex shape. The reasons for these shapes and relationships must be understood at least qualitatively if one is to appreciate the compromises required in engine design, and it is such qualitative understanding that we aim for here. Techniques are available for much more precise treatment of all these matters, some in the literature and some as part of the fund of proprietary information held by each engine manufacturer, but neither the space nor the author's understanding is sufficient for a quantitative treatment here.

Many of the unique features of engine structures stem from the requirements of high-speed turbomachinery, so the structural characteristics dictated by these requirements will be discussed first. The high operating temperatures and the resulting thermal stresses produce a second set of characteristics. Finally, the need to support the rotating components of an aircraft engine in the proper spatial relationships against the thrust, pressure, and inertial loads requires specialized static structural members or frames.

7.1 Centrifugal Stresses

A brief introduction to the problem of centrifugal stress was given in section 1.10, where the stress in a rotating bar of constant cross section was computed. Now consider a rotor, such as that sketched in figure 7.1. Typically it consists of a disc of variable thickness, with inner and outer rims and blades attached to the outer rim. The reason for variation of the disc's thickness will become clear as we proceed. An analysis of this structure for arbitrary shape can be done within the framework of the theory of elasticity; however, to develop the rationale for the shape and an understanding of its design, we will proceed by making some simplifying assumptions and building the rotor up from its component parts. These are the disc, the inner and outer rims, and the blades.

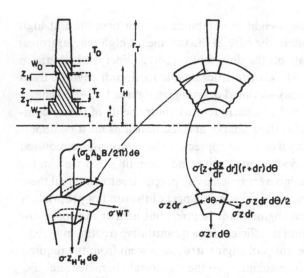

Figure 7.1
Schematic of turbomachine rotor, consisting of disc and blades,
with free-body diagram showing stresses.

7.1.1 Discs

Consider first the disc, assuming that its (axial) thickness can be so
adjusted that the tensile stress is uniform in the disc and the same in all
directions in planes perpendicular to the axis of rotation. Let this stress be
σ and consider the force balance on a small element of the disc as shown in
figure 7.1. The centrifugal force on the element is $\rho\omega^2 r d\theta dr z$, and this is
opposed by the difference between the radial forces acting on the inner and
outer surfaces of the element, plus the radial component of the tangential
forces acting on the two sides. The latter force is $\sigma z dr \sin(d\theta) \approx \sigma z dr d\theta$, so
that

$$\rho z \omega^2 r^2 d\theta dr = \sigma z r d\theta - \sigma(r + dr)\left(z + \frac{dz}{dr}dr\right)d\theta + \sigma z dr d\theta$$

$$= -\sigma r \frac{dz}{dr} dr d\theta$$

and the thickness distribution in r that gives constant σ is

$$\frac{1}{z}\frac{dz}{dr} = -\frac{\rho\omega^2 r}{\sigma} \text{ or } z = \text{const} \times e^{-(\rho\omega^2 r^2)/2\sigma}. \tag{7.1}$$

The thickness decreases toward the edge of the disc. If we think of the disc as an assemblage of concentric rings connected radially by tensile stresses, it is easy to see that part of the centrifugal force in the outer rings is supported by tensile stresses in the inner rings; thus we see that σ can be made as small as we like by increasing the thickness (in z) of the inner rings. But the smaller σ, the larger the consequent variation of z with r. If this variation is too rapid, the assumption of constant σ will no longer be valid, because the stresses will not lie approximately in a plane.

Next suppose we add an outer rim to the disc with blades attached to it as shown in figure 7.1; the rim has width W_O, thickness T_O, and a tangential stress σ equal to the disc stress, so that its circumferential strain will be the same as that of the disc. If we imagine the force exerted on the rim by the blades to be uniformly distributed, then for B blades with root area A_b and stress σ_b the force on an element of rim with angle $d\theta$ is $(\sigma_b BA_b/2\pi)\,d\theta$. The force exerted on the rim by the disc is $\sigma z_H r_H\,d\theta$, and the rim supports itself through the circumferential force $\sigma W_O T_O\,d\theta$. Equating the sum of these forces in the radial direction to the centrifugal force on the rim gives

$$(\rho W_O \omega^2 r_H^2 T_O)\,d\theta + \frac{\sigma_b A_b B}{2\pi}\,d\theta = (\sigma z_H r_H)\,d\theta + (\sigma_O W_O T_O)\,d\theta;$$

this equation can be solved for σ to give

$$\sigma = \frac{\rho\omega^2 r_H^2 + (B/2\pi)\sigma_b(A_b/W_O T_O)}{1 + z_H r_H/W_O T_O}.$$

This result shows that the stress has a contribution from the centrifugal force in the rim itself, $\rho\omega^2 r_H^2$, and a contribution from the blades. It further shows that the disc reduces the stress level, if $z_H r_H/W_O T_O$ is appreciable relative to 1, below the stress for the rim and blades alone.

The blade stress σ_b is related to the tip speed ωr_T and to r_H/r_T, as shown in section 1.10. If the blade has constant cross-sectional area with radius, then

$$\sigma_b = (\rho\omega^2 r_T^2/2)[1 - (r_H/r_T)^2],$$

so we can write the rotor stress as

$$\frac{\sigma}{\rho\omega^2 r_T^2} = \frac{\left(\dfrac{r_H}{r_T}\right)^2 + \left(\dfrac{B}{4\pi}\right)\left(\dfrac{A_b}{W_O T_O}\right)\left[1 - \left(\dfrac{r_H}{r_T}\right)^2\right]}{1 + \dfrac{z_H r_H}{W_O T_O}}. \tag{7.2}$$

A similar argument applies to the inner rim, except that in this case the rim supports the disc, so that

$$\frac{\sigma}{\rho\omega^2 r_T^2} = \frac{(r_I/r_T)^2}{1 - z_I r_I/W_I T_I}. \tag{7.3}$$

The shape of the disc between the two rims is found by evaluating the constant in equation 7.1, say, at the edge near the outer rim. Putting z there equal to z_H gives

$$\frac{z}{z_H} = \exp\left\{\frac{\rho\omega^2 r_T^2}{2\sigma}\left[\left(\frac{r_H}{r_T}\right)^2 - \left(\frac{r}{r_T}\right)^2\right]\right\}. \tag{7.4}$$

A schematic procedure for choosing the disc shape can now be seen as follows. Suppose first that the following are known in equation 7.2: the permissible stress σ; the desired tip speed ωr_T; the hub and tip radii r_H and r_T; and the number B and the root area A_b of the blades and their chord, which determines the required W_O. The quantities in equation 7.2 still to be determined are T_O and z_H. If we choose a rim thickness, then the disc thickness is determined by this equation. Going then to equation 7.4, if we

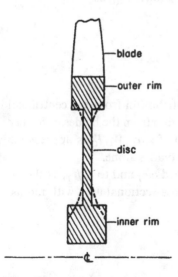

Figure 7.2
Sketch of the disc design described in subsection 7.1.1, with inner and outer rims and tapered disc.

choose r_I/r_T, then z_I/z_H is determined. Finally, $W_I T_i$ is found from equation 7.3.

To illustrate this procedure, consider the following example: $\sigma/\rho\omega^2 r_T = 0.5$, $r_H/r_T = 0.5$, $B = 40$, $A_b/W_O T_O = 0.2$, $r_I/r_T = 0.1$. Equation 7.2 requires $z_H r_H/W_O T_O = 1.91$. For the sake of simplicity suppose $W_O = T_O$, and assume $W_O = 0.1 r_T$. Then $z_H = 0.038 r_T$. From equation 7.4, $z_I = 1.27 z_H$, and from equation 7.3, $W_I/r_T = 0.12$. This disc is shown in cross section in figure 7.2 by the solid lines. Serious stress concentrations would result from the sudden area changes at the junctures of disc and rims, so they would in practice be faired as shown by the dashed lines.

Two general observations can be derived from this simple analysis:

• Geometrically similar rotors of varying size will have similar stress distributions if $\sigma/\rho\omega^2 r_T$ is constant with changing size.
• Other things being equal, the disc stress level increases as the hub/tip radius ratio decreases, because the second term in the numerator of equation 7.2, representing the blade forces, increases. This term contributed about 0.65 of the total stress for the above example.

An important criterion for design of gas turbine discs is the "burst speed," the speed at which the disc will separate into pieces and fly apart. The fragments have so much energy that in such an event total wreckage of a gas turbine and loss of an aircraft is likely to result. A fairly reliable estimate of the burst speed is obtained by regarding the two halves of the disc (separated by a diameter) as bodies held together by a uniform stress. The burst speed is the speed for which this uniform stress equals the material's ultimate strength. This relatively simple model applies because the disc material is ductile enough to yield before fracturing, thus distributing the stress uniformly. Typically, the burst speed should be 1.3 to 1.4 times design speed.

7.1.2 Centrifugal Stresses in Blades

The simple argument given in section 1.10 is readily generalized to account for variations in blade cross-sectional area with radius. If we let $A(r)$ be the area, the mean stress in the cross section at any radius r will be

$$\sigma A(r) = \int_r^{r_T} \rho\omega^2 A(r) r \, dr.$$

For a linear taper from hub to tip—that is, $A(r) = A_H(1 - \alpha r)$—we find

$$\frac{\sigma(r_H)r_H}{\rho\omega^2 r_H{}^2} = \frac{1}{2}\left\{1 - \left(\frac{r_H}{r_T}\right)^2 - \left(\frac{2\alpha r_T}{3}\right)\left[1 - \left(\frac{r_H}{r_T}\right)^3\right]\right\}. \qquad (7.5)$$

If, for example, $r_H/r_T = 0.5$ and $\alpha r_T = 0.5$, the taper reduces the right-hand side from 0.75 to 0.46, a 40 percent decrease in σ, for given ωr_T.

For untapered blades ($\alpha = 0$), equation 7.5 shows that the stress level is proportional to the ratio of annulus flow area to total area, so it is clear that the highest blade stress levels will occur in the first stage of a compressor and in the last stage of a turbine if the tip speed is constant throughout the components.

7.2 Gas Bending Loads on Blades

The gas-dynamic forces acting on a small radial element of a rotating blade are indicated in figure 7.3 to be principally axial and tangential. There is also a radial component, but it is always negligible in comparison with the centrifugal forces. The axial and tangential forces generate a bending moment about the root of the blade, which may be thought of as a cantile-

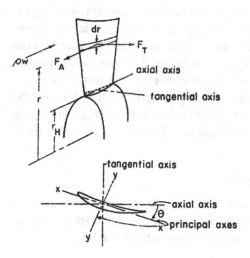

Figure 7.3
Diagram of tangential and axial forces acting on a blade to produce bending stresses at the root about the axial and tangential directions, or about the principal axes x and y.

vered beam. The maximum stress due to the gas forces will be at the root; since the centrifugal stresses are largest there also, it is the critical location.

To estimate the stresses, we can calculate the moment M_z about an axis through the blade root and parallel to the rotational axis and the moment M_θ about a tangential axis, then combine them to find the moments about the principal axes of the blade cross section and hence the bending stress.

For the sake of simplicity let us assume that the fluid streamlines do not change radius as they pass through the rotor, that the fluid is incompressible, and that the axial velocity is constant. The tangential force acting on the element dr of the blade is then simply $dF_\theta = \dot{m}(v_c - v_b)$, where $\dot{m}$ is the mass flow associated with the element dr of the blade and $v_c - v_b$ is the tangential velocity change across the rotor. If B is the number of blades,

$$BdF_\theta = \rho\omega^2\pi r(v_c - v_b)\,dr.$$

The axial force is due to the static pressure rise across the blade row, since we are assuming that the axial velocity is constant. It is therefore

$$BdF_z = (p_c - p_b)(2\pi r dr).$$

The two moments are then

$$BM_z = \int_{r_H}^{r_T} BdF_\theta(r - r_H) = \int_{r_H}^{r_T} \rho w 2\pi r(v_c - v_b)(r - r_H)\,dr$$

and

$$BM_\theta = \int_{r_H}^{r_T} BdF_z(r - r_H) = \int_{r_H}^{r_T} (p_c - p_b)2\pi r(r - r_H)\,dr.$$

To estimate the first integrand in terms of easily interpreted quantities, we may use the Euler equation 5.7, which states that

$$v_c - v_b = \frac{c_p(T_{tc} - T_{tb})}{\omega r} = \frac{c_p T_{tb}(\tau_s - 1)}{\omega r}.$$

The pressure difference can be found by noting that the stagnation pressure of the fluid relative to the rotor is constant across the rotor, so

$$p_c - p_b = \rho[(V_b')^2 - (V_c')^2]/2$$

in the incompressible limit, and this can be written

$$p_c - p_b = \left(\frac{\rho w^2}{2}\right)(\tan^2 \beta'_b - \tan^2 \beta'_c)$$

$$= \left(\frac{\rho w^2}{2}\right)(\tan^2 \beta'_b - \tan^2 \beta'_c)(\tan^2 \beta'_b + \tan^2 \beta'_c).$$

However, for constant axial velocity,

$$\tan \beta'_b - \tan \beta'_c = (v_c - v_b)/w,$$

so that

$$p_c - p_b = \tfrac{1}{2}\rho w (v_c - v_b)(\tan \beta'_b + \tan \beta'_c)$$

$$= \frac{c_p T_{tb}(\tau_s - 1)}{\omega r}\tfrac{1}{2}\rho w(\tan \beta'_b + \tan \beta'_c).$$

Finally, the moments are

$$BM_z = \int_{r_H}^{r_T} \frac{1}{\omega}[2\pi\rho wc_p T_{tb}(\tau_s - 1)](r - r_H)\,dr$$

and

$$BM_\theta = \int_{r_H}^{r_T} \frac{1}{\omega}[\pi\rho w(\tan \beta'_b + \tan \beta'_c)c_p T_{tb}(\tau_s - 1)](r - r_H)\,dr.$$

If τ_s is independent of r, the first integral can be evaluated without further assumptions; it becomes

$$BM_z = \frac{1}{\omega}\rho wc_p T_{tb}(\tau_s - 1)(\pi r_T^2)\left\{\left[1 - \left(\frac{r_H}{r_T}\right)^2\right]\left(1 - \frac{2r_H}{r_T + r_H}\right)\right\}. \qquad (7.6)$$

The second integrand is modified by the factor $(\tan \beta'_b + \tan \beta'_c)/2$, which is just the mean stagger angle. Because most of the contribution to the integral comes from the tip, where $\tan \beta'_c$ is not much different from $\tan \beta'_b$, we approximate the factor by $\tan \beta'_b$. Then, in the notation of subsection 5.1.2,

$$\tan \beta'_b = [(\omega r_T/w) - A](r/r_T),$$

where A expresses the turning of the inlet guide vane. The tangential moment is then

$$BM_\theta = \frac{1}{\omega}\rho wc_p T_{tb}(\tau_s - 1)(\pi r_T^2)\left(\frac{\omega r_T}{w} - A\right)\left[\frac{2}{3} - \frac{r_H}{r_T} + \frac{1}{3}\left(\frac{r_H}{r_T}\right)^3\right]. \qquad (7.7)$$

The moments about the principal axes of the blade cross section are related to M_z and M_θ by

$$M_{xx} = M_z \cos\theta + M_\theta \sin\theta, \quad M_{yy} = M_\theta \cos\theta - M_z \sin\theta. \tag{7.8}$$

The maximum bending stress is then

$$\sigma_{\text{bend}} = \frac{M_{xx} y_{\text{max}}}{I_{xx}} + \frac{M_{yy} x_{\text{max}}}{I_{yy}}, \tag{7.9}$$

where y_{max} and x_{max} are the maximum distances from the axes and I_{xx} and I_{yy} are the principal moments of inertia.

To estimate the magnitude of the bending stresses, consider just the factor in front of the braces in equation 7.6 and that in front of the brackets in equation 7.7; the factors involving r_H/r_T are of order unity. Take $\sigma_{\text{bend}} \approx M_y/I$; $y \approx t$, the blade thickness, and $I \approx ct^3$, where c is the blade chord; then we find that

$$\frac{\sigma}{p} \approx \left(\frac{w}{\omega r_T}\right)(\tau_s - 1)\left(\frac{s}{2c}\right)\left(\frac{r_T}{t}\right)^2 \tag{7.10}$$

apart from a numerical factor of order unity. This result shows the following scaling rules:

• The stress is proportional to the air pressure and to the temperature rise of the blade row.
• The stress is inversely proportional to the solidity and inversely proportional to the square of the ratio of blade thickness to tip radius.

Taking for example

$w/\omega r_T = \frac{1}{2}$,

$\tau_s - 1 = \frac{1}{4}$,

$c/s = 2$,

and

$r_T/t = (r_T/c)(c/t) = (6)(10) = 60$,

we find $\sigma/p = 112$, and for $p = 1$ atm, $\sigma \approx 112$ atm. This is a modest stress level; however, for thinner blades and higher aspect ratios $[(r_T - r_H)/c]$ the bending stress can become an appreciable part of the total stress.

7.3   Thermal Stresses

Thermal stresses result from the tendency of materials to expand with increasing temperatures. In a turbine blade, for example, if the leading and trailing edges are hotter than the middle part of the airfoil, they tend to grow radially. But to grow they must stretch the center of the blade. The result is a compressive stress in the leading and trailing edges and tension in the center. This tensile stress would add to the centrifugal stress already present.

The thermal stress distribution can be analyzed as follows. Suppose first that the blade cross section were divided into a large number of small elements, as suggested in figure 7.4, each independent of the others mechanically, so each could expand radially in accordance with its temperature and its centrifugal stress. Suppose also that the temperature distribution in the blade were known. The total radial strain (fractional elongation) would then be

$$\varepsilon = \frac{\sigma_c}{E} + \alpha(T - \overline{T}),\tag{7.11}$$

where α is the thermal coefficient of expansion, σ_c is the centrifugal stress, E is the elastic modulus of the material, and $\overline{T}$ is the area-averaged temperature across the blade cross section. If the blade is long and has the assumed

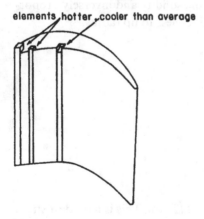

elements hotter, cooler than average

Figure 7.4
Schematic representation of the mechanism that generates thermal stresses in blades due to chordwise temperature gradient.

temperature distribution all along its length, then the total strain e must be the same in each of the elements. The thermal stress is the stress that must be added to each individual fiber so that its strain equals the average strain. That is, if $\bar{\varepsilon}$ is the average strain,

$$\frac{\sigma_T}{E} = \bar{\varepsilon} - \frac{\sigma_c}{E} - \alpha(T - \bar{T}).$$

Since the thermal stresses are all within the blade, they must sum to zero over the cross section, so that $\bar{\varepsilon} = \sigma_c/E$, and the thermal stress is just $\sigma T/E = -\alpha(T - \bar{T})$. The total stress is

$$\frac{\sigma}{E} = \frac{\sigma_c}{E} + \alpha(T - \bar{T}). \tag{7.12}$$

The centrifugal stress, being tensile, is negative, so equation 7.12 indicates that the tensile stress is highest in the lower-temperature elements. A typical value for α is about $0.9 \times 10^{-5}\ ^\circ K^{-1}$. E is about 2×10^6 atm for steels, so αE is 18 atm/$^\circ$K, and we see that a temperature difference of 100°K will produce a thermal stress of 1800 atm!

A better appreciation for the importance of thermal stresses can be had by relating the temperature difference to heat fluxes required in internally cooled blades. Consider the skin of thickness t of the blade in figure 6.9. For this case, $q_w = k(\Delta T/t)$, where ΔT is the temperature drop across the skin. If we put $T - \bar{T}$ in equation 7.12 equal to $\Delta T/2$, then

$$\sigma - \sigma_c = \alpha E(\Delta T/2) = \alpha E q_w t/2k.$$

Recall from equation 6.8 that $q_w \approx \rho u c_p (T_r - T_w) St$. Combining these facts leads to

$$\frac{\sigma - \sigma_c}{p} = \frac{\alpha E}{k}(ut)\frac{\gamma}{2(\gamma - 1)}[1 + \tfrac{1}{2}(\gamma - 1)M_{T_r}{}^2]\left(1 - \frac{T_w}{T_r}\right)St. \tag{7.13}$$

The first factor, $\alpha E/k$, expresses the material's susceptibility to thermal stresses. For high-alloy steels it is of order 54,000 sec m^{-2}. The second factor is the flow velocity over the blade times the thickness of the blade skin; for consistency it would have dimensions of m^2/sec. As an example, suppose $u = 600$ m/sec, $t = 1.5$ mm, and $St = 0.003$; then

$$(\sigma - \sigma_c)/p \approx 300(1 - T_w/T_r).$$

In a high-pressure-ratio engine, $p \approx 20$ atm, so

$$\sigma - \sigma_c \approx 6000(1 - T_w/T_r) \text{ atm.}$$

Thus, cooling just 10 percent below the gas temperature gives a thermal stress of 600 atm under such circumstances in a 1.5-mm-thick skin. In other words, increasing T_r increases the blade thermal stress, other things being equal. The thermal stress for the above example is 3000 atm, where T_r is twice T_w. This is one reason that film cooling must be used for turbines with large ratios of gas temperature to metal temperature. From the standpoint of thermal stress, the blade thickness must be maintained small as engine size is increased; this is an example of the inapplicability of geometric scaling in going from small to large engines.

The importance of precision in cooling designs can readily be appreciated from these estimates. With a heat-transfer distribution such as in figure 6.9, the internal cooling and the film cooling must be arranged to make the metal temperature T_w as nearly uniform as possible, at a level acceptable to the material (currently about 1250°K maximum for aircraft applications and nearer 1150°K for ground applications). A difference of perhaps 15°K in local temperature can result in a factor-of-2 change in turbine life at the 1250°K level. Now, if the blade is held at 1250°K in a 1600°K stream, with cooling air at 850°K, the adiabatic effectiveness must be about 0.5, and it must be known within about 5 percent to hold the temperature variations to 15°K. Currently, the limit on turbine inlet temperature is set by the precision of blade-cooling design techniques.

There can be serious thermal stresses in turbine discs also. When hot gases impinge on the outer rim of the disc or when heat is conducted into it from the blades, it tends to expand relative to the cooler inner portions of the disc. This reduces the tensile stress in the rim and increases that in the disc, so care must be exercised in cooling the disc and blade mounts.

7.4   Critical Speeds and Vibration

In figure 7.5, a gas turbine engine is idealized for the purpose of vibration analysis as a flexible rotor consisting of a lumped mass mounted on a flexible shaft mounted in flexible bearings. This model illustrates some of the more important vibrational characteristics of engines. In addition, the individual blades and discs can vibrate both alone and cooperatively, but different models will be required to represent these modes.

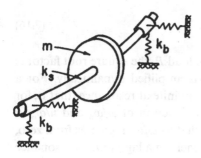

Figure 7.5
Model of a turbomachine rotor of mass m, with shaft spring constant k_s and bearing spring constant k_b.

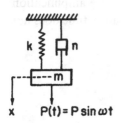

Figure 7.6
Single-degree-of-freedom mass-spring system with damping.

Frequent reference will be made to the forced oscillation of a mass-spring system with one degree of freedom, such as that sketched in figure 7.6. The differential equation describing its motion for a harmonic driving force is (reference 7.1)

$$\ddot{x} + 2n\dot{x} + \omega_n{}^2 x = \left(\frac{P}{m}\right)\sin\omega t, \qquad (7.14)$$

where $\omega_n{}^2 = k/m$ is the natural frequency, n is a damping coefficient, and P is the amplitude of the exciting force. The solution for the forced steady-state vibration is $x = A\sin(\omega t - \phi)$, where the amplitude A is given by

$$A = \frac{P}{k}\frac{1}{\sqrt{[1 - (\omega^2/\omega_n{}^2)]^2 + 4n^2\omega^2/\omega_n{}^4}} \qquad (7.15)$$

and the phase angle ϕ is given by

$$\tan\phi = \frac{2n\omega}{\omega_n^2 - \omega^2}.\qquad(7.16)$$

Since P/k is the static deflection under the load P, the square root factor is the ratio by which the static deflection is amplified dynamically. For a system with no damping it would become infinite at resonance (that is, for $\omega = \omega_n$. The amplification is shown as a function of ω/ω_n and n/ω_n in figure 7.7, as is the phase angle ϕ. We see that as ω/ω_n increases from zero, ϕ increases; the motion lags the forcing, reaching a lag of $\pi/2$ at resonance and approaching π as $\omega/\omega_n \to \infty$.

The behavior near resonance is critical for gas turbines since they must at times pass through resonant speeds to reach their normal operating speed. This steady-state solution indicates that a very large amplification will occur at resonance for low damping, but this large amplitude does not

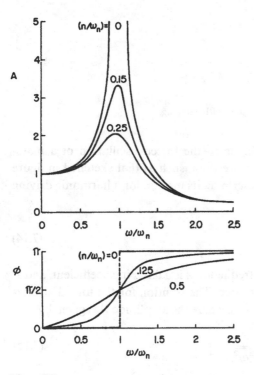

Figure 7.7
Amplification factor A and phase lag ϕ for the single-degree-of-freedom vibrating system of figure 7.6.

occur instantly when the system is driven at resonant frequency, because it is initially at rest. The amplitude in fact increases linearly in time; this may be seen by constructing the solution for $n = 0$ with the initial condition $x(0) = 0$. The result is

$$x = -\frac{P}{k}\left(\frac{\omega_n t}{2}\right)\cos\omega_n t, \tag{7.17}$$

so the amplification factor becomes unity in two natural periods. Damping would reduce the rate of growth.

7.4.1 Shaft Critical Speed

Let us now apply these results to the rotor system of figure 7.5, supposing that excitation is caused by a displacement of the center of mass c of the disc by a distance e from the center s of the shaft. Viewed along the axis of rotation, the relation of points c and s might be as in figure 7.8. The center of mass c is at x,y, and in general s could be at any point distant e from c.

The deflections of the shaft in x and y are $x - e\cos\theta$ and $y - e\sin\theta$; thus, the two equations of motion are

$$m\ddot{x} = -k_s(x - e\cos\theta), \qquad m\ddot{y} = -k_s(y - e\sin\theta). \tag{7.18}$$

There are three unknowns here, x, y, and θ, but we note that a solution exists for $\theta = \theta_0 + \omega t$, where s and c lie on a common radial line that rotates at angular velocity ω. For this case

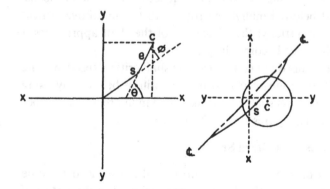

Figure 7.8
Relation of center of mass c to axis of rotation $x = 0$, $y = 0$ and centerline-deflected shaft s for disc-shaft assembly.

$$m\ddot{x} + k_{\mathrm{s}}x = e\cos(\omega t + \theta_0), \quad m\ddot{y} + k_{\mathrm{s}}y = e\sin(\omega t + \theta_0), \qquad (7.19)$$

so the motion in y is just that in x shifted $\pi/2$ in ωt, and we do not need the third equation, which is required for the general case of arbitrary ϕ. Either of the equations is mathematically the same as equation 7.14 with $n = 0$; thus, the solution has the same form. (P/m has become e/m, however.) Then we have

$$x = \frac{e}{1 - (\omega/\omega_{\mathrm{ns}})^2}\cos\omega t,$$

where $\omega_{\mathrm{ns}}^2 = k_{\mathrm{s}}/m$ and where the phase angle ϕ is zero for $\omega < \omega_{\mathrm{ns}}$ but is π for $\omega > \omega_{\mathrm{ns}}$. That is, for $\omega < \omega_{\mathrm{ns}}$ the center of mass lies outside the shaft center, and as $\omega \to \omega_{\mathrm{n}}$ from below, x/e tends toward ∞. For $\omega > \omega_{\mathrm{ns}}$, however, the center of mass lies inside the shaft center and $x \to -0$ as $\omega/\omega_{\mathrm{ns}} \to \infty$, so the disc tends to rotate about its center of mass. The first condition is termed *below critical speed*, the second *above critical speed*. Note that the critical angular speed of the rotor is just equal to the vibrational angular frequency in transverse vibrations it would exhibit if not rotating.

There are several points of practical importance here. First, the amplitude of vibration becomes large if operation is prolonged at speeds near critical speed. Second, there is little damping within the rotor, because the geometry of the rotor is fixed as it spins in this mode. Thus, if a rotor is to be run above critical speed, the passage through critical speed must be rapid. The passage through critical speed can be eased somewhat by introducing damping in the bearings, through oil squeeze films or other means, as will be explained below. Finally, the rotor can be run above critical speed with a small eccentricity; the deflection of the shaft approaches a limiting value e as the speed becomes large.

Many gas turbine rotors do in fact run above shaft critical speed, but the modern tendency is to design rotors to be very stiff in bending by using large-diameter conical shafts such as that sketched in figure 6.8. A number of examples can be seen in figures 1.15–1.20 as well.

7.4.2   Structural Loads and Critical Speeds

No rotor is perfectly balanced, so the rotation will always lead to some shaft deflection and to a fluctuating load on the bearings. For the system of figure 7.5, with $k_{\mathrm{b}} \gg k_{\mathrm{s}}$, the load would be half the spring force of the shaft per bearing, or $-k_{\mathrm{s}}(x - e\cos\omega t)$ for the two bearings. With the expression for x, the force on the bearings is then

$$P(t) = -k_s e \left(\frac{(\omega/\omega_{ns})^2}{1 - (\omega/\omega_{ns})^2} \right) \cos\omega t.$$

The oscillating load on the bearings first increases as ω^2, tending to infinity as $\omega \to \omega_{ns}$, then switches sign and decreases from infinity, approaching $k_s e \cos\omega t$ as $\omega/\omega_{ns} \to \infty$. Thus, the "flexible shaft" rotor will have moderate bearing loads at high rotor speeds, but a stiff rotor would produce very large loads in rigid bearings such as have been assumed thus far. This problem is overcome by making the bearing mounts flexible. If we think of the whole rotor system for $\omega_{ns} \gg \omega$ as a mass mounted on a spring k_b representing the bearing stiffness, the assembly is directly analogous to the mass-spring system of figure 7.6 and will have the same response to the excitation, which is now $P(t) = -me\omega^2 \cos\omega t$. The motion of the shaft at the bearing will be $x_b = A_b \cos\omega t$, where

$$A_b = \frac{-e}{\sqrt{(\omega_{nb}^2/\omega^2 - 1)^2 + 4n_b^2/\omega^2}}; \tag{7.20}$$

this shows that $A_b/e \to 0$ for $(\omega_{nb}/\omega)^2 \to \infty$, as expected; but further, for $(\omega_{nb}/\omega)^2 \to 0$ (i.e., for very soft bearings), we have

$$\frac{A_b}{e} \to \frac{-1}{\sqrt{[1 + 4n_b^2/\omega^2]}},$$

which is always less than 1. The force transmitted to the engine structure through the bearing will be

$$P_b(t) = x_b k_b = x_b m\omega_{nb}^2 = \frac{em\omega_{nb}^2}{\sqrt{1 + 4n_b^2/\omega^2}}. \tag{7.21}$$

This result shows that reducing the stiffness and increasing the damping of the bearing mounts will greatly reduce the vibrational loads on the engine structure without resulting in shaft displacements in excess of e provided $\omega^2 \gg \omega_{nb}^2$.

Many modern aircraft gas turbines use very soft oil-damped bearings in which the outer race is surrounded in its housing by pressurized oil in such a way that the oil must be squeezed out of a radial clearance space when the bearing's outer race moves radially. This provides the high level of

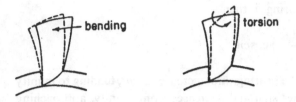

Figure 7.9
Bending and torsional modes of vibration of a cantilevered compressor blade.

damping desired. A soft mechanical spring may be included to center the outer race at low speeds.

It should be clear that vibrations can arise from deflections of the non-rotating structure of the engine, as well as from deflections of the rotating components. Since there is a high premium on low weight in these components, they tend to be flexible, and critical speeds can arise if one or more resonant modes of the structure lie in the operating range of the engine. As we shall see in the discussion of engine layout, these considerations have a large influence on the overall arrangement of the components in an engine.

7.4.3   Blade and Disc Vibration

Each of the turbomachine blades individually, and the assembly of blades and their supporting disc collectively, are capable of vibrating in a number of modes. If the blades are attached only at their roots, then in first approximation they may be thought of as cantilevered beams, as sketched in figure 7.9.

Such a blade can vibrate in bending and in torsion. The density of most turbomachine blades is so large relative to that of the fluid that the modes of vibration are very nearly those for the blade in a vacuum. That is, the effect of the fluid is only to provide excitation or damping; it does not change appreciably the mode shape or frequency. The frequencies are influenced by temperature through its effect on the elastic modulus and by the speed of rotation N, the centrifugal force raising the frequency. The possibilities for excitation of a given blade can be represented by a "Campbell diagram" in which the frequencies of the various modes are plotted versus N, as in figure 7.10, and lines representing multiples of N are superimposed. Any intersection of the latter with a blade frequency curve within the normal speed range of the machine is a potential source of destructive vibration. Not all can be avoided, of course. The strongest excitation usu-

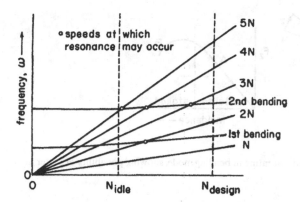

Figure 7.10
Frequency diagram or "Campbell diagram" representing possible resonant excitation points for a blade in bending vibration.

ally results from resonances with N, $2N$, $3N$, and perhaps $4N$ excitations, so it is usual to avoid these.

Often blades are coupled at their tips or at an intermediate radius by "shrouds" which at operating speed effectively join into a solid ring. Then more complex blade-shroud modes are possible (reference 7.2).

The discs by themselves are capable of vibration in various plate-type modes. In the "umbrella" mode, the disc center oscillates axially against the rim. A series of modes occur in which the nodes are diameters, and the disc rim assumes a wavy axial displacement between nodes. These are particularly likely to be excited in cooperation with a bending vibration of the blades.

7.5 Blade Flutter

The vibrations discussed in section 7.4 all result from some forced excitation resulting from the rotation of the machine, due either to structural inertia forces or to fluctuating fluid-mechanical forces in which the time dependence is derived from blades passing other blades, struts, and so on. Another class of vibrations resulting from instabilities can arise when fluid does work on a vibrating blade (or other part) to amplify or maintain the vibration. Blade flutter is such a vibration. It is a very complex subject, so it will not be discussed in any comprehensive way here. But unlike the forced

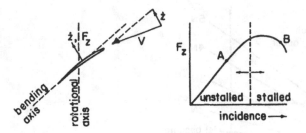

Figure 7.11
Force diagram for isolated blade vibrating in bending mode, showing possibility of flutter in stalled range of incidence.

vibration discussed above, the mechanisms of excitation of flutter are not so readily apparent from experience with mechanical systems, so physical descriptions of some of these mechanisms will be given.

7.5.1 Bending Flutter

Consider first the circumstances in which an isolated blade may flutter in bending. In the cascade representation of this blade, bending results in a vibrating velocity $\dot{z}$ more or less perpendicular to its chord, as shown in figure 7.11. As it vibrates, the direction of the incident flow is changed as indicated by the dashed vector. From the curve of blade force F_z versus incidence, at the right of figure 7.11, we see that if the blade is operating in the unstalled range of incidence, say at point A, the force F_z is reduced by the motion $\dot{z}$, and this will retard or damp the motion. On the other hand, if the blade is operating in the stalled range of incidence, F_z is increased by $\dot{z}$, increasing the force in the direction that will cause the motion to grow. Thus we conclude that flutter is possible whenever the blade is at or beyond stalling incidence.

This argument suggests that unstalled blades will not flutter. This is true for isolated blades but not for cascades, as may be seen by examining the influence of neighboring blades of the cascade on any particular blade (reference 7.3). Consider the cascade sketched in figure 7.12, where the blade on which we wish to find the forces is denoted by 0, its first neighbor in the positive z direction by $+1$, and so on. Now, in an infinite cascade, the blades must all flutter with the same amplitude, because nothing distinguishes one from another, but there can be a phase shift from one to the next. In figure 7.12, the $+1$ and -1 blades are drawn dashed as they would

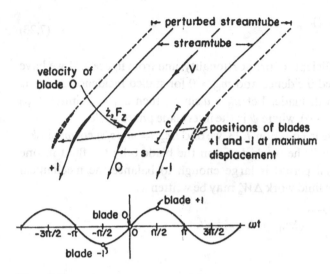

Figure 7.12
Diagram of compressor cascade in bending flutter.

appear if the phase shift from blade 0 to blade $+1$ were $\pi/2$. That is, when blade 0 has maximum positive velocity $\dot{z}$ and zero displacement, blade $+1$ will have maximum positive displacement, and blade -1 will have maximum negative displacement. The sketch shows that the streamtube between blades -1 and $+1$ is then wider than for the nonvibrating cascade. If the turning across the cascade is not decreased much by the vibration, the widened streamtube will result in an increased force on blade 0 in the direction of its motion; hence, the motion will be amplified. We conclude that an unstalled cascade can flutter in bending for an interblade phase angle $\psi = \pi/2$ if these exciting forces are not dominated by damping.

Now suppose the force perturbation ΔF_z can be represented as the sum of ΔF_{zip}, which would result if all blades had uniform spacing (if they vibrated in phase), and $\Delta F_{z\psi}$, which is due to the phase shift between blades. Both force perturbations should be proportional to the dynamic pressure $\rho V^2/2$, so we put $\Delta F_z = \Delta c_z(\rho V^2/2)(ch)$, where c is the blade chord and h is the blade height. By the above physical argument,

$$\Delta c_{zip} = \frac{m_{ip}\dot{z}_0}{V} \tag{7.22}$$

and

$$\Delta c_{z\psi} = \frac{m_\psi(z_{+1} - z_{-1})}{2s}. \tag{7.23}$$

m_{ip} and m_ψ are coefficients of proportionality, and from figure 7.11 we have $m_{ip} < 0$ for unstalled incidence and $m_{ip} > 0$ for stalled incidence. $\dot{z}_0$ is the velocity of the zeroth blade. Let $z_0 = a \sin\omega t$; then $z_{+1} = a \sin(\omega t + \psi)$ and $z_{-1} = a \sin(\omega t - \psi)$, where ψ is the interblade phase angle.

Substituting these in equation 7.23, we find $\Delta c_{z\psi} = m_\psi(a/s) \cos\omega t \sin\psi$.

Flutter will occur if the work done on the blade 0 by the fluid in one complete vibrational period is large enough to balance the mechanical damping work. The fluid work ΔW_a may be written as

$$\Delta W_a = \tfrac{1}{2}\rho V^2(ch) \int_0^{2\pi/\omega} (\Delta c_{zip} + \Delta c_{z\psi})\dot{z}_0 dt,$$

which yields

$$\Delta W_a = \tfrac{1}{2}\rho V^2(ch)\pi a \left[\left(\frac{m_{ip}}{V}\right) a + m_\psi \left(\frac{a}{s}\right) \sin\psi \right].$$

If we define a mechanical logarithmic decrement δm as the fractional decrease in vibrational energy per cycle due to mechanical damping, then the damping work per cycle, ΔW_m, is equal to $-\delta_m m_b(\omega^2 a^2/2)$, where m_b is the effective mass of the blade. The condition for marginal stability is then $\Delta W_a + \Delta W_m = 0$. If we define a reduced velocity $\bar{V} \equiv V/c\omega$, this condition becomes

$$\mathrm{m}_\psi \sin\psi \bar{V}^2 + \frac{m_{ip}s}{c}\bar{V} - \frac{1}{4}\left(\frac{s}{c}\right)\mu\delta_m = 0, \tag{7.24}$$

where $\mu \equiv 4m_b/\pi c^2\rho h$ is a blade density relative to the fluid density.

This result exhibits the main parameters controlling at least one form of blade flutter. Both m_ψ and m_{ip} are functions of Mach number, incidence, cascade geometry, and sometimes Reynolds number. Equation 7.24 shows that flutter will occur when the first term, which is proportional to $\bar{V}^2$, becomes large enough to offset both the last term, which is always negative, and the second term, which is usually negative for unstalled operation. Since ψ is arbitrary, it can assume whatever value maximizes $|m_\psi \sin\psi|$. The argument above suggests this value is $\psi = \pi/2$ for compressors. This is roughly confirmed by detailed analysis (reference 7.4).

7.5.2   Flutter Clearance

It is usual to represent the flutter behavior of a compressor by indicating the regions of the compressor map in which flutter may be encountered. Figure 7.13 shows the various possible flutter boundaries for a high-speed fan, the region in which flutter may occur being shaded. Several types of flutter have been identified, more or less by the regions of the compressor map in which they tend to occur. Supersonic unstalled flutter occurs as corrected speed is increased along the normal operating line, so it can set an upper limit to operating speed. Stall flutter may occur as the pressure ratio is increased at fixed corrected speed, and is likely to be a bending flutter of the type discussed. Because it occurs in a region of the map where the compressor is also susceptible to rotating stall, it may be confused with this during compressor testing. Choke flutter occurs when the compressor is operating with low backpressure and high flow speeds over the blades.

Establishing that flutter instability regions do not overlap the normal operating range of the machine is termed *flutter clearance*. This is a particularly difficult procedure because the flutter behavior of the compressor is

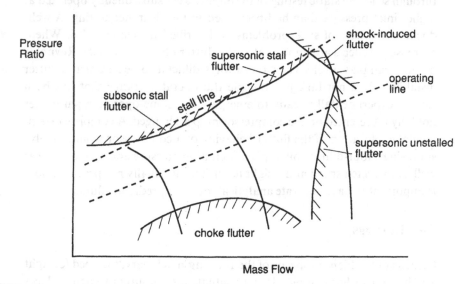

Figure 7.13
Regions of the compressor map in which flutter is encountered. (Adapted from reference 7.5.)

not fully represented on the conventional compressor map. Whereas the performance of the compressor can be represented with some confidence in terms of the corrected speed and weight flow (and sometimes the Reynolds number), the flutter phenomena introduce at least the additional parameter of reduced velocity, and sometimes the μ parameter, representing the blade density relative to air density. The reduced velocity compares the flow time to the vibrational period of the relevant blade mode, so it does not depend only on the flow Mach number. For example, as air temperature is increased at fixed corrected speed and weight flow (i.e., at a fixed point on the compressor map), the reduced velocity increases, so one might find a flutter instability boundary at a particular corrected speed when the inlet temperature is high that does not exist when the inlet temperature is low. Similarly, the air density can influence the stability through the Reynolds number and also through the ratio of air and blade densities. Ordinarily, the first two terms of equation 7.24 are larger than the last, so the limiting value of $\overline{V}$ is almost independent of fluid density, which appears only in the last term. It has happened, however, that this last term has been of controlling importance when engines that had been experimentally "cleared" of flutter problems through rig tests of the components or through sea-level static testing of the engine were subsequently operated at higher inlet pressure than had been used in the clearance testing. A well-documented case of such problems is described in reference 7.6. Whenever viscous effects are important in the flutter phenomenon, the Reynolds number can play a role. Thus, it is always difficult to be sure that a flutter instability does not lurk just outside the operating range that has been cleared experimentally, ready to emerge when some operating parameter not fully represented on the compressor map is changed. A complete experimental exploration of the flutter behavior of a compressor would involve surveying a four-dimensional space (including pressure and temperature as well as corrected speed and weight flow). This is usually not practical, so it is important to have accurate analytical tools for predicting flutter.

7.6 Bearings

Because of the high tip speeds of the rotating machinery, the need for light weight, and rather complex loading situations, an aircraft engine places rather difficult requirements on the bearings used to support the rotating

assemblies. Whereas fluid-film bearings are used on many types of stationary high-speed turbomachinery and magnetic bearings are now in the experimental phase, thus far aircraft engines have used rolling-element (ball
or roller) bearings exclusively. As we shall see in section 7.7, each of the
rotating elements is supported by two or more bearings. One of these is a
ball bearing of the single-row or Conrad type, which can accept both radial
and axial loads. It serves to position the rotating element axially as well as
to absorb the radial loads at its end of the shaft. The other bearings on this
rotating element are cylindrical roller bearings, which absorb radial loads
while allowing axial movement to permit unconstrained differential thermal expansions between the rotating element and its supporting structure.

As was noted in section 7.4, it is usually desirable to make the rotors stiff
in bending, and, other things being equal, this suggests the largest possible
diameter. The bearings limit the rotor diameters at the support points
through their limited capability to tolerate high tangential velocities of
their balls or rollers. This is conventionally expressed in terms of a limit on

DN = (Bore diameter in mm) × (Speed in rpm).

For commercial high-precision bearings such as are used on high-speed
grinding spindles, the DN limit is usually set at about 10^6 (mm) (rpm). In
aircraft engines, more sophisticated designs, better materials, and higher-
capacity oil cooling permit values of DN up to about 2.2×10^6 (mm) (rpm).

It is helpful to translate the DN limit into terms comparable to those
used to describe the aerodynamics of the compressor or turbine. Thus, in
terms of a tangential velocity at the radius,

$r_s = D/2,$

and in more conventional units,

$$(DN)_{max} \to \omega r_s = 2.0 \times 10^6 (10^{-3}) \frac{2\pi}{60} \frac{1}{2} = 105 \text{ m/sec}.$$

Thus, the maximum shaft radius is related to the tip radius of the rotor by

$$\frac{(r_s)_{max}}{r_T} = \frac{105}{\omega r_T}. \tag{7.25}$$

If, for example, the tip speed is 500 m/sec (roughly 1500 ft/sec), $(r_s)_{max}/r_T =$
0.21. Examination of the engine cutaway drawings in chapter 1 will show
that the bearings on the high-speed rotors are fairly close to this limit.

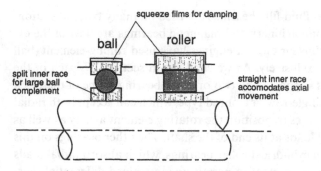

Figure 7.14
Cross sections of typical main-shaft ball and roller bearings.

Cross sections of typical main-shaft bearings are shown in figure 7.14. The ball bearing has a one-piece outer race and a divided inner race; the latter allows a larger ball complement, and hence a greater load capacity, than could be included with the one-piece inner race used in most commercial bearings. Cooling is by oil injected through the inner race, or by means of jets impinging on the balls from the side.

To introduce damping, the outer race may be mounted in an oversized cavity, into which oil is injected, so that the bearing can move radially by forcing the "squeeze film" to flow circumferentially in the housing.

In the roller bearing, the rollers are positioned axially by either the outer or the inner race, and the other race is cylindrical, to allow axial movement.

Both ball and roller bearings sometimes exhibit complex behaviors. The balls or rollers may "slip" or "skid," instead of rolling, under circumstances where they lose contact with the rotating (usually inner) race. This can lead to fretting damage to the balls or rollers, or to the races. It can be avoided for the ball bearings by ensuring that there is always some axial load. Roller-bearing races are sometimes deliberately made slightly elliptical, to ensure contact without uncontrolled radial loads due to thermal expansion. This also helps to prevent "skewing" of the rollers, i.e., their axes' deviating from parallel to the axis of rotation. See reference 7.7 for a discussion of these phenomena. As was noted above, consideration has recently been given to the use of magnetic bearings in aircraft engines. They have been used in high-speed stationary machinery, such as turbomolecular pumps (see, for example, reference 7.8), but are in the experimental phase for aircraft engines. The operating principles of such bearings are illus-

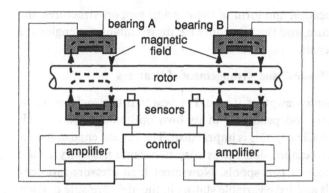

Figure 7.15
Schematic of magnetic bearing system, showing two radial bearings. (Adapted from reference 7.9.)

trated in figure 7.15. Since there is no stable location in a steady magnetic field for a magnetic object such as the shaft, the system depends for stability on feedback from position sensors. The current in the magnets is varied so as to position the shaft in the plane of the drawing and also perpendicular to it, at the desired locations. An actual bearing might use as many as eight poles for each bearing. The axial position of the shaft would be controlled similarly by magnets working on the face of a flange.

Magnetic bearings offer many benefits to aircraft engines. They eliminate the *DN* limitation, thus making possible higher rotational speeds and stiffer rotating structures. They may eliminate the oil system entirely, and the associated requirement for cooled bearing sumps and the associated multiple seals. Active control of the bearings also opens up the possibility of using the control system to damp vibrations, to eliminate the effects of unbalance, and to actively position the rotor for any of a number of reasons. These benefits are such that engines will very likely incorporate magnetic bearings in the future.

7.7 Engine Arrangement and Static Structure

There is room for ingenuity in arranging the basic components of the engine, namely the fan, the compressor, the combustor, and the turbines. Among the major decisions which the designer must make are the number of independent rotating assemblies ("spools"), the number and location of the bearings, the location and form of the bearing support structures, the type of rotor structure, and the procedure for disassembing the engine for maintenance and repair.

7.7.1   Number of "Spools" and Arrangement of Bearings

From the aerodynamic viewpoint, it would be best to rotate each compressor and turbine stage independently at its own optimum speed at each operating condition. Clearly, this is impractical. The earliest engines used a single spool, but as compression ratios rose it became necessary to use either variable stators or two spools. Now most high-pressure-ratio engines use two spools and have variable stators in the high-pressure portion of the compressor. An important exception is the Rolls-Royce RB-211 series of engines, which use three spools but no variable stators.

The arrangements of two very successful high-bypass engines, the Pratt & Whitney JT9D and the General Electric CF-6, are shown in figure 7.16. These diagrams owe their elegance to Mr. Donald Jordan, formerly of

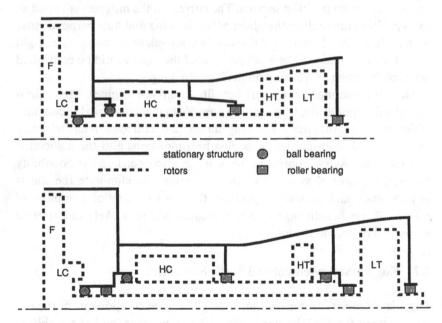

Figure 7.16
Arrangements of two high-bypass turbofans: the P&W JT9D (top) and the GE CF-6.

Pratt & Whitney Aircraft, who gave a series of MIT classes the benefit of his comprehensive understanding of aircraft engines in the period 1981–1983.

From the arrangements, one can see that the two manufacturers adopted quite different strategies. The JT9D was designed as a short engine, with the minimum number of bearings (just two on each rotor). Because inlet guide vanes were ruled out by noise considerations, the fan is cantilevered in each of these engines, but in the JT9D this dictates a large-diameter, stiff, low-speed shaft to absorb the gyroscopic loads from the fan. This requires a relatively large bearing diameter for the high-speed spool, and hence a large DN. In the CF-6, the fan is carried on two bearings, one ball and the other roller, allowing (or perhaps resulting from) the use of a long, relatively flexible, low-speed shaft.

In both engines, the bearings of the fan and those of the forward compressor are supported by a single structure, which penetrates the compressor inlet airstream through a set of streamlined struts. The aft bearing of the high-pressure compressor is supported in the JT9D by a structure that extends aft, inside the combustor, from this forward support. This avoids the problems of carrying a support structure through the high-temperature region of the turbine. In the CF-6, this support is between the high-pressure turbine and the low-pressure turbine. In both cases there is a bearing aft of the low-pressure turbine.

As was noted, both of these engine arrangements have been very successful. Each has both advantages and disadvantages. The short flow path of the JT9D restrains the aerodynamic choices, while the CF-6 is more complex mechanically. Newer engine designs by both companies tend to compromise between the extremes of the JT9D and the CF-6. For example, the Energy Efficient Engine (E^3) arrangements proposed by P&W and GE and shown in figure 7.17 both used five bearings, with two-bearing high-speed spools, two bearings on the cantilevered fan, and one bearing for each turbine. They differed mainly in the support of the turbine bearings—P&W proposed using the conventional support between the high- and low-pressure turbines, while GE proposed supporting the low-speed spool from behind the low-pressure turbine and using an "intershaft" bearing to support the high pressure turbine from the low. In the GE design both the inner and outer races rotate. There has been some experience with such arrangments, not all of it favorable.

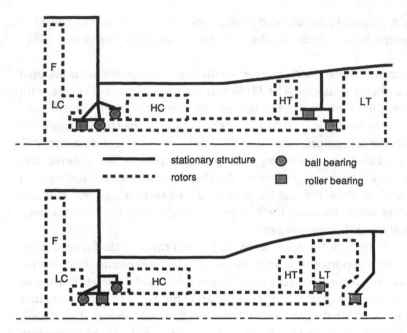

Figure 7.17
Arrangements proposed for Energy Efficient Engines by Pratt & Whitney (top) and General
Electric.

7.7.2 Rotor Structures

Within the dictates of the centrifugally generated stresses discussed in section 7.1, there are two general approaches to the structural design of compressor rotors. One is to assemble the rotor from discs, tied together by
axial tension members. The blades are then generally inserted into slots
broached more or less axially into the rim of each disc. A second approach
is to use a drum, with peripheral slots for the blades. Depending on the
peripheral speed of the drum, it may be necessary to support it with disc-
like ribs on the internal surface. In either case, the rotor is made stiff
in bending by using large-diameter conical shapes, rather than small-
diameter shafts, whenever possible. If the rotor is made up of discs, they
are separated by spacers which in assembly make up the conical shapes.
Examination of the cutaway drawings in chapter 1 will reveal both types of
rotors in the compressors, but only disc structures in the turbines, where
the stress levels are higher.

An important consideration is assembly and disassembly. Turbines are generally assembled element by element, axially. Some compressors with disc-type rotors have been assembled ("stacked") axially as well. With the drum rotors it is necessary to use a case split in a plane through the axis of rotation, as is also conventional in stationary gas turbines.

References

7.1 S. Timoshenko, *Vibration Problems in Engineering.* Van Nostrand, 1937.

7.2 F. O. Carta, "Coupled Blade-Disc-Shroud Flutter Instabilities in Turbine Engine Rotors." *Journal of Engineering for Power*, July 1967: 419.

7.3 S. I. Ginsburg, "Calculation of the Boundary of Excitation of Flexural Flutter of a Homogeneous Array in a Quasistationary Approximation." *Strength of Materials* 8 (1974): 46–51.

7.4 D. S. Whitehead, Bending Flutter of Unstalled Cascade Blades at Finite Deflection. ARC R & M No. 3386, Her Majesty's Stationery Office, 1965.

7.5 A. A. Mikolajczak, R. A. Arnoldi, L. E. Snyder, and H. Stargardter, "Advances in Fan and Compressor Blade Flutter Analysis and Prediction." *Journal of Aircraft* 12 (1975): 325–332.

7.6 J. D. Jeffers II and C. E. Meece, Jr., "F100 Fan Stall Flutter Problem Review and Solution." *Journal of Aircraft* 12 (1975): 350–357.

7.7 B. A. Tassone, "Roller Bearing Slip and Skidding Damage." *Journal of Aircraft* 12 (1975): 281–287.

7.8 K. Katayama et al., Development of Totally Active Magnetic Bearings. *Mitsubishi Heavy Industries Technical Review* 26, no. 1 (1989).

7.9 G. Schweitzer, "Magnetic Bearings—Applications, Concepts and Theory." International Conference on Advanced Mechatronics, Tokyo, 1989.

Problems

7.1 Show that, according to the discussion of subsection 7.1.1, the burst speed for a disc (without blades) is given by

$$\omega^2 = \left(\frac{\sigma_{\text{ultimate}}}{\rho}\right) \frac{\int_{r_I}^{r_H} z(r)\, dr}{\int_{r_I}^{r_H} r^2 z(r)\, dr},$$

where r_I and r_H are the inner and outer radii and $z(r)$ is the axial thickness. Generalize this expression to include the effect of blades mounted on the outer radius of the disc.

7.2 A popular form of construction for modern aircraft engine compressors is the "drum rotor," in which the blades of all stages are mounted on a thin-walled drum having a radius equal to the hub radius of the blading. A disc with inner rim such as that shown in figure 7.2 supports the drum from the inside, carrying the centrifugal load of the blades. Following the method of subsection 7.1.1, develop a procedure for designing such a structure for constant stress.

7.3   In film cooling, the cooling air is passed through small holes on the skin of the blade. If the cooling air is colder than the blade skin, it will cool the interior surface of these small holes, generating a tensile stress. Estimate this stress by modeling the cooling hole as a long hole in a block of metal whose temperature is uniform far from the hole. Use equation 6.13 to estimate the heat flux at the wall of the hole.

7.4   A turbine mounted between two bearings as in figure 7.5 has a total mass (shaft plus discs plus blades) of 100 kg. It is operating at 10,000 rpm, which is three-fourths of its shaft critical speed, with a tangential velocity of 500 m/sec, when one of its 50 blades, with a mass of 0.5 kg, detaches from the rim. Estimate the bearing force that results for very stiff bearings and the shaft deflection at the bearings for very soft bearing mounts with $n_b/\omega_n = 0.5$.

8 Component Matching and Engine Performance

This chapter briefly explores how the performance and the behavior of an actual engine are related to the characteristics of its components. Our view is now different from that taken in chapters 2 and 3, where the components were characterized by thermodynamic parameters π, τ, η with no reference to the shape or size of the machine. Here we begin with the characteristics of actual components, expressed by performance maps for the compressor (figure 5.33), the turbine (figure 6.17), the inlet (figure 4.22), the fan (figure 5.31), and perhaps other components if necessary. These maps represent the behavior of real devices whose geometry is fully specified. Our task is to determine how they interact when combined into an engine or a propulsion system.

The task is most conveniently divided into two parts. The first (termed *component matching*) involves applying the constraints that result from the need for the components to work together; the second involves predicting the performance of the resulting assembly.

8.1 Compressor-Turbine Matching: The Gas Generator

Suppose that we have a compressor with the map of figure 5.33, and a turbine with the map of figure 6.17, mounted on a single shaft with a combustor between them to form a gas generator, which is the heart of any gas turbine engine. With an inlet and a nozzle, a gas generator becomes a turbojet engine (figure 1.4). With a fan and a fan drive turbine, it becomes a turbofan engine (figure 1.6). With a power turbine, it becomes a shaft turbine (figure 1.7), or it can be used as a direct source of hot, high-pressure air for innumerable other applications.

Matching of the compressor, combustor, and turbine implies the following for the station numbers of figure 1.4:

$$N_t = N_c \quad \text{or} \quad \frac{N_c}{\sqrt{\theta_2}} = \frac{N_t}{\sqrt{\theta_4}}\sqrt{\frac{T_{t4}}{T_{t2}}}, \tag{8.1}$$

$$W_4 = (1+f)W_2 \text{ or}$$

$$\frac{W_4\sqrt{\theta_4}}{\delta_4 A_4} = (1+f)\frac{W_2\sqrt{\theta_2}}{\delta_2 A_2}\frac{p_{t2}}{p_{t4}}\sqrt{\frac{T_{t4}}{T_{t2}}}\frac{A_2}{A_4}, \tag{8.2}$$

$$W_2 c_{pc}(T_{t3} - T_{t2}) = W_4 c_{pt}(T_{t4} - T_{t5}) \text{ or}$$

$$1 - \frac{T_{t5}}{T_{t4}} = \frac{c_{pc}}{(1+f)c_{pt}}\frac{T_{t2}}{T_{t4}}\left(\frac{T_{t3}}{T_{t2}} - 1\right). \tag{8.3}$$

The connection among f, T_{t4}, and T_{t3} is given by

$$\frac{hf}{\bar{c}_p T_{t2}} = \frac{T_{t4}}{T_{t2}} - \frac{T_{t3}}{T_{t2}}. \tag{8.4}$$

If we specify two independent variables (usually $N_c/\sqrt{\theta_2}$ and T_{t4}/T_{t2}), then with these relations, the compressor and turbine maps, and a pressure-drop relation such as equation 4.37 for the burner we can determine the mass flow $W_2\sqrt{\theta_2}/\delta_2$, the pressure ratio p_{t5}/p_{t2}, the temperature ratio T_{t5}/T_{t2}, and the fuel flow $hf/c_p T_{t2}$ for the gas generator.

In general the process is involved and tedious, but it is straightforward when the turbine nozzles are choked. Because this is the usual situation at full power, we will deal only with it. If the turbine nozzles are choked, then $W_4\sqrt{\theta_4}/A_4\delta_4$ has a unique value, determined by the geometry of the turbine nozzles, and equation 8.2 becomes an explicit expression for p_{t3}/p_{t2} as a function of $W_2\sqrt{\theta_2}/A_2\delta_2$ and T_{t4}/T_{t2}:

$$\frac{p_{t3}}{p_{t2}} = \left[\frac{(1+f)(A_2/A_4)}{\pi_b(W_4\sqrt{\theta_4}/A_4\delta_4)}\right] \frac{W_2\sqrt{\theta_2}}{A_2\delta_2} \sqrt{\frac{T_{t4}}{T_{t2}}}. \tag{8.5}$$

The quantities in the brackets are constant in first approximation. This is the equation of a straight line through the origin on the compressor map, as shown in figure 5.33. Its slope depends on the values of T_{t4}/T_{t2} and A_4/A_2; thus, if we select a design value of T_{t4}/T_{t2}, the turbine nozzle area must be selected relative to compressor area A_2 to put the operating line in the most advantageous portion of the compressor map.

Once the turbine geometry has been set, equation 8.5 gives p_{t3}/p_{t2} as a function of $N/\sqrt{\theta_2}$ and T_{t4}/T_{t2}. For a given $N/\sqrt{\theta_2}$, raising T_{t4}/T_{t2} increases p_{t3}/p_{t2}; in excess this will stall the compressor. Now, if we specify T_{t4}/T_{t2} and $N/\sqrt{\theta_2}$, the operating point on the compressor map is determined; hence p_{t3}/p_{t2}, η_c, and therefore T_{t3}/T_{t2} are determined. From equation 8.4, $hf/\bar{c}_p T_{t2}$ is determined, but f is calculable only if T_{t2} is specified. If we neglect f compared to unity, T_{t5}/T_{t4} from equation 8.3 depends only on $N/\sqrt{\theta_2}$ and T_{t4}/T_{t2}. Finally, η_t is determined by $N/\sqrt{\theta_4}$ and T_{t5}/T_{t4}, so p_{t5}/p_{t4} can be found. Thus, except for the determination of f, which requires T_{t2}, we can find $W_2\sqrt{\theta_2}/\delta_2$, p_{t5}/p_{t2}, T_{t5}/T_{t2}, and $hf/\bar{c}_p T_{t2}$ as functions of $N/\sqrt{\theta_2}$ and T_{t4}/T_{t2}. These are called the *pumping characteristics* of the gas generator.

The pumping characteristics for a gas generator using the compressor of figure 5.33, with $T_{t4}/T_{t2} = 6$ and turbine efficiency $\eta_t = 0.90$, are shown in

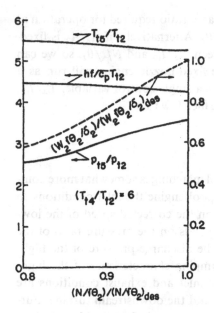

Figure 8.1
Pumping characteristics for a gas generator based on the compressor of figure 5.33, with $T_{t4}/T_{t2} = 6$.

figure 8.1. We see that at 100 percent corrected speed the gas generator produces a stagnation pressure ratio of 3.30 with a stagnation temperature ratio of 5.20. A set of curves such as those in figure 8.1 can be found for each value of T_{t4}/T_{t2}.

8.2 Matching the Gas Generator and the Nozzle

Two independent parameters, $N/\sqrt{\theta_2}$ and T_{t4}/T_{t2}, are required to specify the operating point of the gas generator. But if we specify the size of the exit nozzle, A_n, this is reduced to one parameter. This is seen by writing the statement that the nozzle mass flow equals the turbine mass flow, or $W_n = (1 + f)W_2$, so

$$\frac{W_n\sqrt{\theta_n}}{A_n\delta_n} = (1 + f)\frac{W_2\sqrt{\theta_2}}{A_2\delta_2}\frac{A_2}{A_n}\frac{p_{t2}}{p_{t5}}\sqrt{\frac{T_{t5}}{T_{t2}}}. \tag{8.6}$$

If the nozzle is choked, $W_n\sqrt{\theta_n}/A_n\delta_n$ has a specific value. If it is not choked, $W_n\sqrt{\theta_n}/A_n\delta_n$ depends on p_{t5}/p_0. In either case, we can regard equation 8.6

as determining the nozzle/compressor area ratio required for operation at any particular set of T_{t4}/T_{t2} and $N/\sqrt{\theta_2}$. Alternatively, if A_n/A_2 is fixed, equation 8.6 gives a relationship between T_{t4}/T_{t2} and $N/\sqrt{\theta_2}$, so we can find the corrected speed at which the fixed-nozzle engine will run as a function of T_{t4}/T_{t2}. The engine now has a single control variable, T_{t4}/T_{t2}, which is of course directly related to the fuel flow rate.

8.3 Multi-Spool Matching

In two- or three-spool engines, the spool matching is somewhat more complex, because (for example) in a two-spool engine the inlet conditions to the high-pressure compressor depend on the corrected speed of the low-pressure compressor. This, in turn, depends on the pressure ratio of the low-pressure turbine, which is set by the discharge pressure of the high-pressure turbine. Nevertheless, the pumping characteristics of the high-pressure spool remain as before; their inlet and exhaust conditions are modified by the upstream compressor and the downstream turbine. Furthermore, the matching of the low-pressure spool is described by relations entirely analogous to equations 8.1–8.5, except that the combustor is replaced by the gas generator.

The approach may be illustrated by the example of a turbofan engine with separate fan and core nozzles. It is necessary to add to figure 1.6 a station number between the high-pressure and low-pressure turbines, which will be denoted 4.5, and one behind the fan in the core airflow, denoted 2.5. Then the analogue of equation 8.1 is

$$N_{lt} = N_{lc} \quad \text{or} \quad \frac{N_{lc}}{\sqrt{\theta_2}} = \frac{N_{lt}}{\sqrt{\theta_{4.5}}} \sqrt{\frac{T_{t4.5}}{T_{t2}}}. \tag{8.7}$$

The analogue of equation 8.2 is

$$\frac{W_{4.5}\sqrt{\theta_{4.5}}}{\delta_{4.5}A_{4.5}} = (1+f)\frac{W_{2.5}\sqrt{\theta_2}}{\delta_2 A_{2.5}}\frac{p_{t2}}{p_{t4.5}}\sqrt{\frac{T_{t4.5}}{T_{t2}}}\frac{A_{2.5}}{A_{4.5}}, \tag{8.8}$$

where $W_{2.5}$ and $A_{2.5}$ are the weight flow and the area associated with the core flow in the case of the turbofan engine. For a two-spool turbojet they would be equal to W_2 and A_2; for the turbofan, they differ because of the bypass, so that $1 + \alpha = W_2/W_{2.5}$.

The power match becomes

$$W_2 c_{pc}(T_{t2.5} - T_{t2}) = W_{4.5} c_{pt}(T_{t4.5} - T_{t5}),$$

or

$$1 - \frac{T_{t5}}{T_{t4.5}} = \frac{(1+\alpha)c_{pc}}{(1+f)c_{pt}} \frac{T_{t2}}{T_{t4.5}} \left(\frac{T_{t2.5}}{T_{t2}} - 1 \right). \qquad (8.9)$$

To close the calculation it is necessary to specify either the fan nozzle area or the pressure ratio of the low-pressure compressor, and the nozzle area of the low-pressure turbine. A calculation might then proceed as follows:

(1) Choose a value of $T_{t4}/T_{t2.5}$ and a trial value of $N_c/\sqrt{\theta_{2.5}}$ for the high-pressure spool. The pumping characteristics then yield $W_{2.5}\sqrt{\theta_{2.5}}/\delta_{2.5}$, $p_{t4.5}/p_{t2.5}$, $T_{t4.5}/T_{t2.5}$, and $hf/c_p T_{t2.5}$. These must be consistent with equation 8.8; thus, we can iterate until, for a fixed $T_{t4}/T_{t2.5}$, we find a value of $N_c/\sqrt{\theta_{2.5}}$ that satisfies equation 8.8.

(2) Specify $N_{lc}/\sqrt{\theta_2}$ and $p_{t2.5}/p_{t2}$, thus setting the operating point of the low-pressure compressor, including its mass flow. From the mass flow one can find the required nozzle area, as well as all values at station 2.5.

(3) From equation 8.9, find $T_{t5}/T_{t4.5}$; from the turbine efficiency, calculate $p_{t5}/p_{t4.5}$. The core nozzle area that is consistent with the assumed operating condition of the low-pressure compressor can then be found from an expression analogous to equation 8.8.

(4) Varying the choice of the operating point of the low-compressor in step 3 yields a range of fan exhaust areas, and also a range of core nozzle areas. From these, the operating point of a fixed-geometry engine can be found as a function of T_{t4}/T_{t2} alone.

8.4 Engine-Inlet Matching and Distortion

The mass flow of the engine must also match that of the inlet. Since the corrected speed $N/\sqrt{\theta_2}$ and hence the corrected weight flow $W_2\sqrt{\theta_2}/\delta_2$ of the compressor are determined by T_{t4}/T_{t2} for the fixed-nozzle engine, the inlet must provide a variable Mach number M_2 at the engine face in response to changes in T_{t4}/T_{t2}, and it must do so over a range of flight Mach numbers M_0. This is the reason for the complex variable-geometry inlets discussed in section 4.2. They have variable throat area, a provision for bypassing air around the engine, and so on, to allow the inlet to operate with good pressure recovery while meeting the corrected weight-flow requirement of the engine.

No discussion of inlet-engine matching would be complete without some reference to inlet distortion, one of the most troublesome problems in modern propulsion systems. The flow the engine receives from the inlet is not uniform either radially or circumferentially. Even in a perfectly axisymmetric inlet there would be regions of low stagnation pressure near the outer walls. If the inlet is at an angle of attack or side-mounted on a fuselage, there may be large circumferential variations. Some inlets have S-bends (e.g. the center inlets on the Boeing 727 and the Lockheed L-1011) that produce strong secondary flows.

The effect of these non-uniformities is to lower the stall margin of the fan or compressor that first receives the flow. If it has sufficient tolerance to distortion, it will reduce the non-uniformities, thus shielding the following components; if not, then its stall may result in a general breakdown of flow throughout the engine.

Because the response of the compressor to distortion depends on the details of the flow, which may be unsteady and have variations in stagnation pressure and temperature both radially and circumferentially, no simple description of inlet-compressor interaction has been successful in accurately predicting when stall will occur. This can be determined only by running the engine with the inlet in place. Approximate techniques have been developed for use in the design process, however. They consist of estimating or measuring the flow expected from the inlet, then representing it as a combination of radial and circumferential distortion patterns such as are sketched in figure 8.2. The patterns are characterized by the radial or circumferential extent of the region of low p_{t2}, and by its amplitude in the

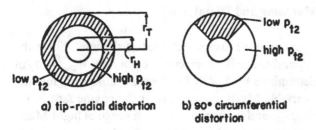

a) tip-radial distortion   b) 90° circumferential
                              distortion

Figure 8.2
Distortion patterns used for evaluating a compressor's response to distortion of the inlet flow.

form

$$\frac{(p_{t2})_{\max} - (p_{t2})_{\min}}{(p_{t2})_{\max}}.$$

Distortion patterns approximating these are generated in the test facility by inserting screens of the required radial or circumferential extent and the required density in the flow upstream of the engine. The stall line is then determined with distortion.

Such steady-state distortion testing is now routine for all new aircraft propulsion systems. Unfortunately, it is not always sufficient—the inlet outflow may be quite unsteady, so that *instantaneously* the distortion may be worse than the mean represented by the steady-state test. Extensive studies have shown that a compressor will stall if a distortion pattern that would lead to steady-state stall persists for a time on the order of the flow time through the compressor or longer (references 8.1, 8.2). Much stronger distortions can be tolerated for very short time periods.

8.5 Overall Performance

The pumping characteristics exemplified by figure 8.1 make it relatively simple to estimate thrust and specific impulse (or power output and specific fuel consumption). Consider the turbojet for simplicity. The thrust is given by equation 1.12, which is best written as

$$\frac{F}{A_n p_0} = \frac{W_2 \sqrt{\theta_2}}{A_2 \delta_2} \frac{\delta_2 u_0}{\sqrt{\theta_2} p_0 g} \frac{A_2}{A_n} \left((1 + f) \frac{u_e}{u_0} - 1 \right) + \frac{A_e}{A_n} \left(\frac{p_e}{p_0} - 1 \right). \qquad (8.10)$$

From the characteristics of the gas generator, we have $W_2 \sqrt{\theta_2}/A_2 \delta_2$, p_{t5}/p_{t2}, and T_{t5}/T_{t2}. The area ratios A_n/A_2 and A_e/A_n must be set. Then u_e/u_0 follows from

$$\frac{u_e}{u_0} = \frac{M_e}{M_0} \left(\sqrt{\frac{\gamma_t R_t T_e}{\gamma_c R_c T_0}} \right)$$

$$= \frac{M_e}{M_0} \left(\sqrt{\frac{\gamma_t R_t T_{te}}{\gamma_c R_c T_{t0}}} \right) \left(\sqrt{\frac{1 + \frac{1}{2}(\gamma_c - 1)M_0^2}{1 + \frac{1}{2}(\gamma_t - 1)M_e^2}} \right); \qquad (8.11)$$

for the nonafterburning turbojet $T_{te}/T_{t0} = T_{t5}/T_{t2}$, and for the afterburning turbojet $T_{te}/T_{t0} = T_{t6}/T_{t2}$ (see figure 1.4). The exit Mach number is determined by the nozzle pressure ratio p_{t6}/p_0 and area ratio A_7/A_n. For a choked convergent nozzle, $M_e = 1$. As A_7/A_n is increased from unity, $M_e = M_7$ increases as shown by figure 4.1, but a γ lower than 1.4 should be used for good accuracy. Having found M_e, one can determine p_e/p_0 by

$$\frac{p_e}{p_0} = \frac{(p_{t6}/p_{t2})(\pi_d \delta_0)}{[1 + \frac{1}{2}(\gamma_t - 1)M_e^2]^{\gamma_t/(\gamma_t-1)}} \tag{8.12}$$

and the expression for $F/A_n p_0$ is complete.

The specific impulse I may be written as

$$I \equiv \frac{F}{\dot{m}_f g} = \frac{\dfrac{F}{A_n p_0}}{\dfrac{W_2\sqrt{\theta_2}}{A_2 \delta_2} \dfrac{hf}{\bar{c}_p T_{t2}}} \frac{A_n p_0}{A_2 \delta_2} \frac{h}{\bar{c}_p T_{t2}} \sqrt{\theta_2}. \tag{8.13}$$

Two interesting observations can be made from equations 8.10 and 8.13. From equation 8.10, for a given flight Mach number M_0, a given θ_2, and a given value of T_{t4}/T_{t2}, the thrust is directly proportional to the ambient pressure. T_0 is nearly constant in the stratosphere, so this proportionality will apply there. From equation 8.13, I is independent of p_0 but varies as $\sqrt{\theta_2}/T_{t2}$, which is proportional to $1/\sqrt{T_{t2}}$, so the specific impulse increases as T_0 decreases at fixed M_0.

Once the performance of an engine is established by test at sea-level static conditions, its performance can be estimated by these means over a wide range of Mach numbers, engine speeds, and altitudes. The estimated thrust, specific fuel consumption, air flow, and bypass ratio for the JT3D-1 are shown in figure 8.3 for an altitude of 35,000 ft.

8.5.1 Performance Trends

Engine performance has improved systematically with time, as component efficiencies have improved, turbine inlet temperatures increased, and weights of the structures decreased. These trends are illustrated by figures 8.4 and 8.5, which show the evolution of the specific fuel consumption and the thrust/weight ratio of commercial turbofan engines over the time period from about 1960 to 1990. Projections to the near future are indicated by the dashed lines.

Some important trends are not immediately evident from these plots. First, the trend toward lower specific fuel consumption is made possible

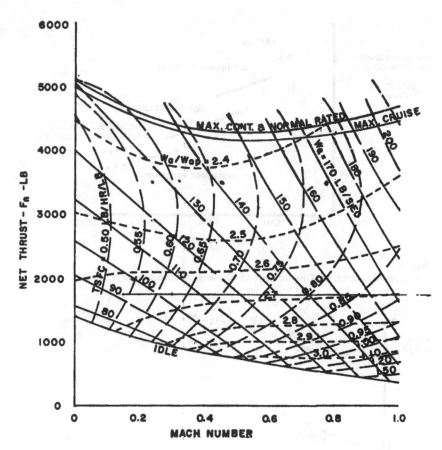

Figure 8.3
Estimated thrust, specific fuel consumption, and airflow of JT3D engine at 35,000 ft altitude.

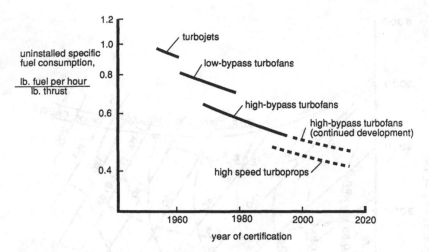

Figure 8.4
Specific fuel consumption of uninstalled turbofan engines of successive generations.

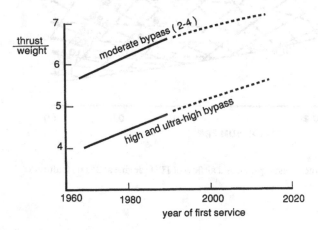

Figure 8.5
Thrust-to-weight ratio for turbofan engines as a function of first year of service. The upper curve is for low to moderate bypass ratios, the lower one for high to very high bypass ratios.

by a systematic increase of compression ratio, which in turn is made possible by increasing turbine inlet temperature. Similarly, the improvement in thrust/weight is made possible by the higher power density of the gas generators (which results from higher turbine inlet temperatures), but also by improved materials, and by higher component aerodynamic performance (which reduces the number of stages required). All these trends are expected to continue for some time, as suggested by the dashed projections.

8.6   Control and Acceleration

Gas turbines require sophisticated control systems because they generally operate at speeds and temperatures close to their limits of durability. Especially in aircraft engines, the range of environmental conditions is large, and so is the number of variables that the control system must deal with. For an afterburning turbofan engine, the following *control variables* might be listed:

primary fuel flow rate
afterburner fuel flow rate
exhaust nozzle area
guide vane and stator angles
bleed valve settings

To control these variables, the control system might sense some or all of the following *measurables*:

inlet temperature T_{t2}
inlet pressure p_{t2}
compressor discharge pressure p_{t3}
turbine blade temperature
turbine outlet temperature T_{t5}
turbine outlet pressure p_{t5}
gas generator speed N_2
fan speed N_1

The pilot interacts with the engine by setting the position of a power lever, which in turn makes an input to the control system, but the pilot does not directly control any of the above control variables. Instead, the power-lever position selects a thrust level, from idle to maximum, and the control manipulates the control variables to give the desired thrust while observing the operating limits of the engine.

The limiting parameters differ from engine to engine, but physically they stem from speed limits set by stresses in the rotating parts, from turbine temperature limits, from compressor or fan pressure-ratio limits set by stall or surge, and from blade Mach number limits set by flutter or vibration. Stress, blade temperature, and the presence of stall or flutter are not sensed directly by the control; they are inferred from the measurements listed above (except in some very recent engines, where the turbine bucket temperature is actually monitored). The control must therefore have a built-in logic that allows it to schedule the variables in accordance with the sensed measurables to observe the actual physical limits.

Further, the control must operate in the hot, vibrating environment of the engine nacelle with high reliability. Until very recently, this has dictated that the controls be hydromechanical, that is, that all the logic be performed with levers, cams, and flowing fuel or air. This has resulted in complex and expensive mechanisms and some real limitations on the functions of the control. For example, in current engines the motion of all compressor stators is ganged as a function of $N/\sqrt{\theta_2}$, whereas better performance could be had by varying the several rows individually.

Recent advances in integrated circuits have made digital electronic control feasible. Now used in all large modern engines, it permits a more complex logical system to be used and results in better performance as well as improved reliability.

The details of controls are diverse; good descriptions are given in references 8.3 and 8.4. The remainder of this discussion will deal with the dynamics of engine acceleration.

In addressing engine dynamics, we must be aware of the various time scales that characterize events in an engine. In order of increasing time they are air flow times (of the order of engine length/flow speed $\approx 10^{-2}$ sec); fuel flow times $\approx 10^{-1}$ sec; guide vane and nozzle response times (limited by control power, $\approx 10^{-1}$ sec); and angular acceleration time (limited by engine power and surge margin, ≈ 1–10 sec.

8.6.1 Acceleration

Because of the long times required for angular acceleration of the engine, we can assume in treating this acceleration process that the components of the engine behave as they would in steady state except for the power matching between the turbine and other components. Consider a turbojet for simplicity. The turbine power is not equal to the compressor power during acceleration, since part of the turbine power goes to increasing the rotational energy of the rotor. We can write the turbine torque as

$$\mathbb{T}_t = \frac{\text{Turbine power}}{\text{Angular velocity}} = \frac{c_p T_{t4}(1 - \tau_t) W_2}{\omega};$$

the compressor torque is

$$\mathbb{T}_c = \frac{c_p T_{t2}(\tau_c - 1) W_2}{\omega}.$$

Thus, if J is the angular moment of inertia of the rotor, the angular acceleration $d\omega/dt$ is given by

$$J\frac{d\omega}{dt} = \mathbb{T}_t - \mathbb{T}_c = \frac{c_p T_{t2} W_2}{\omega}\left(\frac{T_{t4}}{T_{t2}}(1 - \tau_t) - (\tau_c - 1)\right).$$

In terms of corrected parameters,

$$\frac{d(N/\sqrt{\theta_2})}{dt} = \frac{c_p T_0 \delta_2\left(\dfrac{W_2\sqrt{\theta_2}}{\delta_2}\right)}{4\pi^2 J\sqrt{\theta_2}\left(\dfrac{N}{\sqrt{\theta_2}}\right)}\left(\frac{T_{t4}}{T_{t2}}(1 - \tau_t) - (\tau_c - 1)\right). \tag{8.14}$$

Now, for steady-state operation, the quantity

$$g = \frac{T_{t4}}{T_{t2}}(1 - \tau_t) - (\tau_c - 1)$$

is zero. From section 8.1, g can be regarded as a function of $N/\sqrt{\theta_2}$, T_{t4}/T_{t2}, and A_n/A_2; thus, we can represent g by an expansion around the steady-state operating condition:

$$g = \frac{\partial g}{\partial (N/\sqrt{\theta_2})}\left[\frac{N}{\sqrt{\theta_2}} - \left(\frac{N}{\sqrt{\theta_2}}\right)_0\right] + \frac{\partial g}{\partial (T_{t4}/T_{t2})}\left[\frac{T_{t4}}{T_{t2}} - \left(\frac{T_{t4}}{T_{t2}}\right)_0\right]$$

$$+ \frac{\partial g}{\partial (A_n/A_2)}\left[\frac{A_n}{A_2} - \left(\frac{A_n}{A_2}\right)_0\right],$$

where the subscript zero denotes the value of each of the variables at the steady state. Suppose that the engine is running at one steady-state condition and that step changes are made in A_n/A_2 or T_{t4}/T_{t2}, which are then held constant while $N/\sqrt{\theta_2}$ changes to its new steady-state value. During the transient in $N/\sqrt{\theta_2}$, A_n/A_2 and T_{t4}/T_{t2} both have their new steady-state values; thus, the equation for $N/\sqrt{\theta_2}$ becomes

$$\frac{d}{dt}\left[\frac{N}{\sqrt{\theta_2}} - \left(\frac{N}{\sqrt{\theta_2}}\right)\right]_0 = \frac{-1}{t_{acc}}\left[\frac{N}{\sqrt{\theta_2}} - \left(\frac{N}{\sqrt{\theta_2}}\right)_0\right],$$

where

$$t_{acc} \equiv - \frac{4\pi^2 J\sqrt{\theta_2}(N/\sqrt{\theta_2})_0}{c_p T_0 \delta_2 (W_2\sqrt{\theta_2}/\delta_2)_0}\left(\frac{1}{\partial g/\partial(N/\sqrt{\theta_2})}\right)_0$$

is a characteristic time for acceleration. The solution is

$$\frac{N}{\sqrt{\theta_2}} - \left(\frac{N}{\sqrt{\theta_2}}\right)_0 = \text{const} \times e^{-t/t_{acc}},$$

so the corrected speed relaxes exponentially from its original value, as shown in figure 8.6.

The time t_{acc} required for the engine to accelerate is independent of whether the change in $(N/\sqrt{\theta_2})_0$ is due to a change in T_{t4}/T_{t2} or to a change in A_n/A_2.

To obtain the magnitude of t_{acc} we must estimate $\partial g/\partial(N/\sqrt{\theta_2})$. We can do this approximately as follows. We use an asterisk to denote each of the quantities in g divided by its value at the steady state; thus,

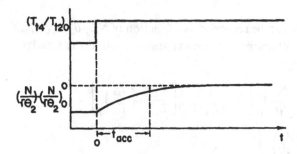

Figure 8.6
Response of the corrected speed to a step change in its steady-state value.

$$\frac{N/\sqrt{\theta_2}}{(N/\sqrt{\theta_2})_0} = (N/\sqrt{\theta_2})^*,$$

$$\frac{\tau_c}{(\tau_c)_0} = \tau_c^*;$$

and we assume that

$$\left(\frac{W_2\sqrt{\theta_2}}{\delta_2}\right)^* = \left(\frac{N}{\sqrt{\theta_2}}\right)^*, \tag{8.15}$$

$$\left(\frac{W_2\sqrt{\theta_4}}{\delta_4}\right)^* = 1 \qquad \text{(choked turbine nozzle)}, \tag{8.16}$$

$$\left(\frac{W_2\sqrt{\theta_5}}{A_n\delta_5}\right)^* = 1 \qquad \text{(choked exit nozzle)}, \tag{8.17}$$

and

$$\eta_t = \eta_c = 1. \tag{8.18}$$

From equation 8.16, $(W_2\sqrt{\theta_2}/\delta_2)^* \, (p_{t2}/p_{t4})^* = 1$; thus, equations 8.18 and 8.15 give

$$(\tau_c^*)^{\gamma/(\gamma-1)} = \pi_c^* \approx (p_{t4}/p_{t2})^* = (N/\sqrt{\theta_2})^* \, \sqrt{(T_{t4}/T_{t2})^*}.$$

From equation 8.17, $(N/\sqrt{\theta_2})^* \, (\sqrt{T_{t5}/T_{t4}})^* \, (1/\pi_c^*\pi_t^*) = 1$, and equation 8.18 gives

$$(\tau_t^*)^{-(\gamma+1)/2(\gamma-1)} = \frac{(\tau_c^*)^{\gamma/(\gamma-1)}}{(N/\sqrt{\theta_2})^*} = 1.$$

The function g thus becomes

$$g = \left(\frac{T_{t4}}{T_{t2}}\right)_0 [1 - (\tau_t)_0] - (\tau_c)_0(N/\sqrt{\theta_2})^{*(\gamma-1)/\gamma} + 1,$$

and

$$\left[\frac{\partial g}{\partial\left(\dfrac{N}{\sqrt{\theta_2}}\right)}\right]_0 = \frac{1}{\left(\dfrac{N}{\sqrt{\theta_2}}\right)_0} \left[\frac{\partial g}{\partial\left(\dfrac{N}{\sqrt{\theta_2}}\right)^*}\right]_0 = -(\tau_c)_0\left(\frac{\gamma-1}{\gamma}\right)\frac{1}{\left(\dfrac{N}{\sqrt{\theta_2}}\right)_0}. \tag{8.19}$$

Thus, the acceleration time becomes

$$t_{acc} = \frac{4\pi^2 J \sqrt{\theta_2} \left(\dfrac{N}{\sqrt{\theta_2}}\right)^2 \dfrac{\gamma}{\gamma - 1}}{c_p T_0 \delta_2 \left(\dfrac{W_2 \sqrt{\theta_2}}{\delta_2}\right)_0 (\tau_c)_0} = \frac{4\pi^2 J N^2}{c_p T_{t2} W_2} \frac{\gamma}{\gamma - 1} \frac{1}{(\tau_c)_0}. \tag{8.20}$$

This result has a relatively simple interpretation: that the numerator represents the rotational energy of the rotor $J\omega^2$, while the denominator represents the available accelerating power, which is clearly proportional to $c_p T_{t2} W_2$.

For a typical turbojet with $W_2 = 50$ kg/sec, $N \approx 200$ sec^{-1}, $\tau_c \approx 2$, $T_{t2} = 290°$K, and $J \approx 5$ kg m^2, we find $t_{acc} \approx 0.8$ sec. At high altitudes, where W_2 which is proportional to p_0) is reduced, t_{acc} can be much longer. If an engine is scaled geometrically, at constant tip speed and with L representing any length scale, J is proportional to L^5, N is proportional to L^{-1}, W_2 is proportional to L^2, and t_{acc} is proportional to L, so a large engine takes longer to accelerate than a small one.

8.6.2   Acceleration Stall Margin

The acceleration process discussed above implies that T_{t4}/T_{t2} is held constant during the acceleration. This may not be possible, because raising T_{t4}/T_{t2} above its steady-state value at any $N/\sqrt{\theta_2}$ also raises the compressor pressure ratio above its steady-state value, driving the compressor toward stall. This is shown schematically in figure 8.7, where $N/\sqrt{\theta_2}$ increases with T_{t4}/T_{t2} as indicated by the points. If the engine is operating at

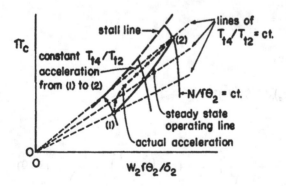

Figure 8.7
The steady-state operating line for a fixed-nozzle engine, and the transient followed in a constant T_{t4}/T_{t2} acceleration, driving the compressor into stall.

point 1, and T_{t4}/T_{t2} is suddenly raised to the value required for steady operation at point 2, π_c rises along the constant $N/\sqrt{\theta_2}$ line and then follows the new T_{t4}/T_{t2} line as the engine accelerates. The situation shown would lead to stall, so it is unacceptable; the control system would restrict the actual increase in T_{t4}/T_{t2} to prevent stall, but then the acceleration time would be greater than t_{acc} as given by equation 8.20. Nevertheless, some excess compression ratio is always required for acceleration. It is called the *acceleration stall margin*.

8.6.3 Other Transients

When the afterburner is lit on a turbojet engine (figure 1.4), the nozzle throat area must be increased simultaneously; otherwise the increased backpressure on the turbine will cause the engine to decelerate. But the timing is not very critical, because of the long time scale for change of $N/\sqrt{\theta_2}$. In an afterburning turbofan, on the other hand, this timing is critical, because the backpressure from an afterburner light is communicated within milliseconds to the fan (figure 2.9) and may cause it to stall. If it does, the compressor may follow and a general breakdown in the flow will result. This problem has existed in all military afterburning turbofans to date. The solution lies in a complex control that minimizes the timing error between afterburner light and nozzle opening and in afterburner fuel systems that give a nearly continuous variation of T_{t6} without sudden steps.

Sudden transients may also result from ingestion of hot gas into the compressor when the compressor suddenly is shifted to a lower corrected speed while the pressure in the combustor is still that corresponding to the higher corrected speed. Stall is very probable if the hot gas persists for a compressor flow time or longer. This situation occurs in military aircraft when guns or rockets are fired, and some aircraft have provision for momentarily reducing fuel flow to the engines when this is done.

References

8.1 G. A. Plourde and B. Brimelow, Pressure Fluctuations Cause Compressor Instability. AFAPL-TR-69-103, 1970.

8.2 B. Brimelow and T. P. Collins, Engine Testing in a Dynamic Environment. AIAA Paper 74-1198, 1974.

8.3 Pratt & Whitney Aircraft, *The Aircraft Gas Turbine Engine and Its Operation* (PWA Operating Instruction 200), June 1952 (revised May 1974).

8.4 I. E. Treager, *Aircraft Gas Turbine Engine Technology.* McGraw-Hill, 1970.

Problems

8.1 Carry out the compressor-turbine matching described in section 8.1 to find all the pumping characteristics for one value of $(N/\sqrt{\theta_2})/(N/\sqrt{\theta_2})_{des}$ on figure 8.1.

8.2 Repeat the calculation leading to figure 8.1 for several values of T_{t4}/T_{t2}, compute the nozzle area A_n/A_2 required to operate the engine at design speed at $T_{t4}/T_{t2} = 6$, and determine the variation of $(N/\sqrt{\theta_2})/(N/\sqrt{\theta_2})_{des}$ with T_{t4}/T_{t2} for this fixed A_n/A_2.

8.3 Using the pumping characteristics of figure 8.1, calculate the thrust and the specific impulse for a turbojet with a convergent nozzle and a mass flow of 100 kg/sec at sea-level static conditions, taking $T_{t4}/T_{t2} = 6$ at the design point. Assume $\pi_d \approx 0.95$.

8.4 When a very rapid increase in thrust is desirable (as in a wave-off from an aircraft carrier) and a variable-area nozzle is available, it might be desirable to keep the (turbojet) engine at full rpm, lowering the thrust for approach by opening the nozzle. By suddenly closing the nozzle, thrust can be recovered without the delay associated with angular acceleration of the engine. Describe what happens in the engine when the nozzle is suddenly closed.

8.5 The acceleration dynamics of a two-shaft turbofan differs from that of a turbojet because the core accelerates independently of, and much faster than, the fan and its associated turbine. Develop a model analogous to that of subsection 8.5.1 for a turbofan. Assume that the core changes speed instantly.

8.6 In a duct-burning turbofan with a fan pressure ratio of 3, the afterburner is inadvertently lit with the nozzle still in the nonafterburning position, raising T_{t6} by a factor of 2. Compute the resulting change in fan outlet pressure. If the same thing occurs in a turbojet, what will be the result?

9 Aircraft Engine Noise

Though long of concern to neighbors of major airports, aircraft noise first became a major problem with the introduction of turbojet-powered commercial aircraft (Boeing 707, Dehavilland Comet) in the late 1950s. These aircraft were powered by turbojet engines originally developed for military aircraft. It was recognized at the time that the noise levels produced by military aircraft would be unacceptable to persons living under the takeoff pattern of major airports such as New York's Kennedy (then Idlewild) and London's Heathrow. Accordingly, much effort was devoted to developing jet noise suppressors (reference 9.1), with some modest success. Takeoff noise restrictions were imposed by some airport managements, notably the Port of New York Authority, and nearly all first-generation turbojet-powered transports were equipped with jet noise suppressors at a significant cost in weight, thrust, and fuel consumption.

The introduction of the turbofan engine (Rolls-Royce Conway, Pratt & Whitney JT3-D), with its lower jet velocity, temporarily alleviated the jet noise problem but increased the high-frequency turbomachinery noise, which became a severe problem on landing approach as well as on takeoff. This noise was reduced somewhat by choosing proper rotor and stator blade numbers and spacing and by using engines of the single-mixed-jet type.

In spite of these efforts, the increasing volume of air traffic resulted in unacceptable noise exposures near major urban airfields in the late 1960s, leading to great public pressure for noise control. This pressure, and advancing technology, led to Federal Aviation Rule Part 36 (FAR-36), which became effective on December 1, 1969, and which set maximum takeoff, landing, and "sideline" noise levels for certification of new turbofan-powered aircraft. It is through the need to satisfy this rule that the noise issue influences the design and operation of aircraft engines. A little more general background on the noise problem may be helpful in establishing the context of engine noise control.

The Federal Aviation Administration (FAA), the aircraft manufacturers, the airline operators, and the airport managers all have roles in the control of aircraft noise, which is complex technically, socially, and legally. Only a brief summary will be given here.

It is the responsibility of the airport operator to ensure that operations from the individual airport do not impose unacceptable noise on the surrounding neighborhoods. This responsibility is set by the legal process, many suits having been filed against airport operators by citizens' groups and individuals. The judgments in these suits by and large prohibit any increases in noise impact on the affected neighborhoods, and imply large economic penalties on airports that violate the limits. The airport management controls the noise impact by its allocation of takeoff and landing rights to airlines. The airlines, in turn, are influenced in their purchases of equipment by the noise characteristics of the aircraft, because they may be allowed more flights, or flights at critical times of the day, if their equipment is quieter than that of competing airlines. All these factors provide incentives for the manufacture of quieter engines and aircraft.

It was in this complex motivational and legal context that the FAA issued FAR-36 (which, as noted, establishes the limits on takeoff, approach, and sideline noise for individual aircraft). The rule has since been revised several times, reflecting both improvements in technology and continuing pressure to reduce noise. As of this writing, FAR-36 is enunciated as three progressive stages of noise certification. The noise limits are stated in terms of measurements at three measuring stations, as shown in figure 9.1: under the approach path one nautical mile or 2000 m before touchdown, under the takeoff path 3.5 miles or 6500 m from the start of the takeoff roll, and at the point of maximum noise along the sides of the runway at a distance of 450 m (0.35 nautical mile for four-engined aircraft). The noise of any given aircraft at the approach and takeoff stations depends both on the engines and on the aircraft's performance, operational procedures, and loading, since the power settings and the altitude of the aircraft may vary. The sideline station is more representative of the intrinsic takeoff noise characteristics of the engine, since the engine is at full throttle and the station is nearly at a fixed distance from the aircraft. The

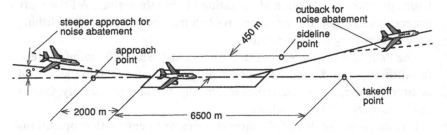

Figure 9.1
Schematic of airport runway showing approach, takeoff, and sideline noise-measurement stations.

actual distance depends on the altitude the aircraft has attained when it produces the maximum noise along the designated measuring line.

The limits prescribed by FAR-36 will be given in subsection 9.4.3, after a discussion of the technical background essential to understanding their significance.

Since FAR-36 and international rules set by the International Civil Aviation Organization (ICAO annex 16, chapter 3) which are generally consistent with it have been in force, airport noise has been a major design criterion for civil aircraft. Thus, an understanding of the mechanisms of noise production and of the techniques for alleviating it is crucial for aircraft propulsion research and development. FAR-36 has been applied only to subsonic transports, because no new supersonic commercial aircraft have been developed since its promulgation. The Concorde Mach 2 aircraft has operated from a limited number of airports under a waiver, having been certified before the rule was in force.

Meeting the present noise regulations of FAR-36 stage 3 is a major challenge to development of a new generation of supersonic transports. There are substantial differences between the noise characteristics of modern high-bypass engines for subsonic aircraft and those of the engines most suitable (from the viewpoint of performance) for supersonic aircraft. As indicated schematically in figure 9.2, the subsonic turbofan radiates noise forward and backward from its large, high-tip-speed fan. It also produces jet noise from both the fan jet and the primary jet, but because a low jet velocity gives good propulsive efficiency in cruise the jet noise can be reduced while improving fuel economy and range. Thus, fan noise is the most critical problem for the subsonic high-bypass turbofan, both on approach and on takeoff.

For supersonic aircraft, high thrust per unit of frontal area is required at cruise and transonic operating conditions in order to minimize drag. This implies high thrust per unit of airflow, which favors a turbojet such as that shown schematically in figure 9.2 or a low-bypass-ratio turbofan, either of which produces a high jet velocity at takeoff conditions. On the positive side, supersonic operation requires a long inlet with a throat which is sonic at supersonic operating conditions and which can be choked on approach, to suppress compressor noise, so that approach noise from the turbomachinery is not a critical problem. A supersonic aircraft also has a high thrust/weight ratio on takeoff (on the order of 0.32, versus 0.25 for a subsonic aircraft) so that after liftoff it can climb very rapidly. This reduces the

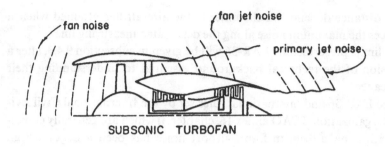

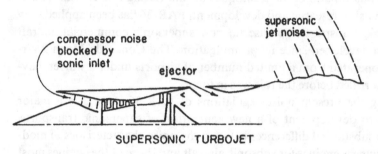

Figure 9.2
Schematic illustration of noise sources from subsonic turbofan engines and supersonic turbojet engines.

noise at the takeoff measuring point, particularly if the engines are throttled back just before that point. But the full noise impact of the high thrust and the high jet velocity is sensed at the sideline station.

Supersonic transports other than the Concorde currently are not covered by FAR-36. The Concorde operates under an exception such that although its noise must be measured at the same measuring stations as for the subsonic aircraft, the actual levels at those stations are limited to "the lowest levels that are economically reasonable, technologically practicable, and appropriate for the Concorde type design" (reference 9.2). These levels are, in fact, considerably above those deemed acceptable by the managements of most airports, so Concorde operations are limited to just a few locations.

It seems likely that, to be acceptable for widespread operations, any new supersonic transport will have to meet noise rules requiring that it impose no worse conditions on the airport neighborhoods than those imposed by large subsonic transports of the same technology generation. Because of the fast climbout of the supersonic transport, it can be argued that its noise impact on the overall community near the airport is no worse than that of a subsonic aircraft, even though the sideline noise exceeds the present rule for subsonic aircraft. However, this argument may not be convincing to the inhabitant of a house near the sideline. Readers with special interest in this matter are advised to follow the attempts of the FAA and the ICAO to deal with it.

At this time it appears that, to provide satisfactory subsonic and supersonic performance and also meet a noise rule equivalent to FAR-36 stage 3, an advanced SST engine will have to be capable of operating with a fairly high bypass ratio on takeoff and then converting to a lower ratio for transonic acceleration and supersonic cruise. Such an engine has been termed a *variable cycle engine*. One possible way of implementing such a design— through the use of a supersonic throughflow fan—was discussed in subsection 4.2.2.8.

The remainder of this chapter will be devoted to the physical phenomena of noise production by jets and by turbomachinery and to some techniques that have been developed for reducing noise emission.

Since this chapter was rewritten, a comprehensive summary of aeroacoustics as applied to aircraft noise has appeared (reference 9.21). It provides an excellent summary of the state of noise-suppression technology as of August 1991. The serious reader should consult it for the most current view of this complex subject.

9.1 Noise Sources: Unsteady Flow

All noise emanates from unsteadiness—time dependence in the flow. In aircraft engines there are three main sources of unsteadiness (figure 9.3): motion of the blading relative to the observer, which if supersonic can give rise to propagation of a sequence of weak shocks, leading to the "buzz saw" noise of high-bypass turbofans; motion of one set of blades relative to another, leading to a pure-tone sound (like that from a siren) which was dominant on approach in early turbojets; and turbulence or other fluid instabilities, which can lead to radiation of sound either through interaction with the turbomachine blading or other surfaces or from the fluid fluctuations themselves, as in jet noise.

These unsteady phenomena can all be described in principle by the equations for compressible fluids. In practice the description is complex,

TURBOMACHINE NOISE

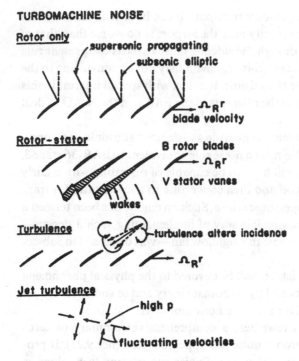

Figure 9.3
Sources of unsteady flow giving rise to noise from aircraft engines.

although it is relatively straightforward from a mathematical viewpoint because the system of perturbation equations is linear. Here we will attempt only a qualitative treatment sufficient for conveying a physical understanding of the phenomena.

Since most readers of this book are expected to have a classical background in fluid mechanics, we begin with the equations of conservation of mass and conservation of momentum (inviscid) in vector form:

$$\frac{D\rho}{Dt} + \rho \nabla \cdot \mathbf{u} = 0 \tag{9.1}$$

and

$$\frac{\rho D\mathbf{u}}{Dt} = -\nabla \rho, \tag{9.2}$$

where $D/Dt \equiv \partial/\partial t = \mathbf{u} \cdot \nabla$ is the "substantial derivative" with respect to time, following the fluid motion. If the fluid is assumed non-heat-conducting as well as inviscid, then the entropy is constant for a fluid element as it is carried along by the flow; that is,

$$\frac{Ds}{Dt} = 0. \tag{9.3}$$

To distinguish sound or noise from other small disturbances in the flow, we first linearize these equations, taking $\mathbf{u} = \mathbf{u}_0 + \mathbf{u}'$, $p = p_0 + p'$, and so on, where $\mathbf{u}_0$ and p_0 are uniform in space and time. Dropping terms quadratic in the small disturbances, we find

$$\frac{Dp'}{Dt} + \rho_0 \nabla \cdot \mathbf{u}' = 0, \tag{9.1a}$$

$$\rho_0 \frac{D\mathbf{u}'}{Dt} = -\nabla p', \tag{9.2a}$$

and

$$\frac{Ds'}{Dt} = 0, \tag{9.3a}$$

where now $D/Dt = \partial/\partial t + \mathbf{u}_0 \cdot \nabla$. In addition, we have the equation of state $p = \rho R T$, which yields

$$\frac{p'}{p_0} = \frac{\rho'}{\rho_0} + \frac{T'}{T_0}, \tag{9.4}$$

and the connection between entropy, pressure, and temperature

$$s - s_0 = c_p \ln \frac{T}{T_0} - R \ln \frac{p}{p_0},$$

which when "linearized" is

$$s' = c_p \frac{T'}{T_0} - R \frac{p'}{p_0} = c_v \frac{p'}{p_0} - c_p \frac{\rho'}{\rho_0}.$$

Differentiating this gives

$$\frac{Ds'}{Dt} = 0 = \frac{c_v}{p_0} \frac{Dp'}{Dt} - \frac{c_p}{\rho_0} \frac{D\rho'}{Dt},$$

so that

$$\frac{D\rho'}{Dt} = \frac{\rho_0}{\gamma p_0} \frac{Dp'}{Dt}. \tag{9.5}$$

After substituting this in equation 9.1a, we have four equations to solve for p' and $\mathbf{u}'$. The velocity can be eliminated by subtracting $\nabla \cdot (9.2a)$ from $D/Dt(9.1a)$ to get

$$\frac{\rho_0}{\gamma p_0} \frac{D^2 p'}{Dt^2} - \nabla^2 p' = 0. \tag{9.6}$$

This is a wave equation for p', with the wave velocity $a_0 = \sqrt{\gamma p_0/\rho_0}$, so we identify this quantity as the *sound velocity*.

Our basic relation for what follows is equation 9.6, but if we take $\nabla \times$ (9.2a), then, since $\nabla \times \nabla \equiv 0$, we have

$$\frac{D}{Dt}(\nabla \times \mathbf{u}') = 0. \tag{9.7}$$

Since $\nabla \times \mathbf{u}' = \omega'$ is the vorticity disturbance, this says that small amplitude vorticity is convected and has associated with it no pressure fluctuations. Equations 9.3a, 9.6, and 9.7 describe three disturbance modes: entropy, sound, and vorticity, which are independent in that they do not interact in the uniform background flow. They do interact in passage through any gradient in the mean flow, however, and especially at shock waves or combustion discontinuities.

Now consider the behavior of sound as described by equation 9.6. For simplicity, we take $\mathbf{u}_0 = 0$. No generality is lost, because we can transform to a coordinate system moving at $\mathbf{u}_0$ without changing the wave behavior. The equation is then

$$\frac{\partial^2 p'}{\partial t^2} - a_0^2 \nabla^2 p' = 0.$$

9.1.1 Waves, Acoustic Power, and Decibels

For one-dimensional or plane waves we have

$$\frac{\partial^2 p'}{\partial t^2} - a_0^2 \frac{\partial^2 p'}{\partial x^2} = 0,$$

which is satisfied by any function having the argument $x \pm a_0 t$, that is, $p' = p'(x \pm a_0 t)$. This simply says that p' is constant along "characteristics" $x = \pm a_0 t$. To determine the actual form and magnitude of p' we must specify some boundary conditions on the solution. Suppose, for example, we want to find the sound radiated by a plane perpendicular to the x axis, vibrating according to $x = A \sin\omega t$, so that its velocity is $\dot{x} = A\omega \cos\omega t$. To find the velocity in the wave we return to equation 9.2a, which is, for this case,

$$\rho_0 \frac{\partial u'}{\partial t} = -\frac{\partial p'}{\partial x}.$$

Now, if we assume a solution $p = P \cos k(x - a_0 t)$, then

$$\frac{\partial u'}{\partial t} = -\frac{1}{\rho_0}\frac{\partial p'}{\partial x} = +\frac{Pk}{\rho_0} \sin k(x - a_0 t),$$

and integrating gives

$$u' = (P/\rho_0 a_0) \cos k(x - a_0 t).$$

Matching this to the boundary condition requires $P/\rho_0 a_0 = A\omega$ and $-ka_0 = \omega$.

The energy transferred across a plane by this wave per cycle per unit area is the force per unit area (pressure) times the velocity integrated over a cycle, or

$$\int_0^{2\pi/\omega} p'u' \, dt = \int_0^{2\pi/ka_0} \frac{P^2}{\rho_0 a_0} \cos^2 k(x - a_0 t) \, dt = \frac{P^2}{\rho_0 a_0} \frac{\pi}{ka_0}. \tag{9.8}$$

The power per unit area, dP/dA, is the number of cycles per unit time, or $ka_0/2\pi$ times this, so

$$dP/dA = P^2/2\rho_0 a_0,$$

where $P^2/2$ will be recognized as the mean square pressure fluctuation.

For a spherical wave such as would emanate from a point source, equation 9.6 is

$$\frac{\partial^2 p'}{\partial t^2} - a_0^2 \frac{1}{r^2}\frac{\partial}{\partial r}\left(r^2 \frac{\partial p'}{\partial r}\right) = 0,$$

and we find

$$p' = P\left(\frac{r_0}{r}\right) \cos k(r - a_0 t), \tag{9.9}$$

and

$$u' = \frac{P}{\rho_0 a_0} \frac{r_0}{r} \cos k(r - a_0 t) - \frac{P}{\rho_0 a_0} \frac{r_0}{kr^2} \sin k(r - a_0 t),$$

while the sound power per unit area is

$$\frac{dP}{dA} = \frac{P^2}{2\rho_0 a_0} \left(\frac{r_0}{r}\right)^2.$$

It dies off as $1/r^2$, the total power radiated over spherical surfaces surrounding the radiator being constant.

Because the human ear's response to sound is such that the perceived sound is proportional to the logarithm of the pressure fluctuation, both the sound pressure level and the sound power level are quoted in decibels (dB), so that

Sound pressure level $= 20 \log_{10}(p'/0.00002)$, dB,

where p' is in Nm^{-2} and

Sound power level $= 10 \log_{10}(P/10^{-12})$, dB,

where P is in watts.

For reference, a pressure of $1 \ Nm^{-2}$ is equal to 94 dB, while one atmosphere $(1.015 \times 10^5 \ Nm^{-2})$ equals 194.1 dB.

9.1.2   Monopoles, Dipoles, Quadrupoles

We can think of the p' given by equation 9.9 as resulting from the harmonic expansion and contraction of a small sphere of radius r_0, the amplitude of pressure fluctuation at its surface being P. This vibrating sphere might model the acoustical effect of a pulsating jet, as indicated in figure 9.4. In the limit as $r_0 \to 0$ with Pr_0 held constant, equation 9.9 is the solution for a point source, or monopole, with acoustic power

$$\mathbf{P_m} = 2\pi \frac{(Pr_0)^2}{\rho_0 a_0}. \tag{9.10}$$

From this fundamental solution of equation 9.6, other solutions can be constructed by superposition. Two of special importance here are the dipole and the quadrupole.

The dipole is composed of two monopoles of opposite signs, or shifted in phase by π if harmonic, separated by some distance d as shown in figure

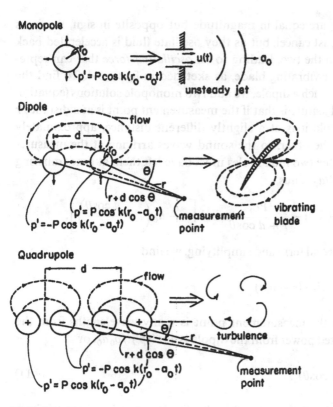

Figure 9.4
Monopole, dipole, and quadrupole sources and some fluid disturbances they can represent.

9.4. Because they are equal in magnitude but opposite in sign, the mass flows of the two just cancel; but as they oscillate fluid is accelerated back and forth between the two, leading to an *oscillating force* that can represent the effect of a vibrating blade, as sketched in figure 9.4. To find the sound radiated by such a dipole, we add two monopole solutions (equation 9.9). The essential feature is that if the measurement point is at a distance r from one monopole it is at a slightly different distance, approximately $r + d \cos\theta$, from the other; so the sound waves arriving at the measurement point from the two monopoles must have left them at times differing by roughly $d \cos\theta/a_0$. Thus,

$$p' = \frac{Pr_0}{r} \cos k(r - a_0 t) - \frac{Pr_0}{r + d \cos\theta} \cos k\left[r - a_0\left(t - \frac{d \cos\theta}{a_0}\right)\right].$$

Expanding the second term and simplifying, we find

$$p' \approx \frac{Pr_0}{r}(kd \cos\theta) \sin k(r - a_0 t)$$

if $kr \gg 1$, so that the measurement point is many wavelengths from the dipole. The radiated power from the dipole is $4\pi r^2 (p')^2/\rho_0 a_0$, or

$$\mathbf{P_d} = \frac{2\pi(Pr_0)^2}{\rho_0 a_0}(kd \cos\theta)^2. \tag{9.11}$$

Comparing equation 9.10, we see that the dipole power is $(kd \cos\theta)^2$ times the monopole power. The directional characteristic is embodied in $\cos^2\theta$. The factor $kd = 2\pi d/\lambda$ is the ratio of spacing between the monopoles to the wavelength λ of the radiated sound, so for a given spacing d set by the geometry of the blading which the dipole represents, the acoustical power of the dipole goes down as the frequency is lowered.

If we combine two dipoles so that the force they exert on the fluid cancels, we have the simplest representation of a fluid disturbance involving *no mass injection and no externally applied force*, but only fluid pressures and accelerations, which are in balance according to equation 9.2. Such a disturbance can still radiate sound, as we see by repeating the argument used to construct the dipole, separating two dipoles by a distance d to form the (linear) quadrupole sketched in figure 9.4. Then we find the quadrupole power

$$\mathbf{P_q} = \frac{2\pi(Pr_0)^2}{\rho_0 a_0}(kd)^4, \tag{9.12}$$

where the directional behavior has been neglected. Thus, for the same level of pressure fluctuation, the quadrupole radiates $(kd)^2$ as much power as the dipole, and $(kd)^4$ as much as the monopole.

9.2   Jet Noise

When fluid issues as a jet into a stagnant or more slowly moving background fluid, the shear between the moving and stationary fluids results in a fluid-mechanical instability that causes the interface to break up into vortical structures as indicated in figure 9.5. (See references 9.3, 9.4, and also 9.5; the last presents many photographs of jet flows.) The vortices travel downstream at a velocity which is between those of the high and low speed flows, and the characteristics of the noise generated by the jet depend on whether this propagation velocity is subsonic or supersonic with respect to the external flow. We consider first the case where it is subsonic, as is certainly the case for subsonic jets.

9.2.1   Noise from Subsonic Jets

For the subsonic jets, Lighthill (reference 9.6) has argued that the turbulence in the jet can be viewed as a distribution of quadrupoles. This is consistent with the facts that there are no sources of fluid and no bodies to generate forces, so the lowest-order acoustical source is the quadrupole. The argument for estimating the quadrupole magnitude is complex, but

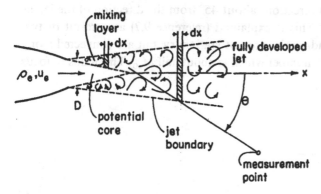

Figure 9.5
A subsonic jet mixing with ambient air, showing the mixing layer followed by the fully developed jet.

physically it is equivalent to the following: The scale of the turbulent fluctuations and hence of the quadrupoles is D (the jet diameter), and the magnitude of the pressure fluctuations is $\rho_e u_e^2$, so $(\mathbf{P}r_0)^2$ becomes $(\rho_e u_e^2 D)^2$. The wave number $k = \omega/a_0$, and $\omega \approx u_e/D$, so $kd \approx u_e/a_0$. Substituting these in equation 9.12 and taking $\rho_e \approx \rho_0$, we find the jet acoustic power

$$\mathbf{P}_j = \frac{2\pi\rho_0 u_e^8 D^2}{a_0^5};\tag{9.13}$$

since the argument is dimensional not quantitative, the factor of 2π is not to be regarded seriously. The striking feature of this relation is the dependence on u_e^8, which was first derived by Lighthill (reference 9.6). It has been verified as correct for subsonic jets over a wide range of velocities.

It is instructive to rewrite equation 9.13 as

$$\mathbf{P}_j = D^2(\rho_e u_e^3)\frac{\rho_e}{\rho_0}\left(\frac{u_e}{a_0}\right)^5,\tag{9.14}$$

from which we identify $D^2(\rho_e u_e^3)$ as the jet kinetic power and $(u_e/a_0)^5$ as the fifth power of jet Mach number based on the speed of sound in the ambient air. Thus, we can write the jet's "acoustic efficiency" as

$$\eta_{\text{jet noise}} \propto \left(\frac{\rho_e}{\rho_0}\right)M_{e0}^5.\tag{9.15}$$

This expression will be quantified later by comparison to data on jet noise.

An important feature of jet noise is its directional characteristics. The intensity is largest in directions about 45° from the direction of the jet, as shown in figure 9.6. This is explained (reference 9.7) as a result of two effects. First, the quadrupoles are not stationary, but are convected along by the flow at a Mach number which is about half M_{e0}. This tends to focus

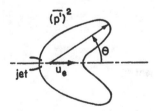

Figure 9.6
Directional characteristics of the noise from a subsonic jet.

their radiation downstream. The second effect is refraction. As the sound propagates out of the jet into the still air, it is turned away from the axis, leaving a quiet zone directly behind the jet.

The frequency distribution of the sound from a jet can be explained by thinking of the jet as divided axially into a series of slices of thickness dx (reference 9.7) (see figure 9.5), each emitting sound at a frequency $\omega \approx u/d$, where u is the local jet velocity and d is the local height of the mixing region or the size of the vortical structures indicated in figure 9.5. The initial portion of the jet comprises an inviscid core and a linearly growing mixing region; here we take $u = u_e/2$, the volume element $dV \approx xD\,dx$, and $\omega \approx u/d = u_e/2x$. From equation 9.13, the acoustic power per unit of jet volume is about $\rho_0 u^8/La_0^5$, where L is interpreted as the distance along the jet to the radiating element, so the contribution of the elements in the mixing region, where $L \approx x$, is

$$dP = \left(\frac{\rho_0 u_e^8 D}{a_0^5}\right) dx$$

and for these elements $\omega \approx u_e/2x$.

In the fully developed part of the jet, u decreases so that $ux \approx u_e D$, and $\omega \approx u/d \approx u_e D/x^2$ while the volume element is $x^2\,dx$, and again $L \approx x$, so we have

$$dP \approx \frac{\rho_0 u_e^8 D}{a_0^5}\left(\frac{D}{x}\right)^7 dx$$

and for these elements $\omega \approx u_e D/x^2$.

We see that the mixing layer contributes to the high frequencies, and the fully developed jet to the lower frequencies. To estimate the frequency spectrum we note that $dP/d\omega = (dP/dx)(dx/d\omega)$ and find

$$\frac{dP}{d\omega} \approx \left(\frac{\rho_0 u_e^9 D}{a_0^5}\right)\omega^{-2} \quad \text{(mixing region—high frequencies)}$$

and   (9.16)

$$\frac{dP}{d\omega} \approx \left(\frac{\rho_0 u_e^5 D^5}{a_0^5}\right)\omega^2 \quad \text{(developed jet—low frequencies)}.$$

Thus, $dP/d\omega$ increases as ω^2 for low frequencies and decreases as ω^{-2} for high frequencies, as sketched in figure 9.7. The peak occurs for a frequency

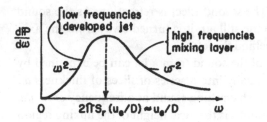

Figure 9.7
Frequency spectrum of jet noise, showing contributions of mixing layer and developed jet (from reference 9.7).

$$\frac{\omega}{2\pi} \approx \mathrm{Sr}\left(\frac{u_e}{D}\right), \tag{9.17}$$

where the Strouhal number Sr is between 0.15 and 0.20 for subsonic jets.

9.2.2 Supersonic Jets

This description must be modified for high-velocity jets where $M_{e0} > 2$. First of all, it is clear that the jet's acoustic efficiency as given by equation 9.15 cannot exceed unity, so the $M_{e0}{}^5$ law must fail for large M_{e0}. In fact, experiments show that for large M_{e0} the jet acoustic power varies as $u_e{}^3$, not as $u_e{}^8$.

Two mechanisms exist for noise production in supersonic jets that are not present in subsonic jets. They are shown schematically in figure 9.8. First, the large vortical structures in the supersonic shear layer radiate Mach waves when the rate of convection of the eddies is more than the speed of sound in the ambient air (when $M_{e0} \geq 2$). Second, if the jet is not perfectly expanded, then shocks occur in the adjustment to ambient pressure. Since these shocks are formed by the deflection of the jet boundary, they will be unsteady if the boundary is, and they can add considerably to the noise. This can add 3–5 dB to the noise level for jets that are not perfectly expanded (reference 9.8). The effect is shown in figure 9.9, where data from a wide range of jet conditions are plotted. Here, to account for effects of jet Mach number, a modified jet acoustic efficiency has been used, $P/\rho_0 A_e a_0{}^3 M_e{}^3$, where M_e is not equal to M_{e0} but is rather equal to u_e/a_e. It seems clear that M_e must govern the jet structure, not M_{e0}. The same factor is put in the abscissa, so that if $P/\rho_0 A_e a_0{}^3 M_e{}^3$ were actually proportional to $(u_e/a_0)^8 M_e{}^{-3}$ the data would fall on a straight line (the Lighthill predic-

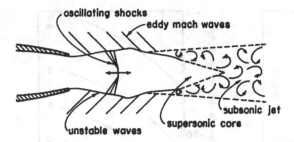

Figure 9.8
Schematic of supersonic jet, showing shock structure due to imperfect expansion and
unstable waves in supersonic mixing layer.

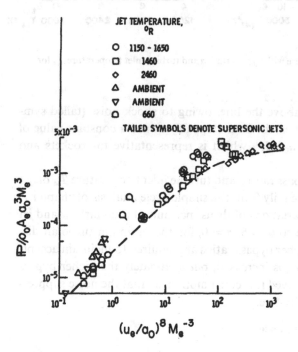

Figure 9.9
Jet acoustic power relative to jet kinetic power as a function of
$(u_e/a_0)^8 M_e^{-3} = (u_e/a_0)^8 (a_e/u_e)^3$ (from reference 9.8).

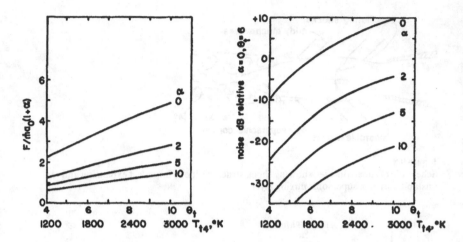

Figure 9.10
Variations of thrust and jet noise with bypass ratio α and turbine inlet temperature T_{t4} for turbofan engines.

tion). Some points fall above the line, owing to shock noise (tailed symbols). At large jet velocities, the data slope off toward a constant value of about 0.003 for $P/\rho_0 A_e a_0^3 M_e^3$, which is representative for rockets and which implies $P \propto u_e^3$.

The effects of both bypass ratio α and turbine inlet temperature T_{t4} on jet noise can be estimated readily from the simple cycle analysis of chapter 2. Figure 9.10 shows the variations of thrust per unit of total airflow and jet noise, the latter relative to $\theta_t = 6$, $\alpha = 0$, for the optimum turbofan discussed in section 2.5. Higher bypass ratios are required to maintain acceptable jet noise levels as T_{t4} is increased, but fortunately the higher bypass ratio also results in improved fuel consumption, so that the noise suppression is in this sense penalty-free.

9.2.3 Suppression of Jet Noise

Methods for suppressing jet noise have exploited the characteristics of the jet itself and those of the human observer. For a given total noise power, the human impact is less if the frequency is very high, as the ear is less sensitive at high frequencies. A shift to high frequency can be achieved by replacing one large nozzle with many small ones, since $\omega \approx u_e/D$, where D is the jet diameter. This was one basis for the early turbojet engine suppressors (reference 9.1). Reduction of the jet velocity can have a powerful effect since $\mathbf{P}$ is proportional to the jet velocity raised to a power varying from 8 to 3, depending on the magnitude of u_e. The multiple small nozzles reduced the mean jet velocity somewhat by promoting entrainment of the surrounding air into the jet. Some attempts have been made to augment this effect by enclosing the multinozzle in a shroud, so that the ambient air is drawn into the shroud in the way described in subsection 4.3.2.

Certainly the most effective of jet noise suppressors has been the turbofan engine, which in effect distributes the power of the exhaust jet over a larger airflow, thus reducing the mean jet velocity.

In judging the overall usefulness of any jet noise reduction system, several factors must be considered in addition to the amount of noise reduction. Among these factors are loss of thrust, addition of weight, and increased fuel consumption.

Consider the tradeoff between thrust and noise suppression. At takeoff the thrust F is proportional to $\dot{m}_e u_e$. If $M_{e0} < 2$ (as in turbofans), we conclude from equation 9.15 that $\mathbf{P_j} \propto \dot{m}_e u_e{}^7$. Three interesting special cases can be identified:

• *With an ideal ejector*, in which the jet power is held constant as the mass flow is increased, $\dot{m}u_e{}^2$ is constant, and F, which is proportional to $(\dot{m}u_e{}^2)/u_e$, varies as $1/u_e$; thus, as u_e is lowered F increases and $\mathbf{P_j}$ decreases like $u_e{}^5$, or $\mathbf{P} \propto F^{-5}$. The noise level in dB is $10 \log_{10}\mathbf{P}$, so the noise reduction in decibels that results from changing from thrust F_0 to F by changing u_e is

$$\Delta \text{dB} = -50 \log_{10}\left(\frac{F}{F_0}\right) \quad \text{(ideal ejector)}. \tag{9.18}$$

This is plotted as $F/F_0 = 10^{-\Delta \text{dB}/50}$ in figure 9.11. The only practical approach to the ideal ejector thus far is the turbofan engine. For such engines, if $\dot{m}_0$ represents the mass flow at thrust F_0 (turbojet), then

$$\frac{\dot{m}}{\dot{m}_0} = \left(\frac{u_{e0}}{u_e}\right)^2 = \left(\frac{F}{F_0}\right)^2 = 1 + \alpha,$$

where α is the bypass ratio. The resulting values of α are shown in figure 9.11, where we see that a bypass ratio of 2 gives a bit more than 10 dB reduction in jet noise. This was about the gain made in going to the first widely used turbofan engine (i.e., from the Pratt & Whitney JT-3 turbojet to the JT-3D).

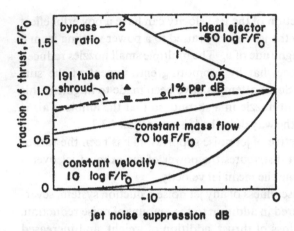

Figure 9.11
Variation of thrust with jet noise reduction for ideal-ejector (turbofan), constant-mass-flow, and constant-exhaust-velocity engines, compared with experimental results.

• *Smaller engines* with the same jet velocity would give $\mathbf{P} \propto \dot{m} \propto F$, so

$$\Delta dB = 10 \log_{10}\left(\frac{F}{F_0}\right) \quad \text{(reduced engine size)}, \tag{9.19}$$

which is plotted as the lowest curve on figure 9.11. Clearly, this is a bad solution.

• *Constant-mass flow*, with a reduction in u_e to effect the noise reduction, would give $\mathbf{P} \propto F^7$ and

$$\Delta dB = 70 \log_{10}\left(\frac{F}{F_0}\right) \quad \text{(constant-mass flow)}, \tag{9.20}$$

which is plotted as the middle full line in figure 9.11. Such a noise suppression could be obtained in a turbojet engine by operating at full corrected speed (full $W_2 \sqrt{\theta_2}/\delta_2$) with the exhaust nozzle opened as T_{t4}/T_{t2} is reduced to reduce u_e.

Measured against these simple cases, the results of actual noise-suppression programs (reference 9.9) as of the date of that reference (1968) —after major efforts at both subsonic and supersonic noise suppression —were better than the constant-mass-flow case, but far worse than the ideal-ejector case in terms of the penalty suffered to achieve a given level of noise suppression. Most suppressors cost about 1 percent loss in thrust per dB of noise reduction, including the very elaborate multi-tube ejectors studied for the US SST (Boeing 2707).

At this writing (1991) it appears that airport noise is still a limiting factor in the feasibility of supersonic civil transports. The problem and the prospects as of 1988 are very well summarized in reference 9.10, where the authors find little reason to believe that jet noise suppression can do better than the thrust loss of about 1% per dB of noise reduction quoted in connection with figure 9.11. This reference summarizes the experience gained in attempting to reduce the noise of the Concorde, and also describes later attempts to devise noise-suppression schemes.

A number of such noise-suppression schemes have been studied, mainly for turbofan engines of one sort or another. These include inverted-temperature-profile nozzles, in which a hot outer flow surrounds a cooler core flow, and mixer-ejector nozzles. In the first of these, the effect is to reduce the overall noise level from that which would be generated if the hot flow were on the inside as in a conventional turbofan, apparently because disturbances propagating at the interface between the cold inner and hot outer jets are subsonic with respect to the outer hot gas. This idea can be implemented either with a duct burner on a conventional turbofan or with a nozzle that interchanges the core and duct flows, carrying the latter to the inside and the former to the outside. In the mixer-ejector nozzle, the idea is to reduce the mean jet velocity by ingesting additional airflow through a combination of the ejector nozzles discussed in subsection 4.4.2 and the chute-type mixer discussed in subsection 4.4.3. Fairly high mass flow ratios can be attained with such arrangements, at the expense of considerable weight.

The most promising solution, however, is some form of "variable cycle" engine that operates with a higher bypass ratio on takeoff and in subsonic flight than at the supersonic cruise condition. This can be achieved to some degree with multi-spool engines by varying the speed of some of the spools to change their mass flow, and at the same time manipulating throttle areas. Another approach is to use a tandem-parallel compressor arrangement, where two compressors operate in parallel at takeoff and subsonically, and in series at supersonic conditions. The possibilities are discussed in reference 9.10.

As of this writing, the sideline noise problem of the supersonic transport must be regarded as unsolved in the sense that no concept has been shown to meet the existing noise regulations with clearly acceptable weight and performance. This is a challenging area for research.

9.3 Turbomachinery Noise

Turbomachinery generates noise by producing time-dependent pressure fluctuations, which can be thought of in first approximation as dipoles since they result from fluctuations in force on the blades or from passage of lifting blades past the observer. As dipoles the blades radiate in accord with equation 9.11, and if we carry out the argument that led to equation 9.13 for jet noise, we find that for the blade noise

$$\mathbf{P}_{\text{blade}} \approx \rho u^6 c^2/a^3 \approx \rho_2 w_2^{\,3} c^2 M_T^{\,3}, \tag{9.21}$$

where c is the blade chord or other characteristic dimension, and $u/a \approx M_T$, the tangential Mach number, leading to the second expression in the notation of chapter 5. By the argument of section 1.9 and subsection 5.1.2, the power exchanged with the fluid by the blade is of order $\rho_2 w_2^{\,3} c^2 M_T^{\,2}$, so

$$\eta_{\text{blade noise}} \propto M_T \tag{9.22}$$

(where $\eta_{\text{blade noise}}$ is the *acoustic efficiency*).

However, the situation is more complex than this. In contrast to the exhaust jet, the rotating blade is not free to radiate sound to the observer; instead the sound waves must propagate out of the engine inlet or exhaust before they can be radiated. Because this propagation has a controlling influence on turbomachinery noise, it will be discussed here at a physical level. A much more complete discussion can be found in reference 9.11.

9.3.1 Duct Modes

Suppose that a rotor or a rotor-stator pair is located at $z = 0$ in an annular duct (figure 9.12) and provides a source of acoustic excitation such that there is a pressure pattern

$$p'(r, \theta, 0) = g(r)e^{i(m\theta - \omega t)}. \tag{9.23}$$

If the pattern is from a rotor with angular velocity Ω_R and m blades, then $\omega = m\Omega_R$. We want to determine how this pressure pattern behaves in the annular duct; in particular, we want to know whether it propagates or is attenuated. We assume that there is a uniform, purely axial velocity W in the duct, or, equivalently, a Mach number M.

The pressure field is governed by equation 9.6, which when written in the cylindrical coordinates appropriate to the duct geometry becomes

Aircraft Engine Noise

387

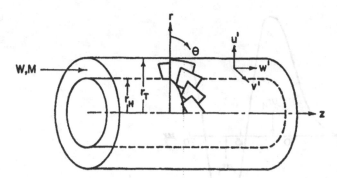

Figure 9.12
Schematic of a turbomachine duct with rotating blade row, illustrating notation for duct
propagation analysis.

$$\frac{1}{a_0^2}\frac{\partial^2 p'}{\partial t^2} + \frac{2M}{a_0}\frac{\partial^2 p'}{\partial z \partial t} + (M^2 - 1)\frac{\partial^2 p'}{\partial z^2} - \frac{1}{r}\frac{\partial}{\partial r}\left(r\frac{\partial p'}{\partial r}\right) - \frac{1}{r^2}\frac{\partial^2 p}{\partial \theta^2} = 0.$$

If we suppose the solution is of the form

$$p' = f(r)e^{i(m\theta + kz - \omega t)},$$

the radial dependence is governed by

$$\frac{1}{r}\frac{d}{dr}\left(r\frac{df}{dr}\right) + \left[-(1 - M^2)k^2 - \frac{2M\omega k}{a_0} + \left(\frac{\omega}{a_0}\right)^2 - \frac{m^2}{r^2}\right]f = 0,$$

or, more compactly,

$$r^2\frac{d^2 f}{dr^2} + r\frac{df}{dr} + (\mu^2 r^2 - m^2)f = 0, \tag{9.24}$$

where $\mu^2 = -(1 - M^2)k^2 - 2M\omega k/a_0 + (\omega/a_0)^2$. This is a Bessel equation with solutions (see reference 9.12)

$$f(\mu r) = c_1 J_m(\mu r) + c_2 Y_m(\mu r), \tag{9.25}$$

where $J_m(\mu r)$ is the Bessel function of the first kind of order m, and $Y_m(\mu r)$ is the Bessel function of the second kind. Both functions are tabulated and plotted in reference 9.13. A plot of $J_{10}(\mu r)$ is given in figure 9.13. $Y_m(\mu r)$ behaves similarly for large μr, with a phase shift, but is infinite at $\mu r = 0$.

The solution for the present problem must satisfy the boundary condition that the radial velocity perturbation be zero at the inner and outer

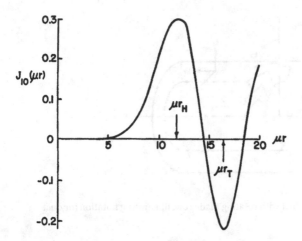

Figure 9.13
Bessel function of first kind of order 10.

walls, $r = r_H, r_T$. From the radial component of equation 9.2a,

$$\rho_0 \frac{\partial u'}{\partial t} = -\frac{\partial p'}{\partial r}$$

or

$$-i\rho_0 \omega u(r) = -\frac{dp(r)}{dr} = -\frac{df}{dr},$$

so the boundary condition is that $df/dr = 0$ at $r = r_H$, r_T; with equation 9.25 we have two relations,

$$c_1 J_m'(\mu r_H) + c_2 Y_m'(\mu r_H) = 0$$

and   (9.26)

$$c_1 J_m'(\mu r_T) + c_2 Y_m'(\mu r_T) = 0,$$

which for a given value of r_H/r_T may be considered to determine μr_T and c_2/c_1. The solution can be carried out numerically for any prescribed r_H/r_T, but to illustrate the results by means of an example without becoming involved in the numerics we can here regard r_H/r_T as a variable to be determined. Since the point marked μr_T in figure 9.13 satisfies equation 9.26 for $c_2 = 0$, we will have a solution with just J_{10} if r_H/r_T is such that μr_H is the second point indicated. This gives

$$\frac{r_H}{r_T} = \frac{\mu r_H}{\mu r_T} = \frac{11.9}{16.5} = 0.72$$

for $\mu r_T = 16.5$ and $m = 10$.

Going back to the definition of μ, we can now write

$$(\mu r_T)^2 = -(1 - M^2)(kr_T)^2 - \frac{2M\omega r_T}{a_0}(kr_T) + \left(\frac{\omega r_T}{a_0}\right)^2;$$

since μr_T is known, this becomes a quadratic relation for the axial wave number kr_T. If we further put

$$\frac{\omega r_T}{a_0} = \frac{m\Omega_R r_T}{a_0} = mM_T,$$

where M_T is the tangential Mach number of the blade tip, we get

$$kr_T = \frac{-MM_T m \pm \sqrt{m^2 M_T^2 - (\mu r_T)^2(1 - M^2)}}{1 - M^2}. \tag{9.27}$$

Now, if kr_T is real, the pressure disturbance is harmonic of the form $e^{ik_R z}$ in z; if kr_T has an imaginary part, however, then p' is of the form $e^{-k_1 z + ik_R z}$ and hence is attenuated in z. This attenuation is called *cutoff* in acoustics. The condition for propagation or "noncutoff" is then

$$m^2 M_T^2 - (\mu r_T)^2(1 - M^2) > 0, \text{ or}$$

$$M_T^2 > \left(\frac{\mu r_T}{m}\right)^2 (1 - M^2). \tag{9.28}$$

For the $m = 10$ mode in the above example, this becomes

$$M_T^2 > (1.65)^2(1 - M^2).$$

Thus, if $M = 0.5$, for example, $M_T > 1.43$ is required for propagation.

But this is not the lowest-order mode, as one combining J_{10} and Y_{10} has a lower $\mu r_T/m$. A solution exists for practical purposes for $\mu r_T = 11.9$ and $\mu r_H < 5$, that is, for $r_H/r_T < 5/11.9 = 0.42$, because J' is nearly zero for these small values of μr_H. For this mode $M_T^2 > (1.19)^2(1 - M^2)$, and $M_T > 1.031$ is required for propagation. As m becomes large, $\mu r_T/m \to 1$, and the condition for propagation becomes simply $M_T^2 + M^2 > 1$; *the relative Mach number to the blade tip must exceed unity for excitation of a propagating mode.* Cutoff tangential Mach number is plotted as a function of r_H/r_T and m for $M = 0$ in figure 9.14.

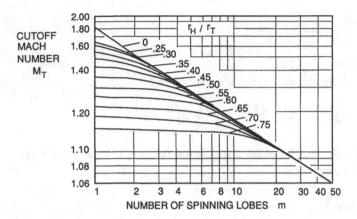

Figure 9.14
Cutoff tangential Mach number as a function of peripheral harmonic number m and hub/tip ratio r_H/r_T (from reference 9.14).

If a mode is appreciably below cutoff, the imaginary part of kr_T, $k_1 r_T$, is of order m, so the mode attenuates as $e^{-m(z/r_T)}$. For large m, then, the attenuation is rapid, and for practical purposes the cutoff modes do not radiate from the duct. It has been observed experimentally (reference 9.9) that the fan speed at which any given frequency first appears in the radiated sound agrees very well with equation 9.28.

9.3.2 Rotor-Stator Interaction

It would appear at first from this argument that compressors or fans should not radiate sound due to blade motion unless the blade tip speed is supersonic, but even low-speed turbomachines do in fact produce a great deal of noise at the blade passing frequencies. This phenomenon was first explained by Tyler and Sofrin (reference 9.15). A simplified version is as follows: Consider a rotor with B blades in close proximity to a stator with V vanes, as indicated in figure 9.3. The stator blades will produce a lift force and hence a pressure field p' dependent on their incidence. Whatever the pressure field, it must be periodic with stator spacing, so let

$$p'_{\text{stator}} \propto (\text{incidence})e^{iV\theta}$$

so that when θ passes from 0 to 2π, p'_{stator} passes through V periods. The disturbance produced by the rotor is periodic with rotor spacing; in rotor coordinates, (incidence) $\propto e^{iB(\theta - \Omega_R t)}$.

It follows that

$p'_{\text{stator}} \propto e^{i[(V-B)\theta - B\Omega_R t]}$.

This can be regarded as a pressure pattern rotating with angular velocity

$$\Omega_{\text{eff}} = \frac{d\theta}{dt} = \frac{B\Omega_R}{V - B}, \qquad (9.29)$$

which implies that the combination of rotor and stator will produce pressure patterns rotating faster than the rotor by the factor $B(V - B)$. This can be large if V and B are close together, as they were in most early engines. In modern turbofans, $V > 2B$, so that rotor-stator interaction is no more likely to excite propagating modes than is the rotor rotation itself. No inlet guide vanes are used in large modern turbofans, in order to eliminate the interaction between their wakes and the rotor. A large axial gap between the rotor and its downstream stator decreases the interaction between them by allowing the rotor wakes to decay.

9.3.3 "Buzz Saw" or Combination-Tone Noise

This analysis applied to an isolated rotor operating at an M_T above cutoff would predict a noise consisting of the blade-passing tone plus all its higher harmonics. No excitation of frequencies lower than blade-passing would be expected from a well-balanced rotor in which each blade is identical. What is in fact observed is a more or less random excitation of most of the harmonics of shaft rotational frequency up to and above the blade-passing frequency, as shown in figure 9.15. The large content of low frequencies is what leads to the characteristic "buzz saw" noise of high-bypass turbofans on takeoff.

There are at least two possible explanations for this fortunate occurrence—fortunate because if the turbofans did radiate at blade-passing, they would be very bad neighbors, as the blade-passing tone is near the peak of the ear's sensitivity. As indicated in figure 9.3, each blade near the tip generates a weak shock that propagates forward into the inlet annulus, but if the blades are not all identical then some of these shocks will be stronger than others and hence will propagate faster, overtaking their weaker upstream neighbors. When two shocks so interfere, they coalesce to form a stronger one, while a gap is left by the advancing shock. This

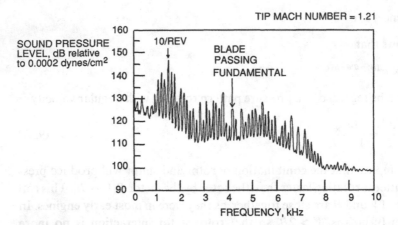

Figure 9.15
Harmonic analysis of "buzz saw" noise from large turbofan engine (from reference 9.14).

mechanism can lead to the formation of a shock pattern that is a replica, in a sense, of the mechanical imperfections of the rotor and hence would contain all harmonics of shaft rotational frequency (reference 9.16).

Although the above is the accepted explanation of the combination tone or "buzz saw" noise from transonic fans, another possible explanation lies in the behavior of the rotor blade wakes downstream of the rotor. It can be argued (reference 9.17) that the wake structure is unstable because of the strong mean swirl of the flow behind the rotor, and that this instability leads to formation of disturbances downstream of the rotor with periodicity less than blade passing. The pressure field of these disturbances could influence the shock structure on the rotor, giving rise in part to the "buzz saw" tones.

9.3.4 Duct Linings

The treatment of propagation in subsection 9.3.1 assumed that the walls of the duct were rigid, so that the radial velocity u was zero there. One consequence of this is that work can be done on the wall by the sound wave, so the wall takes no energy out of the wave. If the wall is modified so that pressure fluctuations at the wall result in a normal velocity component there, then energy can be extracted from the sound field, causing attenuation in the axial direction z.

This can be implemented in a number of ways; one is to line the duct with a porous sheet covering a series of small cavities, as shown in figure

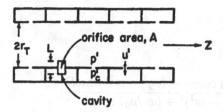

Figure 9.16
Schematic of acoustic damping cavities in an engine duct. The size of the resonators is
exaggerated relative to the duct diameter.

9.16. Each small cavity functions as a "Helmholtz resonator"; the pressure
in the cavity acts as a spring upon which the flow through the orifice
oscillates in response to pressure fluctuations outside the orifice. Thus, if p'_c
is the pressure in the cavity and L is the effective length of the orifice,

$$\rho_0 L \frac{du'}{dt} = p' - p'_c - cu'.$$

c is a viscous drag coefficient, and p'_c is connected to u' by

$$V\left(\frac{d\rho'_c}{dt}\right) = \rho_0 Au' = \frac{V}{a_0} \frac{dp'_c}{dt},$$

where A is the orifice area. Now, taking $p' = Pe^{i\omega t}$, we find

$$\frac{d^2 u'}{dt^2} + \frac{c}{\rho_0 L} \frac{du'}{dt} + \frac{A a_0^2}{LV} u' = i\omega P e^{i\omega t};$$

this has the solution $u' = Ue^{i\omega t}$, where

$$U = \frac{i\omega P/\rho_0 L}{\omega_n^2 - \omega^2 + i\omega c/\rho_0 L}.$$

Here $\omega_n^2 \equiv a_0^2 A/VL$ is the resonant frequency of the cavity. The ratio of
u' to $p'/\rho_0 a_0$ is usually termed the *acoustical admittance*, and is written as

$$\eta \equiv \frac{\rho_0 a_0 u'}{p'} = \frac{i\varepsilon\omega a_0/L}{\omega_n^2 - \omega^2 + i\omega c/\rho_0 L}, \tag{9.30}$$

where ε is the ratio of orifice area to wall area.

The effect on propagation is most easily seen by estimating the energy
extracted from the duct disturbance by the resonators. The power ex-

tracted per unit of wall area is

$$\frac{d\mathbf{P}}{dA_w} = \frac{\omega}{2\pi} \int_0^{2\pi/\omega} \operatorname{Re}(p') \operatorname{Re}(u') \, dt$$

$$= \frac{1}{2\pi} \int_0^{2\pi} \frac{\varepsilon P^2 \cos\omega t \; \omega[(\omega c/\rho_0 L) \cos\omega t - (\omega_n^2 - \omega^2) \sin\omega t]}{\rho_0 L \qquad (\omega_n^2 - \omega^2)^2 + (\omega c/\rho_0 L)^2}$$

$$= \frac{\varepsilon P^2 \omega^2 c}{2(\rho_0 L)^2 [(\omega_n^2 - \omega^2)^2 + (\omega c/\rho_0 L)^2]}.$$

The power in the wave propagating down the duct of diameter $2r_T$ is $\mathbf{P} = \pi r_T^2 P^2 / 2\rho_0 a_0$ (see section 9.1), and $dA_w/dz = 2\pi r_T$, so

$$\frac{1}{\mathbf{P}} \frac{d\mathbf{P}}{dz} = -\frac{\omega^2 a_0 c \varepsilon}{\rho_0 r_T L^2 [(\omega_n^2 - \omega^2)^2 + (\omega c/\rho_0 L)^2]}. \tag{9.31}$$

To estimate the attenuation length, suppose $cu' \approx \rho_0(u')^2/2$ (all the dynamic head of flow through the orifices is lost). Then $c \approx \rho_0 u'/2$, which we estimate as $\rho_0 a_0/2$. The decay length divided by the duct diameter is then

$$\frac{\mathbf{P}}{2r_T} \frac{dz}{d\mathbf{P}} = \frac{L^2 \omega^2}{\varepsilon a_0^2} \left[\left(\frac{\omega_n^2}{\omega^2} - 1 \right)^2 + \left(\frac{a_0}{2\omega L} \right)^2 \right].$$

Near resonance $\omega \approx \omega_n$ and this reduces to simply

$$\frac{\mathbf{P}}{2r_T} \frac{dz}{d\mathbf{P}} \approx \frac{1}{4\varepsilon}, \tag{9.32}$$

so that to attenuate the resonant frequency by a factor of $1/e$ in one duct diameter requires $\varepsilon = \frac{1}{4}$, that is, 25 percent open area in the duct lining.

The length scale of the openings is set by the frequency. We have $\omega_n^2 = a_0^2 A/VL$, and if we take $A \approx \varepsilon V/L$ then $\omega_n^2 \approx a_0^2 \varepsilon / L^2$. Then, for a rotor with B blades and angular velocity Ω_R,

$$\frac{L}{r_T} \approx \frac{\sqrt{\varepsilon}}{BM_T}.$$

A honeycomb structure covered by a perforated plate is often used.

This analysis is meant only to illustrate the principles of duct attenuation. Detailed calculations and extensive experimentation are required to determine the attenuation properties of liners with engineering accuracy. A good summary of data is given in reference 9.18.

9.4 Noise Measurement and Rules

Human response sets the limits on aircraft engine noise. Although the logarithmic relationship represented by the scale of decibels is a first approximation to human perception of noise levels, it is not nearly quantitative enough for either systems optimization or regulation. Much effort has gone into the development of quantitative indices of noise. A review was given in reference 9.19. All that will be attempted here is a brief description of the major factors involved and an explanation of the calculation procedures of Federal Aviation Rule, Part 36, which in a sense represents the official consensus on the measure of noisiness of an individual aircraft.

9.4.1 Noise Effectiveness Forecast (NEF)

It is not the noise output of an aircraft per se that raises objections from the neighborhood of a major airport, but the total noise impact of the airport's operations, which depends on takeoff patterns, frequencies of operation at different times of the day, population densities, and a host of less obvious things. There have been proposals to limit the total noise impact of airports, and in effect legal actions have done so for the most heavily used ones.

One widely accepted measure of noise impact is the Noise Effectiveness Forecast (NEF), which is arrived at as follows for any location near an airport:

1. For each event, compute the Effective Perceived Noise Level (EPNL) by the methods of FAR-36, described below. (The certification limits set by FAR-36 are intended to set an upper limit on this value.)
2. For events occurring between 10 P.M. and 7 A.M., add 10 to the EPNdB.
3. Then $NEF = 10 \log_{10} \Sigma_i \log_{10}^{-1}(EPNDB/10)_i - 82$, where the sum is taken over all events in a 24-hour period. A little ciphering will show that this last calculation is equivalent to adding the products of sound intensity times time for all the events, then taking the dB equivalent of this. The subtractor 82 is arbitrary.

As an example, suppose a point near the airport experiences 50 flyovers, each imposing an EPNdB level of 100; then

$$NEF = 10 \log_{10}[50 \log_{10}^{-1}(100/10) - 82] = 35.$$

Major complaints have occurred when NEF has exceeded about 30. As an

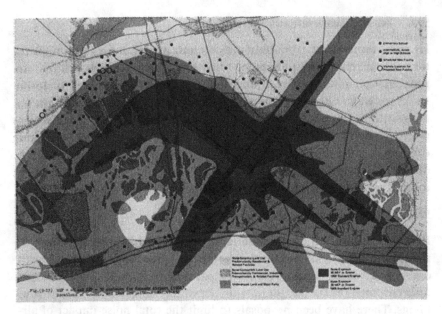

Figure 9.17
NEF = 40 and NEF = 30 contours for John F. Kennedy Airport (1968), locations of schools, and land-use patterns (from reference 9.20).

indication of the magnitude of the problem, consider figure 9.17. Reference 9.20 contains a comprehensive discussion of the noise issues at John F. Kennedy Airport. Although it is a bit dated now, it still reflects the important issues.

9.4.2 Effective Perceived Noise Level (EPNL)

The perceived noisiness of an aircraft flyover depends on the frequency content, relative to the ear's response, and on the duration. The perceived noisiness is measured in NOYs (units of perceived noisiness) and is plotted as a function of sound pressure level and frequency for random noise in figure 9.18. Note the great sensitivity in the range of frequencies from 2000 to 5000 Hz. Pure tones (that is, frequencies with pressure levels much higher than that of the neighboring random noise in the sound spectrum) are judged to be more annoying than an equal sound pressure in random noise, so a "tone correction" is added to their perceived noise level. A "duration correction" represents the idea that the total noise impact depends on the integral of sound intensity over time for a given event.

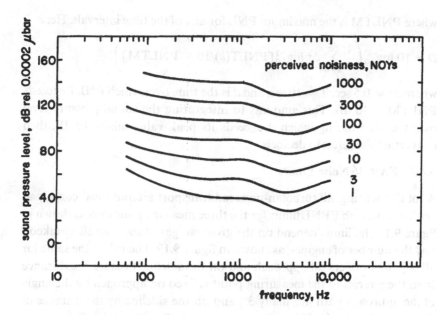

Figure 9.18
Perceived noisiness as a function of frequency and sound pressure level.

The raw data for an EPNL calculation would consist of time histories of the sound pressure (SPL) for each of the one-third-octave frequency bands from 50 Hz to 10,000 Hz. Conceptually, the calculation of EPNL involves the following steps; for the exact prescription see reference 9.2.

1. Determine the NOY level for each band from figure 9.18 or the equivalent, and sum them by the relation

$$N(k) = 0.85n(k) + 0.15 \sum_{i=1}^{24} n_i(k),$$

where k denotes an interval in time, i denotes the several frequency bands, and $n(k)$ is the NOY level of the noisiest band. This reflects the "masking" of lesser bands by the noisiest.
2. The total PNL is then $PNL(k) = 40 + 33.3 \log_{10} N(k)$.
3. Apply a tone correction $c(k)$ by identifying the pure tones and adding to PNL an amount ranging from 0 to 6.6 dB, depending on the frequency of the tone and its amplitude relative to neighboring bands.
4. Apply a duration correction according to EPNL = PNLTM + D,

where PNLTM is the maximum PNL for any of the time intervals. Here

$$D = 10 \log_{10}\left(\frac{1}{T} \sum_{k=0}^{d/t} \Delta t \, \log_{10}^{-1}[\mathrm{PNLT}(k)/10 - \mathrm{PNLTM}]\right),$$

where $\Delta t = 0.5$ sec, $T = 10$ sec, and d is the time over which PNLT exceeds PNLTM $-$ 10 dB. This amounts to integrating the sound-pressure level over the time during which it exceeds its peak value minus 10 dB, then converting the result to decibels.

9.4.3   FAR-36 Noise Limits

As of this writing, all turbofan-powered transport aircraft must comply at certification with EPNL limits for the three measuring stations as shown in figure 9.1. The limits depend on the gross weight of the aircraft at takeoff and the number of engines, as shown in figure 9.19. The rule is the same for all engine numbers on approach and on the sideline because the distance from the aircraft to the measuring point is fixed on approach by the angle of the approach path (normally 3°) and on the sideline by the distance of the measuring station from the runway centerline. On takeoff, however,

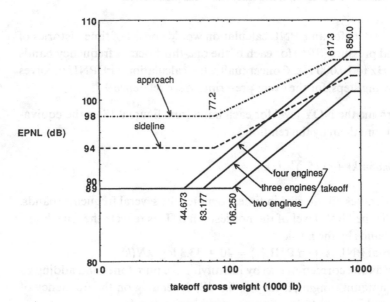

Figure 9.19
Noise limits imposed by FAR-36 for certification of aircraft.

aircraft with fewer engines climb out faster, so they are higher above the measuring point. Here the "reasonable and economically practicable" principle comes in to dictate that three-engine and two-engine aircraft have lower noise levels at the takeoff noise station than four-engine aircraft.

There is some flexibility in the rule, in that the noise levels shown can be exceeded by up to 2 EPNdB at any station provided the sum of the exceedances is not over 3 EPNdB and that the exceedances are completely offset by reductions at other measuring stations.

References

9.1 F. B. Greatrex and R. Bridge, "The Evolution of the Engine Noise Problem." *Aircraft Engineering*, February 1967.

9.2 Federal Aviation Regulations Part 36 , Subpart B, Paragraph 36.101 and Appendix A. US Government Printing Office, 1989.

9.3 G. L. Brown and A. Roshko, *Journal of Fluid Mechanics* 64 (1974): 775–816.

9.4 P. E. Dimotakis, Turbulent Free Shear Layer Mixing and Combustion. ISABE 89-7006, 1989. (See also AIAA Paper 89-0262, 1989.)

9.5 M. Van Dyke, *An Album of Fluid Motion*. Parabolic Press, 1982.

9.6 M. J. Lighthill, "On Sound Generated Aerodynamically, I: General Theory." *Proceedings of the Royal Society, Series A*, 211 (1952): 564–587. Also: M. J. Lighthill, "Jet Noise." *AIAA Journal* 1 (1963): 1507–1517.

9.7 H. S. Ribner, "The Noise of Aircraft." In Proceedings of the Fourth Congress of the International Council of Aeronautical Sciences, Paris, 1964. Also: University of Toronto Institute of Aerospace Sciences, Report 24.

9.8 U. H. von Glahn, Correlation of Total Sound Power and Peak Sideline OASPL from Jet Exhausts. AIAA Paper 72-643, 1972.

9.9 G. S. Schairer, J. V. O'Keefe, and P. E. Johnson, Perspective of the SST Aircraft Noise Problem. AIAA Paper 68-1023, 1968.

9.10 M. J. T. Smith, B. W. Lowrie, J. R. Brooks, and K. W. Bushell, Future Supersonic Transport Noise—Lessons from the Past. AIAA-88-2989, Joint Propulsion Conference, Boston, 1988.

9.11 N. A. Cumpsty, "Engine Noise." In *The Aerothermodynamics of Aircraft Gas Turbine Engines* (AFAPL TR-78-52), ed. G. C. Oates. Air Force Aero Propulsion Laboratory, Wright-Patterson Air Force Base, Ohio.

9.12 F. B. Hildebrand, *Advanced Calculus for Applications*. Prentice-Hall, 1962.

9.13 E. Jahnke and F. Emde, *Tables of Functions and Curves*. Dover, 1945.

9.14 M. J. Benzakein, "Research on Fan Noise Generation." *Journal of the Acoustical Society of America* 51, no. 5, part I (1972): 1427–1438.

9.15 J. M. Tyler and T. G. Sofrin, "Axial Compressor Noise Studies." *SAE Transactions* 70 (1962): 309–332.

9.16 M. R. Fink, Shock Wave Behavior in Transonic Compressor Noise Generation. ASME Paper 71-GT-7, 1971.

9.17 W. T. Thompkins and J. L. Kerrebrock, Exit Flow from a Transonic Compressor Rotor. MIT GTL Report 123, 1975. Also: Unsteady Phenomena in Turbomachinery (AGARD Conference Proceedings 177), 1975.

9.18 Aircraft Engine Noise Reduction. NASA SP 311, NASA Lewis Research Center, 1972.

9.19 J. W. Little and J. E. Mabry, Human Reaction to Aircraft Engine Noise. AIAA Paper 68-548, 1968.

9.20 National Academy of Sciences and National Academy of Engineering, *Jamaica Bay and Kennedy Airport: A Multidisciplinary Study*, vol. 2, 1971.

9.21 Aeroacoustics of Flight Vehicles (NASA Reference Publication 1258 and WRDC Technical Report 90-3052, 1990): volume 1 (Noise Sources) and volume 2 (Noise Control).

Problems

9.1 Consider a wall with shape given by $y = A \sin kx$ ($kA \ll 1$) moving with velocity U in the x direction in a compressible fluid with speed of sound a. Show that if $M \equiv U/a > 1$ sound waves are radiated to $y = +\infty$, whereas if $M < 1$ the pressure disturbance is attenuated in y. Compare these results to those of subsection 9.3.1.

9.2 Following the argument of subsection 9.1.2, work out the directional radiation characteristics of the linear quadrupole shown in figure 9.4, then repeat the argument for a rectangular or square quadrupole having sources of alternate sign on each side.

9.3 Taking the dipole model indicated in figure 9.4 for a vibrating blade, develop an approximate expression for the sound radiated by an airfoil of chord c oscillating with amplitude A in the direction perpendicular to its chord in an airflow of velocity U_0, pressure p_0, and temperature T_0.

9.4 A turbojet engine has a mass flow of 300 kg/sec and a thrust of 300,000 N at takeoff. Estimate (a) its total radiated jet noise power, (b) the sound pressure level in dB at a distance of 0.35 mile, and (c) the frequency of peak sound intensity.

9.5 For the optimum turbofan of section 2.5, find the variations of thrust per unit of airflow and jet noise with bypass ratio for $\theta_t = 6$, and check the results of figure 9.10.

9.6 A transonic fan without inlet guide vanes has a blade stagger angle (β'_b) at the tip of 60°, a hub/tip radius ratio of 0.4, and a tip radius of 1 m. Determine the rotative speed at which the rotor would be expected to first generate a propagating mode in the upstream duct at takeoff.

9.7 Consider two possible classes of subsonic transports, one of 600,000 lb gross weight and the other of 200,000 lb gross weight, but otherwise identical in aerodynamic performance, structural to gross weight ratio, engine performance, and so on, so that the smaller aircraft will carry one-third the number of passengers that the larger one does. For a given total number of passenger movements per day, how do the NEF levels for the two aircraft compare?

10 Hypersonic Engines

Current air-breathing propulsion systems operate routinely at flight Mach numbers up to 3 and at altitudes as high as 90,000 ft. Between these and the orbital operating conditions of spacecraft is a huge range of speed and altitude over which the air-breathing system—because it draws its oxygen from the atmosphere—is in principal capable of much higher specific impulse than is rocket propulsion. Some idea of the potential of high-speed air-breathing propulsion can be gained from figure 10.1, where values of specific impulse are plotted for a number of engine types over a wide range of M_0 and where the parameters of each engine are selected as they would optimize for its usual mission. The values for the turbojet, the turbofan, and the ramjet were taken from chapter 2; those for the supersonic combustion ramjet were estimated by techniques to be explained below and are a bit speculative. But the main point, the potentially high specific impulse of air-breathing relative to rocket propulsion, seems clear. As the enabling technology for the National Aerospace Plane (NASP) concept, the supersonic combustion ramjet (scramjet) has received a good deal of attention since the first edition of this book was published. First seriously considered for propulsion of transatmospheric vehicles in the 1960s, it was studied at a low level of effort during the 1970s and 1980s, and has since been examined intensively in the context of NASP. (For reviews of the earlier work see references 10.1–10.3; for the more recent work, see references 10.4–10.7.) Other types of high-speed air-breathing engines, such as the air-turborocket (ATR) and the liquid air collection (LACE) system, have received some attention and will be discussed in somewhat less detail than the scramjet.

The technical differences between air-breathing engines operating at speeds above $M_0 = 3$ and those for lower speeds stem mainly from the high stagnation temperature levels at large M_0. $T_{t0} = T_0[1 + \frac{1}{2}(\gamma - 1)M_0^2]$ is shown in figure 10.1.

At Mach numbers of 6 and above, conversion of even a part of the air's kinetic energy to thermal energy can raise the temperature to the level where the air dissociates, changing its properties considerably. Even more important, the temperature can be so high that the temperature rises little when fuel is added to the air, because what we think of as the normal combustion products (H_2O and CO_2) are strongly dissociated. This occurs above about 2500°K. In addition, there are the temperature limitations of structural materials, some of which are indicated in figure 10.1. Above $M_0 = 6$, no material will endure the stagnation conditions, so all structures must be cooled, either radiatively or by the fuel or ablatively.

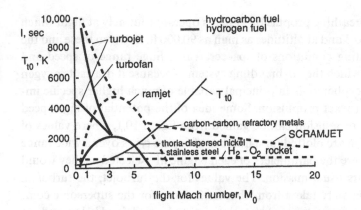

Figure 10.1
Specific impulse versus M_0 for various engines using hydrocarbon and hydrogen fuels.

The factors that limit M_0 are different for the various types of engines. For the turbojet, the limit is set by the turbine temperature limits at about $M_0 = 3$ or 4. With advanced cooling techniques this could be raised; however, as figure 10.1 shows, the ramjet is a better choice above $M_0 \approx 3.5$. In fact the SR-71 aircraft, which has operated at Mach numbers above 3, is powered by a turbo-ramjet engine, in which, at high flight Mach numbers, some of the airflow passes from the inlet and from the fourth compressor stage directly to the afterburner, bypassing the turbine. This system has the characteristics of a ramjet at high Mach numbers and those of a turbojet at lower Mach numbers.

The H_2-fueled turbojet has a higher specific impulse primarily because the heating value of H_2 per unit mass is some 2.3 times that of jet fuel. Because of the greater cooling capability of the liquid H_2 fuel, the H_2 turbojet could probably be operated to $M_0 = 5$. Conventional ramjets, in which the air is slowed to $M_0 \approx 0.2$ before combustion, operate most efficiently for $2 < M_0 < 6$, as indicated in figure 1.2; above 6 dissociation of the combustion products limits the temperature rise, upon which the engine depends for efficient operation. With H_2 fuel, the limit is a bit higher, but it is still in the range of $M_0 \approx 7$.

The supersonic combustion ramjet, shown schematically in figure 10.2, was conceived to minimize the problem of dissociation. Since much of the recent effort on hypersonic propulsion has focused on this concept, it will be given the most attention here. In this concept diffusion is carried out in

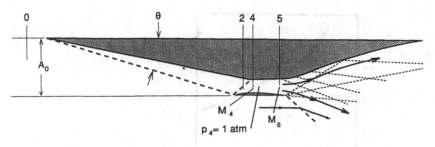

Figure 10.2
Schematic of hypersonic scramjet engine showing inlet, combustor and nozzle.

the inlet from M_0 down to only $M_2 \approx 3$ or higher, depending on the value of M_0, so that the air temperature does not approach T_{t0} prior to combustion. The combustion occurs in the supersonic flow, raising T_t and also T, after which the combustion products are expanded through the nozzle. Even though the diffusion is partial, the thermal efficiency can in principle be high because the temperature ratio in the diffuser is still close to 10, giving an ideal Brayton cycle efficiency of 0.90. Combustion in a supersonic flow can be achieved with H_2, though probably not with hydrocarbons, because of the much higher flame speed and wider flammability limits of H_2.

One of the dominant characteristics of air-breathing engines for hypersonic flight, implied by figure 10.2, is that the engine is mostly inlet and nozzle, and efficient compression and expansion processes are very critical to the engine's performance. This can be seen more quantitatively by estimating the velocity ratios u_e/u_0 that are implied by the values of I given in figure 10.1. We know that

$$I = \frac{F}{\dot{m}f} \approx u_0 \frac{u_e/u_0 - 1}{f}.$$

Above $M_0 \approx 5$ all engines operate approximately stoichiometrically, so f is nearly a constant that works out to $f = 0.0293$ for H_2. Hence,

$$\frac{u_e}{u_0} - 1 = \frac{0.0293 I g}{a_0 M_0} \approx \frac{0.00096}{M_0} I.$$

This equation is plotted in figure 10.3, which shows that the fractional velocity change across the engine is very small indeed for $M_0 > 6$, so that a

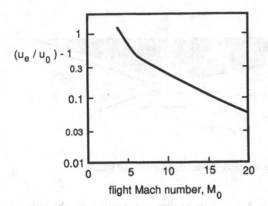

Figure 10.3
Fractional velocity change in hypersonic engines.

small (unexpected) inefficiency in the inlet or in the nozzle could have large consequences.

These arguments serve to define the critical elements of feasibility for scramjet engines:

• Inlets and nozzles of high efficiency are essential.
• Fuel-air mixing and combustion must be carried out at high (supersonic) velocities, instead of at near-stagnation, as in gas turbines.
• Cooling of the engine's structure by fuel or radiation is required.

In the following sections each of these elements will be discussed, first in a qualitative way that brings out the physical phenomena important in high-speed propulsion. That will be followed by a quantitative discussion of the potential performance of scramjets. The chapter will be concluded with a discussion of other potential high-speed air-breathing propulsion systems.

10.1   Hypersonic Inlets

The performance of hypersonic inlets is best expressed in terms of a kinetic energy efficiency, which is defined by

$$\eta_d \equiv \frac{\text{Available kinetic energy after diffusion}}{\text{Available kinetic energy before diffusion}} = \frac{u_2'^2}{u_0^2}. \tag{10.1}$$

It is understood that the flow is expanded to ambient pressure p_0 in both the numerator and the denominator, and the definition assumes that the flow in the diffusor is adiabatic. The advantage of η_d is that it can be nearly constant over a wide range of M_0 for well-designed inlets. In the notation of figure 10.2 and chapter 2,

$$p_{t0} = p_0\left(1 + \frac{\frac{1}{2}(\gamma - 1)u_0{}^2}{\gamma R T_0}\right)^{\gamma/(\gamma-1)}$$

and

$$p_{t2} = p_2\left(1 + \frac{\frac{1}{2}(\gamma - 1)u_2{}^2}{\gamma R T_2}\right)^{\gamma/(\gamma-1)} \equiv p_0\left(1 + \frac{\frac{1}{2}(\gamma - 1)u_2'{}^2}{\gamma R T_2'}\right)^{\gamma/(\gamma-1)}.$$

The last expression is the definition of the available kinetic energy $u_2'{}^2/2$. By the definition of η_d, we have $u_2'{}^2 = \eta_d u_0{}^2$; thus, by conservation of total enthalpy,

$$T_2' = T_0 + \frac{u_0{}^2}{2c_p}(1 - \eta_d),$$

and we find that in terms of η_d we can write $\pi_d = p_{t2}/p_{t0}$ as

$$\pi_d = \left(1 + (1 - \eta_d)\frac{\gamma - 1}{2}M_0{}^2\right)^{-\gamma/(\gamma-1)}. \tag{10.2}$$

Given a value of η_d (0.97 is realistic), we can estimate π_d as a function of M_0 from this relation. Some values are given in table 10.1. Even for such a high efficiency, π_d becomes quite small at large M_0. But the values of π_d for a normal shock at the same M_0 are very much smaller, so the hypersonic inlet must avoid strong shocks in order to achieve $\eta_d = 0.97$.

One way to look at this requirement is to ask what Mach number normal to a single oblique shock is implied by $\eta_d = 0.97$. These values, which are just the normal shock Mach numbers giving $p_{t2}/p_{t0} = \pi_d$, are listed as M_n. Since the shock loss is only a part of the total loss in the inlet, the

Table 10.1
Values of π_d as a function of M_0.

	M_0				
	1	3	6	10	20
$\pi_d(\eta_d = 0.97)$	0.97	0.830	0.505	0.193	0.0137
π_d (normal shock)	1.0	0.328	0.0297	0.00305	0.000108
M_n (single oblique shock)	1.29	1.76	2.48	3.60	7
M_n/M_0		0.59	0.41	0.36	0.35

shocks must be considerably weaker than those found here if the inlet is to achieve $\eta_d = 0.97$, and in fact (as we shall see) there is a strong incentive to achieve values of η_d considerably in excess of 0.97, so that the inlet design must be quite refined, with at least two and probably three or more oblique shocks rather than one. Since the principles involved are not different from those elaborated in chapter 4, this aspect will not be pursued further here. It is important to note, however, that for reasons which will become clear later the scramjet will operate at conditions such that the Reynolds number on the inlet ramps is in the transitional range, and the engine performance can be strongly influenced by the location along the inlet of the transition. Since transition is not well understood at hypersonic conditions, this poses an uncertainty—one that can be resolved only through research on flight vehicles at the relevant conditions, because it is not feasible to simulate these conditions in test facilities.

Finally, the ratio $M_n/M_0 = \sin\theta$, θ being the wave angle of figure 10.2, is tabulated. The interesting point is that M_n/M_0 is nearly constant above $M_0 = 6$, so an inlet of nearly fixed geometry might be able to operate without much spill over this wide range of Mach number.

10.2 Heat Addition in High-Speed Flow

Next consider the effects of heat addition on the flow in the combustor where $M > 1$. In figure 10.2 the combustor is between stations 4 and 5. The flow length from the end of the inlet (station 2) to station 4 is reserved for the processes associated with injection and mixing of the fuel, although in practice the injection, the mixing, and the combustion would occur concurrently. For the sake of simplicity, we will initially assume that the flow area is constant. Then, in differential form, the equations governing the flow are as follows.

conservation of momentum: $\quad \rho u \dfrac{du}{dx} = -\dfrac{dp}{dx}$ (10.3)

conservation of energy: $\quad \rho u \dfrac{dh}{dx} = u \dfrac{dp}{dx} + \dfrac{dQ}{dx}$ (10.4)

conservation of mass flow: $\quad \rho u = \text{const} = \rho_4 u_4.$ (10.5)

Here dQ/dx is the rate at which energy is added to the gas by chemical reaction.

To display the fluid-dynamic effects of heat addition most clearly, assume for the present that $h = c_p T$, and that c_p and c_v are constant. From the equation of state,

$$p = \rho R T$$

and

$$\frac{dp}{dx} = R\left(\rho\frac{dT}{dx} + T\frac{d\rho}{dx}\right) = R\left(\frac{\rho_4 u_4}{u}\frac{dT}{dx} - \frac{\rho_4 u_4 T}{u^2}\frac{du}{dx}\right),$$

where equation 10.5 has been used twice. Eliminating dp/dx with equation 10.3, we then have

$$(\rho_4 u_4)u\frac{du}{dx} = -R\left(\rho_4 u_4 \frac{dT}{dx} - \frac{\rho_4 u_4 T}{u}\frac{du}{dx}\right),$$

and from equation 10.4 we have

$$(\rho_4 u_4 c_p)\frac{dT}{dx} = R\left(\rho_4 u_4 \frac{dT}{dx} - \frac{\rho_4 u_4 T}{u}\frac{du}{dx}\right) + \frac{dQ}{dx},$$

which can then be solved for either dT/dx or du/dx. Noting that $M = u/\sqrt{\gamma R T}$, we find

$$\frac{dT}{dx} = \frac{1 - \gamma M^2}{1 - M^2}\frac{1}{\rho_4 u_4 c_p}\frac{dQ}{dx}, \qquad (10.6)$$

$$u\frac{du}{dx} = \frac{(\gamma - 1)M^2}{1 - M^2}\frac{1}{\rho_4 u_4}\frac{dQ}{dx}, \qquad (10.7)$$

$$\frac{1}{M}\frac{dM}{dx} = \frac{1 + \gamma M^2}{2(1 - M^2)}\frac{1}{\rho_4 u_4 c_p T}\frac{dQ}{dx}, \qquad (10.8)$$

Equation 10.8 shows that adding heat lowers M if $M > 1$ and raises M if $M < 1$, so it always drives M toward 1. Similarly, equation 10.7 shows that u decreases with heat addition for $M > 1$ and increases for $M < 1$.

Expressions for the stagnation conditions are more useful for the present purposes than equations 10.6 and 10.7. Multiplying equation 10.3 by u and adding it to equation 10.4 yields

$$\rho_4 u_4 \frac{d}{dx}(c_p T + \tfrac{1}{2}u^2) = \rho_4 u_4 c_p \frac{dT_t}{dx} = \frac{dQ}{dx}. \qquad (10.9)$$

The stagnation pressure is

$$p_t = p[1 + \tfrac{1}{2}(\gamma - 1)M^2]^{\gamma/(\gamma-1)},$$

so

$$\frac{1}{p_t}\frac{dp_t}{dx} = \frac{1}{p}\frac{dp}{dx} + \frac{\gamma}{\gamma - 1}\frac{(\gamma - 1)M}{1 + \tfrac{1}{2}(\gamma - 1)M^2}\frac{dM}{dx};$$

using equations 10.3, 10.7, and 10.8 we find

$$\frac{1}{p_t}\frac{dp_t}{dx} = -\frac{\gamma M^2/2}{\rho_4 u_4 c_p T_t}\frac{dQ}{dx}. \tag{10.10}$$

This equation shows clearly that *adding heat always lowers* p_t; it further shows that the decrease of p_t is much larger for $M > 1$ than for the small values found in gas turbine combustors. This is one of the negative aspects of supersonic combustion.

Though equations 10.9 and 10.10 show the effects of heat addition most clearly, it is convenient for purposes of computation to have integrated forms of the equations. Eliminating dQ/dx between equations 10.8 and 10.7 gives

$$\frac{dp_t}{p_t} = \frac{-\gamma M(1 - M^2)}{(1 + \gamma M^2)[1 + \tfrac{1}{2}(\gamma - 1)M^2]}dM,$$

which integrates to

$$\frac{p_{t5}}{p_{t4}} = \frac{1 + \gamma M_4^2}{1 + \gamma M_5^2}\left(\frac{1 + \tfrac{1}{2}(\gamma - 1)M_5^2}{1 + \tfrac{1}{2}(\gamma - 1)M_4^2}\right)^{\gamma/(\gamma-1)}.$$

Similarly,

$$\frac{u_5}{u_4} = \frac{\rho_4}{\rho_5} = \frac{1 + \gamma M_4^2}{1 + \gamma M_5^2}\left(\frac{M_5}{M_4}\right)^2,$$

$$\frac{p_5}{p_4} = \frac{1 + \gamma M_4^2}{1 + \gamma M_5^2},$$

$$\frac{T_5}{T_4} = \left(\frac{1 + \gamma M_4^2}{1 + \gamma M_5^2}\right)^2\left(\frac{M_5}{M_4}\right)^2,$$

$$\frac{T_{t5}}{T_{t4}} = 1 + \frac{Q}{\rho_4 u_4 c_p T_{t4}} = \left(\frac{1 + \gamma M_4^2}{1 + \gamma M_5^2}\right)^2\frac{1 + \tfrac{1}{2}(\gamma - 1)M_5^2}{1 + \tfrac{1}{2}(\gamma - 1)M_4^2}\left(\frac{M_5}{M_4}\right)^2. \tag{10.11}$$

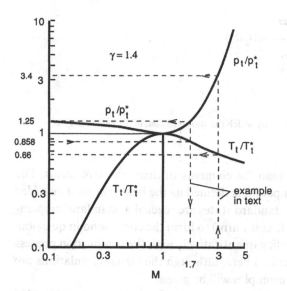

Figure 10.4
Ratios of stagnation pressure and temperature to their values at $M = 1$ in flow with heat addition in a constant-area duct.

Now, if we fix M_5 at 1, we can give ratios of p_{t5}/p_t^*, T_{t5}/T_t^*, and so forth as functions of M_4, just as in subsection 4.1.1. Here we are mainly concerned with the effect of heat addition on p_t, so it is convenient to have p_t/p_t^* and T_t/T_t^* as functions of M. Such functions are plotted in figure 10.4.

Suppose, for example, that enough fuel is burned to increase T_t by 30 percent in a flow with initial Mach number of 3. From equation 10.11 we have $Q/\rho_4 u_4 c_p T_{t4} = 0.3$ and $T_{t5}/T_{t4} = 1.3$. From figure 10.4, for $M = 3$ we have $(T_t/T_t^*)_4 = 0.66$ and $(p_t/p_t^*)_4 = 3.4$. Then $(T_t/T_t^*)_5 = 1.3(T_t/T_t^*)_4 = 0.858$, and from the figure $M_5 = 1.7$, $(p_t/p_t^*)_5 = 1.25$, and finally $p_{t5}/p_{t4} = 1.25/3.4 = 0.37$.

10.3   Heat Release Due to Chemical Reactions

Thus far the combustion process has been represented by an external heat source, Q. In fact, no energy is added to the flowing gas mixture during combustion; energy is only converted from chemical to thermal form. The presence of the chemical energy in the unburned mixture can be represented by the addition of a heat of formation ΔH_f° to the enthalpy of each

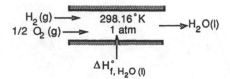

Figure 10.5
Schematic of the steady flow process by which the standard heat of formation ΔH_f° is determined.

chemical substance other than the elements in their standard forms. The heat of formation of a compound is defined as the heat that *must be added* when the elements in their standard states are reacted at standard temperature and pressure (298.16°K and 1 atm) to form the compound in question. A general approach for dealing quantitatively with the combustion process in these terms will be described here, although the actual calculations are so lengthy that only some examples will be given.

Schematically, in the steady flow process of figure 10.5, if 1 mole of H_2 and 0.5 mole of O_2, both gases at 1 atm and 298.16°K, could be reacted at 298.16°K and 1 atm to form 1 mole of liquid H_2O, the heat that would have to be added would be -68.3174 kcal/mole. Actually, since only the initial and final states matter, the temperature can rise during the combustion process and then drop as a result of heat transfer out of the flow channel. In fact this is the way ΔH_f° is measured. Now, if we define a *complete enthalpy*, say H, for the species i by

$$H_i = \int_{T_r}^{T} c_{pi}\,dT + \Delta H_{fi}^\circ,\tag{10.12}$$

where the reference temperature is usually $T_r = 298.16°K$, then the analogue of equation 4.2 is

$$\Sigma x_i H_i + \frac{u^2}{2} = \text{const},\tag{10.13}$$

where x_i is the mass fraction of the species i; that is,

$$x_i = \frac{\rho_i}{\rho},\tag{10.14}$$

where ρ is the total gas density and ρ_i is the density of species i. The statement of conservation of momentum remains

$$\rho u \frac{du}{dx} = -\frac{dp}{dx},$$

or, in integrated form for constant area flow ($\rho u = \rho_4 u_4 = $ const),

$$\rho_4 u_4 (u_5 - u_4) = -(p_5 - p_4). \tag{10.15}$$

The equation of state may be written separately for each species as

$$p_i = \rho_i R_i T, \tag{10.16}$$

where $R_i = R/M_i$, M_i being the molecular weight and R the universal gas constant. The total pressure is simply $p = \Sigma p_i$, and the x_i are related to the p_i by

$$\frac{p_i}{p} = \frac{\rho_i R_i T}{\Sigma p_i} = \frac{\rho_i R_i T}{\Sigma \rho_i R_i T} = \frac{\rho_i R_i}{\rho \Sigma x_i R_i}.$$

Thus, if $R \equiv \Sigma x_i R_i$, then

$$\frac{p_i}{p} = \frac{x_i R_i}{R}. \tag{10.17}$$

If the mass fractions of the various species x_i were known, then the equations of conservation of mass, momentum, and energy could be solved just as in section 10.2 with the $\Delta H_{fi}°$ terms in equation 10.12 replacing the Q. The x_i are determined by chemical reactions, which may be controlled by kinetic phenomena, as outlined in section 4.4. But in many circumstances the chemical reactions proceed essentially to a condition of local chemical equilibrium. This will happen if the reaction rates are large relative to the rate of the phenomenon that is changing the thermodynamic state of the gas (usually a flow process in engines).

When the reactions are near equilibrium, the p_i are governed by the law of mass action, which states that for any set of chemically interacting compounds $A, B, \ldots, L, M, \ldots$ that satisfy the stoichiometric equation

$$aA + bB + cC + \cdots \leftrightarrow lL + mM + nN + \cdots \tag{10.18}$$

there is a function K_p of T alone such that

$$K_p(T) = \frac{p_L^l p_M^m p_N^n \cdots}{p_A^a p_B^b p_C^c \cdots}. \tag{10.19}$$

The set of equations 10.19, plus the statement that elements are conserved in the chemical reactions, serve to relate the pressure of any chemical species to those of the elements of which it is composed, as functions of the temperature and gas pressure.

For example, suppose H_2 and O_2 react to form H_2O, OH, H_2, O_2, H, and O according to

$$H_2 + \alpha O_2 \rightarrow \beta H_2O + \gamma OH + \delta H_2 + \varepsilon O_2 + \zeta H + \eta O,$$

where α is prescribed and we want to find β, γ, ... Usually, α will be given in terms of an "equivalence ratio" ϕ which is the ratio of the fuel flow to that for stoichiometric combustion. Thus, for H_2 and O_2, $\phi = 1/2\alpha$. The stoichiometric reactions by which the K_p are defined are

$$H_2 + \tfrac{1}{2}(O_2) \rightarrow H_2O \quad K_{pH_2O} = p_{H_2O}/p_{H_2}\sqrt{p_{O_2}} = \frac{\beta}{\delta\sqrt{\varepsilon}}\sqrt{\sigma/p},$$

$$H_2 + O_2 \rightarrow 2OH \quad K_{pOH} = p_{OH}^2/p_{H_2}p_{O_2} = \frac{\gamma^2}{\delta\varepsilon},$$

(10.20)

$$H_2 \rightarrow 2H \quad K_{pH} = p_H^2/p_{H_2} = \frac{\zeta^2}{\delta}\frac{p}{\sigma},$$

$$O_2 \rightarrow 2O \quad K_{pO} = p_O^2/p_{O_2} = \frac{\eta^2}{\varepsilon}\frac{p}{\sigma},$$

where $\sigma = \beta + \gamma + \delta + \varepsilon + \zeta + \eta$.

Conservation of elements requires for H and for O

$$2 = 2\beta + \gamma + 2\delta + \zeta,$$

(10.21)

$$2\alpha = \beta + \gamma + 2\varepsilon + \eta.$$

Between equations 10.20 and 10.21 we have six relations from which to solve for the six unknowns β, γ, δ, ε, ζ, and η for given values of T and p.

Because the computation is complex, some suggestions as to methods of approach may be in order. The general problem may be posed as follows. Given ρ_4, u_4, p_4, T_4, M_4, and the mixture ratio, say of H_2 to air (or equivalently α), find the state of the gas at station 3, that is, p_{t5}, T_{t5}, M_5, and the set of x_{i5}.

1. Consider first the simpler case where $M_4 \ll 1$. Then, from equation 10.15, $p_5 \approx p_4$, and in equation 10.13 we may neglect $u^2/2$ compared to

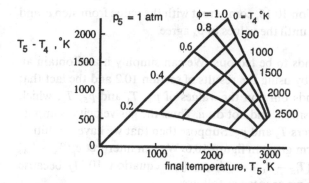

Figure 10.6
Temperature rise in combustion of H_2 with air, as a function of final temperature and equivalence ratio, for pressure of 1 atm and no flow.

$\Sigma x_i H_i$. The solution is obtained as follows:

a. Assume a value of T_5 and compute the x_i from equations 10.20 and 10.21.

b. Compute $(\Sigma x_i H_i)_5$ and compare it with $(\Sigma x_i H_i)_4$.

c. Iterate the choice of T_5 until equation 10.13 is satisfied.

The solution for H_2 and air is given in figure 10.6 for a range of equivalence ratios at a pressure of 1 atm, which is typical of the pressure in the combustor of a scramjet. Note first that as the initial temperature increases, the rise in gas temperature due to combustion becomes smaller, until it is only about 500°K for a stoichiometric mixture for $T_4 = 2500°K$. Since there is no change in the gas' kinetic energy in this case, the chemical energy of the hydrogen-oxygen mixture is mainly in the form of chemical energy of the species formed by dissociation of H_2O.

2. In the general case, where M_4 is not small, we must incorporate the equivalent of the calculation in section 10.2. This can be done as follows:

a. Suppose first that we can specify p_5 rather than p_4. This will be helpful if T_4 is low enough so the $(x_i)_4$ are known independently of p_4. Then we *assume* T_5 and compute the $(x_i)_5$.

b. From equation 10.13, find u_5.

c. From $\rho_5 u_5 = \rho_4 u_4$, find ρ_4 and hence p_4.

d. Find p_4 from equation 10.15, compare it with the value from step c, and iterate the choice of T_5 until the values of p_4 agree.

3. This calculation tends to be tedious. We can simplify it and obtain an approximate solution by using the results of section 10.2 and the fact that the heat release depends only on the values of p_4, T_4 and p_5, T_5, which control the gas composition, and not on u. Thus, the gas velocity is important only in that it lowers T_5 and p_5. Suppose then that we have a solution to problem 1 in the form given in figure 10.6. We can interpret $c_p(T_5 - T_4)$ as an increase in $c_p(T_{t5} - T_{t4}) = Q/\rho_4 u_4$ (see equation 10.11) because $T_t = T$ for case 1. Then we proceed as follows:

a. Assume T_5 and read $c_p(T_{t5} - T_{t4})$ from figure 10.6.

b. Compute $T_{t5}/T_{t4} = 1 + [c_p(T_{t5} - T_{t4})/c_p T_{t4}]$ and find M_5 from figure 10.4.

c. Find T_5 from $T_5 = T_{t5}/[1 + \frac{1}{2}(\gamma - 1)M_5^2]$ and compare it with the assumed value, iterating choices of T_5 until they are equal.

d. Obtain p_{t5}/p_{t4} for final M_5 from figure 10.4.

Though approximate because of the assumption of constant values of c_p and γ in treating the gas dynamics, this method does include the important dissociation phenomena, and it is much easier than method 2 since the calculations leading to figure 10.6 need not be done. The accuracy can be improved by using a value of γ that is an appropriate mean for the hot reacting gases, rather than the value of 1.4 appropriate to cold air.

For accurate calculation it is necessary to carry out the complete calculation as outlined in case 2. This is not difficult, but it involves much detail. An excellent microcomputer-based program prepared for scramjet calculations is presented in reference 10.8 and will be used to develop a quantitative discussion of scramjet performance in section 10.6.

10.4 Nozzle Flow

The flow enters the nozzle in a highly reactive state. As it expands to lower pressure and temperature, chemical reactions will occur toward the completion of combustion, with consequent additional heat release. If the expansion is slow enough that chemical equilibrium is approached, the methods given above can be used to compute the flow. We might then proceed as follows for each point in the nozzle.

a. Choose a pressure $p < p_5$.
b. Assume T and compute the x_i; then compute u from equation 10.13.
c. Compute A from $\rho_5 u_5 A_5 = \rho u A$.
d. From a plot of A/A_5 as a function of p/p_5, construct the variation of p/p_5 for a known nozzle shape.

Having done this *equilibrium calculation*, we could write expressions for the *rates* at which chemical reactions must occur in the nozzle to maintain the assumed equilibrium. By comparing these rates with the actual kinetic rates, as limited by collisional processes between molecules, we can determine at what point in the flow through the nozzle the composition freezes (becomes fixed). A detailed treatment of this process is beyond the scope of this book, but two limiting cases can be treated fairly easily: equilibrium flow, where equilibrium is maintained to the nozzle exit pressure and the method described above is applicable; and frozen flow, where the x_i are assumed to have the values $(x_i)_5$ all the way through the nozzle. The true situation lies between these two cases. If the condition at which freezing will occur can be estimated as outlined, then one can approximate the actual flow by assuming that the flow changes from equilibrium to frozen at the point in the nozzle where these conditions are attained. Usually this condition is stated in terms of a level of pressure.

When the ideal nozzle exit velocity has been computed, viscous effects can be accounted for by the use of either a nozzle velocity coefficient or a kinetic energy efficiency, as defined for the inlet; the former is more usual. Defined simply as

$$C_v = \frac{\text{Actual exhaust velocity}}{\text{Ideal exhaust velocity}},$$

it is on the order of 0.98 for a well-designed nozzle at high Reynolds number.

10.5 Fuel Injection and Mixing

It has been implicitly assumed in the foregoing discussion that the hydrogen fuel can somehow be mixed with the supersonic airflow through the engine uniformly enough so that an acceptable fraction of the theoretically

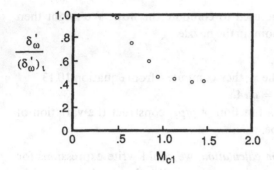

Figure 10.7
Spreading of turbulent shear layers at high Mach numbers. (From reference 10.9.)

available heat release can be achieved. This is in fact one of the most challenging aspects of scramjet engine design, and it has received considerable attention of late. The difficulties stem from a number of factors:

• The severity of the heating and potential shock losses make impractical a closely spaced array of hydrogen injectors in the flow path.
• Because of the high momentum flux and the high Mach number of the engine airflow, it is difficult to obtain thorough penetration of the main stream by jets of hydrogen from the walls.
• Turbulent mixing of hydrogen from the wakes of backward-facing steps is impeded by the low rates of shear layer spreading in hypersonic flows.

The fluid mechanics of these mixing processes is complex and can be given only a cursory treatment here. Because shear layer spreading is likely to be a central issue in any mixing scheme, it will be discussed briefly, and a proposed technique for enhancing mixing by means of shock interaction with the mixing region will be described as an example of the type of strategy that may be required.

At Mach numbers below about 3, turbulent shear layers spread at a nearly constant angle, as was mentioned in subsection 4.4.4. But above about Mach 3, the spreading rate decreases by about half, as described in reference 10.9 and shown in figure 10.7. The ordinate in this figure is the rate of spreading divided by that in the lower speed range. The mechanism for growth of the shear layer is the formation of large vortical structures which entrain fluid from outside the shear layer. These structures are convected at a velocity which is between the velocities of the low- and high-

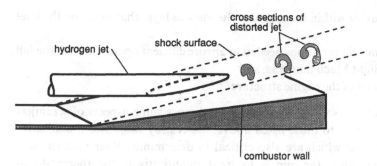

Figure 10.8
Enhancement of mixing by interaction of shock with fuel jet.

speed streams, and the abscissa in figure 10.7 is the Mach number of the high-speed flow relative to the large structures. Evidently the change in spreading rate occurs when the high-speed stream is supersonic with respect to the convected structures. The details of these phenomena are described in reference 10.10, which also addresses their implications for achieving effective heat release in such flows.

A technique for enhancing mixing has been proposed in reference 10.11. Shown schematically in figure 10.8, it consists in generating a shock oriented so that the pressure change across the shock interacts with the density difference between the air and the hydrogen to produce streamwise vorticity, which entrains the hydrogen and carries it into the air stream. Computations and experiments reported in reference 10.11 show that this scheme does enhance the mixing quite effectively, but that it also results in some of the hydrogen being localized in the cores of the streamwise vortices. This is an area of active research, as it is of central importance to scramjets.

10.6   Quantitative Discussion of Scramjet Performance

As a result of intensive recent work in the NASP program, the technical factors that control the feasibility and the performance of the scramjet are rather well known. They have been outlined in a number of publications, including references 10.4–10.7. Probably the most critical are

• the feasibility of mixing the hydrogen fuel into the engine airflow in a flow length that is acceptable from the viewpoint of wall heating and viscous losses, as discussed above,

- the behavior within the engine of the viscous layer that forms on the inlet ramp,
- the geometric variations needed to ensure efficient operation over the full range of flight Mach numbers, and
- the cooling of the engine structure.

The aim of this section, however, is to discuss some aspects of scramjets that, in contrast to those listed above, are readily subject to quantitative treatment, and which are also critical in determining their performance. More specifically, the aim is to treat quantitatively the thermo-fluid-dynamics of the scramjet, including the important effects of real-gas chemistry and to some extent those of combustion and expansion kinetics. It can be argued that an analysis which properly includes these factors provides an optimistic baseline performance estimate, since each of the phenomena listed as critical will in one way or another result in degradation of the performance from that indicated by a thermodynamic analysis of this type. Such an idealized treatment is somewhat trivial in the overall context of NASP; however, because of the large number of design variables and the sensitivity of a scramjet's performance to their choice, the results are useful as a framework for consideration of the more substantive issues.

Estimates of fuel specific impulse as a function of flight Mach number are the primary result of this section. Their implications for the overall performance of scramjet-powered trans-atmospheric vehicles, and in particular the propellant mass ratio required for single-stage-to-orbit vehicles, are discussed in chapter 11.

The scramjet engine is schematically modeled for the present purpose as in figure 10.2; that diagram is scaled for a flight Mach number of 15. The model may be thought of as two-dimensional, but the calculations are in the channel-flow approximation, so that the primary relevance of the geometry is to the determination of the area variation of the flow along the length of the engine. To the extent that the real geometry of a design limits these area variations, it will degrade the performance from that estimated here, since the nozzle will be assumed to be fully expanded and no geometrical restrictions on the inlet are recognized in the calculations. The following subjects will be treated:

- the choice of the combustor inlet Mach number, M_4,
- the effect of combustion kinetics on the choice of the combustor pressure,

which in turn determines the flight altitude as a function of flight velocity,
• the specific impulse that results from these choices.

A brief functional description of the computer program of reference 10.8 is
necessary first, however.

10.6.1   The SCRAMJET Program

This computer program essentially implements the real-gas computational
scheme outlined above, computing the chemical equilibrium composition
of the air in the inlet, and then that of the hydrogen-air mixture and that of
the combustion products at each station in the engine, and evaluating the
thermodynamic properties for the local composition at each point. With
reference to figure 10.2, the functions of the program and the assumptions
made in its execution will be described for each of the elements of the
engine in the order in which the flow sees them, beginning with the inlet.
For a more complete description see reference 10.8.

Inlet   The inlet is modeled in the channel flow approximation by a kinetic-
energy efficiency as discussed in section 10.1. For the bulk of the calculations presented here, an optimistic value of 0.985 has been used. The sensitivity of the results to lower values is explored, however.

Mixer   At the exit of the diffuser and before the combustor, the program
provides a process for mixing the fuel and air from a thermodynamic
standpoint. The temperature and pressure of the fuel prior to mixing can
be specified. The heat necessary to raise the hydrogen fuel from liquid
conditions to the specified state is taken from the combustion gases in the
combustor. There is provision for incomplete mixing of the fuel and air.
For the present calculations, the fuel temperature was set at standard temperature and the fuel pressure at the combustor static pressure, 1 atm.
Complete mixing of the fuel and air is assumed except for some special
cases in which the sensitivity of the results to incomplete mixing is
explored.

Combustor   The combustion conditions are set by the static pressure at
the entrance and by either the pressure ratio across the combustor or the
area ratio. The program then computes the gas composition and state at
the outlet in the channel flow approximation assuming that chemical equilibrium has been reached. For the present calculations, constant-pressure

combustion has been assumed, although the program offers the option of a specified pressure ratio.

Nozzle The nozzle flow is computed in the channel flow approximation according to either of two assumptions, equilibrium chemistry and frozen flow. The transition from equilibrium to frozen behavior can be made at any of 20 stations distributed in the nozzle between the inlet and the outlet. The ratio of exit pressure to ambient pressure can also be specified. Viscous losses are accounted for through a velocity coefficient, as defined in section 10.4. For the calculations presented here, expansion to atmospheric pressure was assumed, with a velocity coefficient of 0.985 for most of the calculations. The freezing location was assigned in such a way as to represent freezing at various pressure levels. For the rationale behind this see the discussion of kinetics below.

10.6.2   Choice of Combustor Inlet Mach Number

As was explained above, it is the supersonic flow into the combustor that defines the scramjet, in distinction to subsonic combustion ramjets; the rationale is that diffusion from the free stream velocity should be limited to an amount that will hold the static temperature to a low enough value so that the heat-producing combustion reactions can proceed nearly to completion. If the static temperature at the combustor entrance is too high, the nominal combustion products such as H_2O are dissociated, so that their chemical energy is not available as thermal energy for conversion to kinetic energy in the nozzle. The question to be dealt with quantitatively is then what static temperature, or what combustor inlet Mach number, is best for any given flight Mach number.

That the existence of such an optimum M_4 depends on finite chemical reaction rates can be seen by comparison of the specific impulse for two cases: one in which chemical equilibrium is assumed throughout the flow, and another in which the flow is assumed to be in equilibrium up to the combustor exit but frozen at that composition during the nozzle expansion. Such calculations are compared in figures 10.9 and 10.10, which are respectively for equilibrium and for frozen nozzle flow.

From figure 10.9 it is evident that there is no optimum M_4 for equilibrium flow, the specific impulse (I_{sp}) increasing continuously as M_4 is decreased (except for an odd behavior near $M_0 = 22$ for $M_4 = 7$ or 8, which seems to be due to the particular atmospheric model used in reference

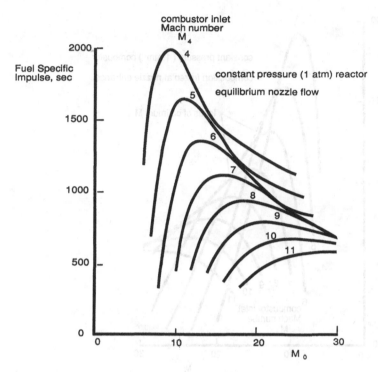

Figure 10.9
Specific impulse for equilibrium nozzle flow. (From reference 10.8.)

10.8). This is not surprising, because for equilibrium flow the chemical energy invested in dissociation is recovered as thermal energy and then kinetic energy as recombination occurs in the nozzle, and the lower the combustor Mach number the lower the entropy increase in the combustor. In contrast, figure 10.10 shows that there is a clear optimum M_4 for flight Mach numbers above about 10. It is defined by the envelope, drawn as a dashed line.

It follows then that the optimum value of M_4 depends on the extent to which recombination occurs in the nozzle, as well as on the degree of dissociation at the combustor exit. This effect is shown in figure 10.11 in terms of the pressure at which freezing occurs in the nozzle. In these calculations as well as those of figures 10.9 and 10.10, the combustor pressure has been fixed at 1 atm.

The top line in figure 10.11 is for freezing at the combustor exit, and corresponds to figure 10.10. As the freezing pressure decreases, the opti-

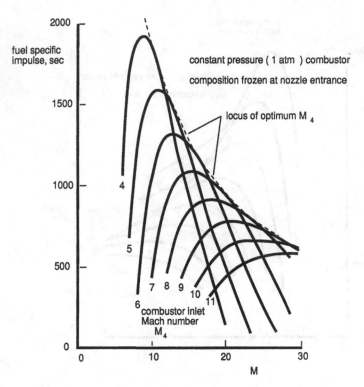

Figure 10.10
Specific impulse for frozen nozzle flow. (From reference 10.8.)

mum M_4 does as well, because with recovery of some of the dissociation energy it is advantageous to allow higher combustor temperatures, which result in less entropy increase in the combustor. For each level of freezing pressure, there is a flight Mach number below which the optimum no longer exists. This value is about $M_0 = 10$ for a freezing pressure of 0.03 atm.

The sensitivity of specific impulse to the choice of M_4 is shown for $M_0 = 15$ in figure 10.12. Figure 10.11 was constructed from such a data set for each of several flight Mach numbers. It can be seen that the penalty for non-optimum choice of M_4 can be rather large.

10.6.3 Effect of Combustion Kinetics

In the calculation discussed above, the combustor pressure was set at 1 atm on the basis that this value should result in fast enough combustion kinet-

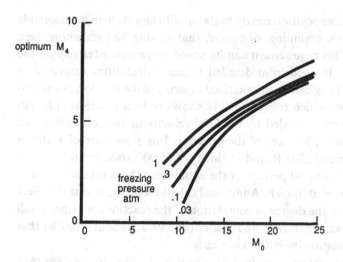

Figure 10.11
Optimum (maximum specific impulse) M_4 as a function of flight Mach number and of the
pressure at which freezing occurs in the nozzle.

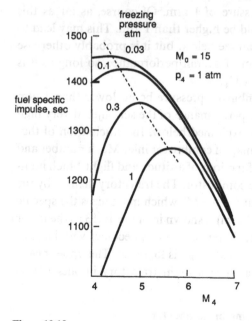

Figure 10.12
Variation of specific impulse with combustor inlet Mach number, M_4, and with freezing
pressure in nozzle, showing optimum M_4 at the dashed line.

ics to bring the composition nearly to its equilibrium state in a reasonable combustor length, assuming, of course, that mixing has somehow been accomplished. This requirement can be posed more quantitatively on the basis of results obtained from detailed kinetic calculations described in reference 10.12. They can be summarized in terms of the fraction of equilibrium heat release, which reference 10.12 shows to be a function of length along the combustor divided by the flow velocity in the combustor and multiplied by the 1.65 power of the pressure. For a pressure of 1 atm, a static temperature of 2500°R, and a velocity of 15,000 ft/sec, reference 10.12 shows that to achieve 90 percent of the equilibrium heat release at 1 atm requires a length of about 3 ft. Additional calculations show that the effect of temperature on the degree of completion of the reactions is rather small so long as the temperature is above a critical value (about 2500°R) that makes the combustion initiation kinetics fast.

Because of the scaling with the 1.65 power of the pressure, we see then that the combustor pressure cannot be much below 1 atm without the energy loss due to incomplete combustion or the combustor length becoming excessive. For this reason, all the performance estimates made here have assumed a combustor pressure of 1 atm. Of course, as far as this argument goes, the pressure could be higher than 1 atm. This may lead to excessive heat transfer, as we shall see below, but it is probably otherwise advantageous from the viewpoint of overall performance, so long as it is not accompanied by a decrease in M_4.

The requirement that the combustor pressure be no lower than 1 atm has serious consequences for the performance of the scramjet at very high flight Mach numbers, as will be explained below. In anticipation of that discussion, figure 10.13 shows a map of combustor inlet Mach number and free stream dynamic pressure as functions of altitude and flight Mach number for a pressure of 1 atm in the combustor. The trajectory defined by the top curve of figure 10.11 (i.e., that for the M_4 which maximizes the specific impulse when freezing occurs at 1 atm) is shown in figure 10.13 as the lower heavy arrow. It indicates that the dynamic pressure becomes very large at the high flight Mach numbers if this strategy is followed. The upper heavy arrow along $q_0 = 0.5$ shows that such a flight trajectory implies lower-than-optimum values of M_4.

10.6.4   Effect of Incomplete Mixing in Combustor

Since the requirement for mixing of the fuel and air in the combustor in an acceptable flow length is considered one of the more difficult to meet, it is

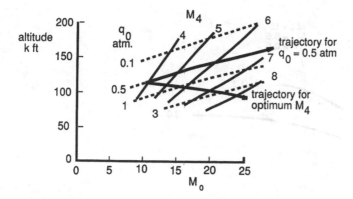

Figure 10.13
Combustor inlet Mach number and free-stream dynamic pressure as functions of altitude and flight Mach number.

important to explore the consequences of not achieving complete mixing. Within the context of the present thermodynamic treatment, this can be done by modeling the flow as two parallel streams, one with a low equivalence ratio and the other with a higher one, their mass-flow weighted average being the overall equivalence ratio. It is relatively easy to show that in this case the effective specific impulse is

$$\bar{I} = \varepsilon\left(\frac{\phi_1}{\phi}\right)I_1 + (1 - \varepsilon)\left(\frac{\phi_2}{\phi}\right)I_2,$$

where ε is the fraction of the airflow in stream 1 and ϕ_1 and I_1 are the equivalence ratio and the specific impulse in stream 1, and ϕ is the overall equivalence ratio.

Results of such calculations for $M_0 = 15$ are shown in figure 10.14 for two overall equivalence ratios. The sensitivity is considerably less for the equivalence ratio of 2 than for the nominal value of 1.2, which has been used in the remainder of the calculations presented here, but so is the base value of I_{sp}.

10.6.5   Specific Impulse for Selected Flight Trajectories

The flight trajectory for a scramjet-powered vehicle is subject to a number of constraints, and actually should be chosen by a systematic process of optimization. The approach that will be followed here is to initially choose

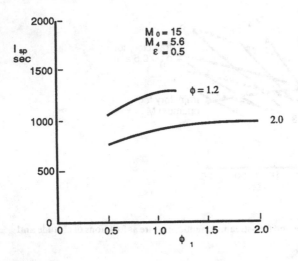

Figure 10.14
Sensitivity of specific impulse to mixing in the combustor.

the engine parameters and the flight profile to maximize the specific impulse at each flight velocity, then to adjust these choices where they prove to be clearly non-optimum from the viewpoint of minimizing the fuel consumption, or where they lead to the violation of some constraint such as that on vehicle dynamic pressure.

Trajectories for the M_4 That Gives Maximum I_{sp}   The nominal values of M_4 are given in figure 10.11 as a function of the flight Mach number and the freezing pressure in the nozzle. Consider first the case where the freezing occurs at 1 atm (i.e., at the combustor exit). Here, there is an optimum M_4 for all flight Mach numbers above 10, and for this M_4 schedule the flight altitude that results in $p_4 = 1$ atm is approximately constant at 100,000 ft. (Figure 10.13 shows that it actually decreases a bit with increasing flight velocity.) At the lowest speeds, near 6000 ft/sec, the resulting vehicle dynamic pressure is quite low, however, and the vehicle acceleration would be correspondingly small, so in this range the altitude has been reduced to that which gives $q_0 = 0.5$ atm. The overall schedule of altitude vs. flight velocity is shown as a dotted line in figure 10.15. Below 6000 ft/sec the trajectory follows $q_0 = 0.5$ atm.

The specific impulse computed for this trajectory is shown in figure 10.16, again as a dotted line.

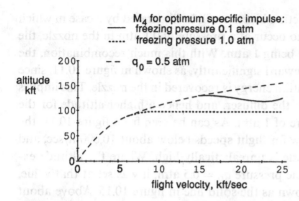

Figure 10.15
Schedule of altitude vs. flight speed for three sets of conditions.

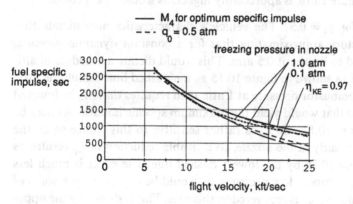

Figure 10.16
Specific impulse for four design choices.

The effect of faster reaction kinetics will be illustrated by a case in which the freezing is assumed to occur at a pressure of 0.1 atm in the nozzle, the combustor pressure still being 1 atm. With this much recombination, the optimum M_4 shifts downward significantly, as shown in figure 10.11, since now some of the dissociation energy is recovered in the nozzle. This implies a lower pressure ratio in the diffuser, and hence a higher altitude for the same combustor pressure of 1 atm. As can be seen from figure 10.11, the optimum M_4 is quite low for flight speeds below about 10,000 ft/sec, and the corresponding altitude is unrealistically high. Where the altitude exceeded that for a dynamic pressure $q_0 = 0.5$ atm, it was set at that value, giving the trajectory shown as the solid line in figure 10.15. Above about 10,000 ft/sec the vehicle flies considerably higher in this case than when the freezing occurred at the nozzle entrance. The specific impulse, shown as the solid line in figure 10.16, is appreciably higher, as would be expected.

Trajectories for $q_0 = 0.5$ The vehicle's characteristics may dictate that it fly a trajectory approximating that for a constant dynamic pressure, usually judged to be about 0.5 atm. This would dictate a schedule of altitude vs. speed as shown in figure 10.15 as the dashed line. Simultaneously holding the combustor pressure at 1 atm then requires that M_4 be lowered from the value that would yield the maximum specific impulse. As may be seen from figure 10.12, the I_{sp} is rather sensitive to this choice when the freezing occurs early in the nozzle, so a sizable reduction in I_{sp} results, as shown in figure 10.16 by the lowest dashed line. The effect is much less when freezing occurs at low pressure, as would be expected since some of the dissociation energy is recovered in this case. This is shown by the upper dashed line, which deviates from the optimum M_4 line only above about $M_0 = 17$.

10.6.6 Effect of Kinetic Energy Efficiency

The kinetic energy efficiency is a key parameter in characterizing the inlet. For all the calculations so far discussed, it has been set at 0.985, a rather optimistic value. To test the sensitivity of the results to this value, the specific impulse has been computed for a lower value of 0.97, for the freezing pressure of 0.1 atm. The result is shown as the dash-dot line in figure 10.16. We see that even for the optimistic assumption about freezing the specific impulse is down to about 300 sec at 25,000 ft/sec. This is well below the value of 450 sec attainable with a hydrogen-oxygen rocket.

10.7 Cooling the Scramjet

Above $M_0 \approx 6$ all of the structure of a scramjet-powered vehicle must be cooled, and only three methods are available: radiation to the environment; heat absorption by the fuel as it is consumed; and heat absorption by the structure of the vehicle, resulting in a rise in its temperature with time. Because the focus of the discussion here is the scramjet engine, its cooling will also be emphasized. In fact, the most severe cooling requirement is presented by the engine combustor, where the mass flux density ρu is a maximum, because the heat flux is proportional to $\rho u c_p (T_t - T_s)$ where T_s is the surface temperature. The heat flux to the inlet ramp and that to the exposed portion of the nozzle are smaller, and it may be that radiation cooling can meet the need there.

To examine the cooling requirement for the engine, we model the engine as a square channel of side D and length L, as sketched in figure 10.17. The walls of the channel are assumed to be cooled by the fuel, through a combination of convective cooling and film cooling. After convectively cooling the wall, the fuel is injected through one or more slots, parallel to the core flow, to act as a film coolant. It is also possible that the primary fuel flow will have been used to convectively cool the wall before injection into the core flow, so that in this scheme the fuel flow available for convective cooling is the primary fuel flow plus that to be used for film cooling.

The temperature of the coolant at injection, T_{tc}, may be below the wall temperature, T_w; indeed it will be if the convective cooling load is less than that which will heat the film cooling fuel to the wall temperature.

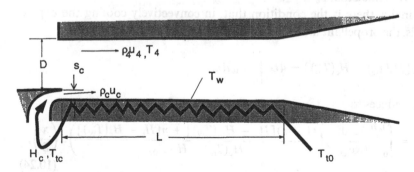

Figure 10.17
Schematic of engine cooling, with convection cooling of wall followed by film cooling.

In the remainder of this section we will examine the restrictions imposed by the fact that the fuel is the only available heat sink, and the only fluid available for film cooling; the heat flux to the wall which must be attained; and the feasibility of and limitations on film cooling with the hydrogen fuel.

10.7.1   Heat and Mass Balances

The total propellant mass flow from the cooling slots for the geometry shown in figure 10.17 is

$$\dot{m}_{cs} = 4D\rho_c u_c s_c,$$

and the engine air mass flow is

$$\dot{m}_4 = D^2 \rho_4 u_4.$$

The stoichiometric fuel-air ratio being 0.0293, we find that the equivalence ratio for cooling is

$$\phi_{cs} = \frac{\dot{m}_{cs}}{0.0293\dot{m}_4} = 136.5\left(\frac{s_c}{D}\right)m, \tag{10.22}$$

where m is the film cooling mass flux ratio ($m \equiv \rho_c u_c / \rho_4 u_4$).

In terms of the geometry and the gas properties,

$$m = \frac{\rho_c u_c}{\rho_4 u_4} = \sqrt{\left[\frac{\gamma_c R_4 T_4}{\gamma_4 R_c T_{tc}}\left(1 + \frac{\gamma - 1}{2}M_c^2\right)\right]\frac{M_c}{M_4}} \tag{10.23}$$

if it is assumed that $p_c = p_4$.

Finally, there is the condition that, in convectively cooling the engine walls, the propellant must absorb the heat transferred to the wall:

$$\dot{m}_{cw}[H_c(T_{tc}) - H_c(T_{t0})] = 4D\int_0^L q_w dx.$$

This reduces to

$$\phi_{cw} = \int_0^{L/D} \frac{2.05}{(\text{Re}_x)^{0.2}}\left(\frac{(1 - \eta)[H_4 - H_{fs}(T_w)] + \eta[H_c - H_c(T_w)]}{H_c(T_{tc}) - H_c(T_{t0})}\right)d\left(\frac{x}{D}\right), \tag{10.24}$$

where η is the "effectiveness" of film cooling, defined as the ratio of the heat

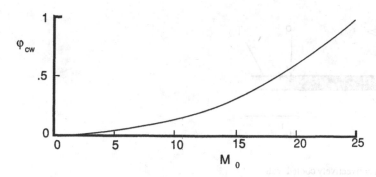

Figure 10.18
Fuel mass flow required to convectively cool the wall, expressed as an equivalence ratio.

flux to the wall with and without film cooling. This value is plotted in figure 10.18 for $\eta = 0$ as a function of M_0 for $p_4 = 1$ atm, $L/D = 15$, and $D = 1$ ft for the values of M_4 that (as discussed in subsection 10.6.2) optimize I_{sp} when the flow is frozen at the nozzle inlet. The dependence on D and p_4 is only through the Reynolds number, hence to the 0.2 power, while the variation with combustor length is $(L/D)^{0.8}$.

In general the result is that the amounts of hydrogen required for cooling the wall, expressed in this way as a change in overall equivalence ratio, are rather small—indeed, within the desired primary fuel flow.

But this does not convey the full cooling requirement, as will be seen from an examination of the heat fluxes implied by convection cooling.

10.7.2 Heat Flux Limits

The magnitude of the heat flux did not enter into the above arguments, yet it is limited by materials and by practical design considerations. In fact, it is this limit that would argue for the use of film cooling. The limit can be understood qualitatively by considering an element of the wall as a thin flat plate with the coolant against its back side, as in figure 10.19.

Then we have the elementary balances:

$$q_w = \rho_4 u_4 \, St(H_{aw} - H_w)$$

$$= k\left(\frac{T_w - T_{wc}}{t}\right)$$

$$= \rho_{H_2} u_{H_2} \, St_c[H_{H_2}(T_{wc}) - H_{H_2}].$$

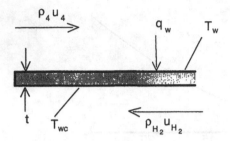

Figure 10.19
Schematic of convectively cooled wall.

Assuming for the moment that the second equality can be satisfied (i.e., that the heat transfer to the hydrogen coolant will support the required heat flux), consider the first equality in terms of the effectiveness required to make possible a practical cooling design. From the definitions of effectiveness (see subsection 10.7.3) and Stanton number,

$$q_W = \rho_4 u_4 \, \text{St} \{ (1 - \eta)[H_4 - H_{fs}(T_W)] + \eta[H_c - H_c(T_W)] \}$$

$$= k \left(\frac{T_W - T_{Wc}}{t} \right),$$

where k is the thermal conductivity of the wall material. To estimate the magnitude of q_W, we put

$$H_4 \approx c_{p \, air} T_4 \left(1 + \frac{\gamma - 1}{2} M_4{}^2 \right), \quad H_{fs} \approx c_{p \, air} T_W, \quad H_c = c_{pH_2} T_{tc},$$

$$H_c(T_W) = c_{pH_2} T_W.$$

Then

$$q_W \approx \frac{\gamma}{\gamma - 1} p_4 u_4 \frac{0.015}{(\text{Re}_x)^{0.2}} \left[(1 - \eta) \left(1 + \frac{\gamma - 1}{2} M_4{}^2 - \frac{T_W}{T_4} \right) \right.$$

$$\left. + \eta \frac{c_{pH_2}}{c_{p \, air}} \left(\frac{T_{tc}}{T_4} - \frac{T_W}{T_4} \right) \right] = k \left(\frac{T_W - T_{Wc}}{t} \right). \quad (10.25)$$

This estimate for q_W normalized by the pressure in atmospheres is shown for $\eta = 0$ in figure 10.20 as a function of flight Mach number and x/D.

In evaluating Re_x, D has been set at 1 ft. At each flight Mach number, the combustor inlet Mach number M_4 has been set at the value that gives the

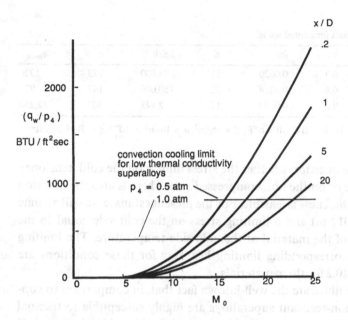

Figure 10.20
Estimate for q_w as a function of flight Mach number and combustor length-to-height ratio, with approximate limits for convective cooling.

highest specific impulse when composition freezing occurs at the nozzle inlet. This provides an upper limit on M_4, and hence a lower limit on q_w.

The second equality in equation 10.25 determines the heat flux that can be transferred through the wall material in terms of the properties of the material. The hot surface of the wall tends to expand relative to the cold one, so the hot surface will develop a compressive stress and the cold one a tensile stress. For a plane-parallel configuration such as that in figure 10.19, the surface stress is

$$\sigma = \frac{\alpha E}{k} q_w t,$$

where α is the coefficient of thermal expansion, k is the thermal conductivity, and E is the modulus of elasticity.

The thermal stress is proportional to the heat flux, to the plate thickness, and to the combination of material properties $\alpha E/k$ (which may be thought of as a thermal stress susceptibility). These are tabulated for three materials in table 10.2.

Table 10.2
Properties of materials for cooled walls.

Material	α	k	E	$\alpha E/k$	σ	q_{limit}
MAR-M 200	6.3	0.0020	31	94,000	123	1320
RENE 41	6.6	0.0014	32	150,000	147	977
Copper	9.2	0.0639	17	2,448	30	12,250

Units: $\alpha = {}^\circ\text{R}^{-1} \times 10^6$; $k = \text{BTU/ft sec } {}^\circ\text{R}$; $E = \text{lb/in}^2$; $\sigma = \text{lb/in}^2 \times 10^{-3}$; $q = \text{BTU/sec ft}^2$

There is some question whether the stress limit is on the cold, tensioned side of the plate or on the hot, compressed side. There is also the question of how small a thickness is feasible. For the present estimate we will assume $t = 0.001$ ft (0.012 in) and a limiting stress on the cold side equal to the tensile stength of the material at the cold-side temperature. The limiting stress and the corresponding limiting heat flux for these conditions are shown in table 10.2 for three materials.

These results illustrate the well-known fact that, in comparison to copper, the oxidation-resistant superalloys are highly susceptible to thermal stress and are not appropriate choices for cooled walls when the principal means of protection of the wall is convective cooling from the back side. They are universally used in gas turbines, where the blade material runs quite hot, high strength is essential, and the environment is highly oxidizing. There, film or convective cooling reduces the blade temperature only moderately below the uncooled adiabatic wall temperature. It appears that for the scramjet combustor a different strategy may be indicated, with a highly conducting material such as copper used for the combustor wall and with the hot side of the wall maintained at a low enough temperature that the strength and oxidation resistance of the copper will be adequate. It may be that some sort of fiber reinforcement of the copper can be used to enhance its strength, and that some form of coating can be devised to improve its oxidation resistance.

High-performance rocket engines provide the experience closest to that projected here. The throat of the nozzle of the Space Shuttle's main engine has a heat flux of about 14,000 BTU/sec ft^2. It is convectively cooled by a scheme very close to that modeled here, with very cold gaseous hydrogen at very high pressure (on the order of 4000 psi). The material of the wall is an alloy of mainly copper, with trace constituents for strengthening. The life of the wall is limited by the thermal strain that occurs in each operating cycle, taking the wall material well into the plastic range. This example shows that—locally, at least—the type of heat flux anticipated in the scramjet can be dealt with, but it may be important to note that there are some differences between the systems requirements of a rocket engine and those of a scramjet. First, the fuel flow of the rocket is much larger relative to the total flow than that of the scramjet. Secondly, the rocket nozzle is very compact compared to the scramjet combustor, which is likely to have a length-to-diameter ratio of 10 or more. Finally, the rocket combustion gases are highly reducing; this may also be true of the scramjet, but that is not entirely clear at this time.

It seems the most that can be concluded from an analysis at this level of specificity is that the design of a convectively cooled combustor for the scramjet will be an interesting challenge. It seems likely that film cooling will be required in at least some areas of the engine, so the elements of this technology are summarized next.

10.7.3 Film Cooling with Hydrogen

The model of film cooling that will be used here is essentially that of reference 10.13 with some simplifications appropriate to the present aims. The film cooling fluid is assumed to be injected from a slot, as in figure 10.17, with mass flux $\rho_c u_c$ at Mach number M_c and with stagnation temperature T_{tc}. The experiments described in reference 10.13 used such a geometry, with a range of gas injection conditions. Measurements included both the hydrogen concentration at the wall and the adiabatic wall temperature, with the injected hydrogen at room temperature (291°K) and the nitrogen free stream at 817°K.

A cooling effectiveness, η, was defined in terms of the hydrogen mass fraction as simply $\eta = c_w$, where c_w is the mass fraction of coolant at the wall (if the free-stream hydrogen mass fraction is zero). This effectiveness then describes the extent of mixing of the free-stream fluid with the injected fluid at the wall. Mass averaging the enthalpies of the free-stream fluid and the injectant, we obtain the adiabatic wall enthalpy,

$$H_{aW} = (1 - \eta)H_4 + \eta H_c,$$

or in perhaps more familiar form,

$$\eta = \frac{H_4 - H_{aW}}{H_4 - H_c}. \tag{10.26}$$

The effectiveness in this form has been correlated as a function of the parameter $\xi = x/m^{0.8}s_c$, where $m \equiv \rho_c u_c/\rho_4 u_4$ and s_c is the slot height (indicated in figure 10.17). The correlations show that there is a distance downstream of the slot, called the *cooled length*, over which $\eta = 1$ to a good approximation, meaning none of the free-stream fluid reaches the wall. This distance is $50 < \xi < 100$. Beyond it the effectiveness falls off more or less linearly on log-log coordinates. From the data of figure 41 in reference 10.13, this behavior is approximated for the present purposes as follows:

$$\eta = 1, \quad \frac{x}{s_c} < 50m^{0.8}$$

$$\eta = \left(\frac{x/s_c}{50m^{0.8}}\right)^{-0.876}, \quad \frac{x}{s_c} > 50m^{0.8}. \tag{10.27}$$

Both the adiabatic wall temperature and the wall heat flux for a given wall temperature are of interest. To estimate either from the effectiveness, note that for the above mixing model the enthalpy of the mixture at the wall, at the wall temperature T_w, is

$$H_w(T_w) = (1 - \eta)H_{fs}(T_w) + \eta H_c(T_w), \tag{10.28}$$

where fs denotes the free-stream fluid, which is air in the present case.

The effects of η and of the relative specific heats of the hydrogen and air on the adiabatic wall temperature may be seen by evaluating equation 10.28 for $T_w = T_{aw}$, so that

$$H_w(T_{aw}) = (1 - \eta)H_{fs}(T_{aw}) + \eta H_c(T_{aw}).$$

Equating this value to H_{aw} from equation 10.26 yields

$$H_{fs}(T_{aw}) + \frac{\eta}{1 - \eta}H_c(T_{aw}) = H_4 + \frac{\eta}{1 - \eta}H_c. \tag{10.29}$$

This can be solved numerically for T_{aw}, but the specific-heat effect is seen more readily by approximating $H \approx c_p T$ to get

$$\frac{T_{aw}}{T_{t4}} = \frac{1 + \dfrac{\eta}{1 - \eta}\dfrac{c_{pH_2}}{c_{p\,air}}\dfrac{T_{tc}}{T_{t4}}}{1 + \dfrac{\eta}{1 - \eta}\dfrac{c_{pH_2}}{c_{p\,air}}}. \tag{10.30}$$

Considering that $c_{pH_2}/c_{p\,air} \approx 14.5$, it is quite clear that as long as η is near unity T_{aw} is strongly driven toward T_{tc}. For example, for $\eta = 0.5$, $T_{tc}/T_{t4} = 0.1$ and $T_{aw}/T_{t4} = 0.16$. A definition of effectiveness in terms of temperatures has sometimes been used; for example,

$$\eta' \equiv \frac{T_{t4} - T_{aw}}{T_{t4} - T_{tc}}.$$

Using equation 10.30, it is easy to see that

$$\eta' = \frac{\eta\, \dfrac{c_{pH_2}}{c_{p\,air}}}{1 + \eta\left(\dfrac{c_{pH_2}}{c_{p\,air}} - 1\right)}, \tag{10.31}$$

a relationship which is plotted for several values of $c_{p\,inj}/c_{p\,fs}$ in figure 10.21. From this figure it can be seen that for hydrogen coolant ($c_{p\,inj}/c_{p\,fs} = 14.5$) a heat-transfer effectiveness of say 0.5 is equivalent to an adiabatic-wall-temperature effectiveness of about 0.9, so the film is much more effective in reducing the adiabatic wall temperature than in reducing the heat transfer. If the materials allow operation at the adiabatic wall temperature, then film cooling with hydrogen can be very effective. But if the resultant temperatures are too high, then the heat-transfer effectiveness is the critical factor, and hydrogen is not an especially effective coolant.

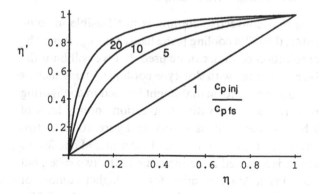

Figure 10.21
Temperature effectiveness vs. heat-transfer effectiveness for several values of the ratio of the specific heats of injected and free-stream gases.

The heat transfer with film cooling is usually estimated by means of the normal boundary-layer heat-transfer relationship,

$$q_\text{w} = \rho_4 u_4 \, \text{St}[H_{\text{aw}} - H_\text{w}(T_\text{w})]. \tag{10.32}$$

Unfortunately, there seem to be no reliable measurements of heat transfer for light-gas film cooling in supersonic flows upon which to base estimates of St. Most of the available data refer to concentration-derived effectiveness, or to measurements of adiabatic wall temperature. There is reason to believe that both the large-scale turbulence and the flow near the wall will be affected by a strong molecular-weight gradient in the layer. Even at the simplest level, if the Stanton number is based on the free-stream quantities in equation 10.32, the light gas lowers the mass flux ρu in the boundary layer while increasing H_{aw} and H_w for a given temperature level. These effects offset one another somewhat, but probably not exactly.

In the absence of data on the Stanton number for light-gas film cooling, the estimates of Van Driest (reference 10.14) for turbulent boundary layers will be used here. For the range of Mach numbers of interest, an adequate approximation in the range of M_4 from 3 to 7 is

$$\text{St} = \frac{0.015}{(\text{Re}_x)^{0.2}}, \tag{10.33}$$

about half the low-Mach-number value.

10.7.4 Application of Film Cooling—An Example

If detailed design shows that convective cooling is not feasible in some portions of the combustor, then film cooling is the likely remedy. It may be that some form of transpiration cooling can be used, but we will limit the discussion here to what can be done with slot-type cooling systems such as that in figure 10.17. To illustrate the logic that might be used in configuring such a cooling system, a rather conservative convection cooling limit of 400 BTU/sec ft^2 will be assumed. This is shown in figure 10.20 as two limiting lines, one for a combustor inlet pressure of 1.0 atm and one for 0.5 atm. The limiting heat flux is assumed to be the same in the two cases, but this heat flux occurs at a lower Mach number for the higher combustor pressure. For this example, film cooling would be required at $x/D = 0.2$ for Mach numbers larger than about 12 for a combustor pressure of 1 atm and for Mach numbers larger than about 15 for a pressure of 0.5 atm.

The approach that will be taken here is to estimate first the "cooled length" (i.e., the value of x/D to which an effectiveness of 1 can be maintained) as a function of flight Mach number and ϕ_c, the equivalence ratio devoted to film cooling. Then, for selected flight conditions, the effectiveness of film cooling will be estimated as a function of x/D.

In developing these estimates, a family of scramjets has been assumed in which the combustor inlet Mach number is optimized to maximize the I_{sp} for each flight Mach number, assuming composition freezing at the nozzle inlet. The combustor inlet pressure, and hence the altitude, are treated parametrically. The coolant injection Mach number is set at half the combustor inlet Mach number, and the pressure in the injection slot is assumed equal to that in the free stream.

The value of m is obtained from equation 10.23, and then s_c/D is calculated from equation 10.22. The value of x/D for which $\eta \approx 1$ is then obtained from the first relation of equation 10.27. It is shown in figure 10.22 as a single curve, since the variations from this curve with flight Mach number and combustor pressure are very small. We see that the combustor lengths that can be completely cooled, in this sense, are rather small for reasonable values of ϕ_c. A film cooling equivalence ratio of 2 is required to cool, with effectiveness of 1, a channel with a length-to-height ratio of unity.

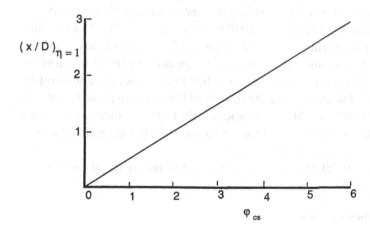

Figure 10.22
Combustor length-to-height ratio downstream of the injection slot, for which the effectiveness is near unity, as a function of fuel mass flow expressed as an equivalence ratio.

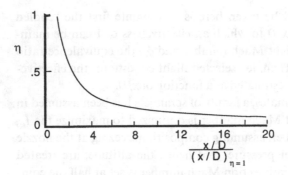

Figure 10.23
Decrease of heat-transfer effectiveness downstream from the point of injection.

The next question concerns the variation of effectiveness from $(x/D)_{\eta=1}$ downstream. This is shown in figure 10.23, again as a single relation for all flight Mach numbers and combustor pressures. Although there is some effectiveness even at distances as high as 20 times $(x/D)_{\eta=1}$, the value of η is only about 0.5 already at twice the "cooled length." From equation 10.25 we see that $\eta = 0.5$ implies about a factor-of-2 reduction in q_W. Thus, in rough terms, an effectiveness of 0.5 or more can be maintained for a length-to-height ratio just about equal to the film cooling equivalence ratio.

So far these results have rather general applicability. To understand their implications for scramjet design, it is necessary to settle on the criteria for determining when film cooling will be required. It is fairly clear that film cooling is expensive, in terms of the excess fuel flow required, so convective or regenerative cooling is the desirable design choice when it is viable. In this example, the limit on convective cooling has been set at a conservative value of 400 BTU/sec ft², and the impact of film cooling requirements on the required fuel flow will be correspondingly large—probably unrealistically large. But the example will serve to illustrate the general features of film cooling in application to scramjets.

From figure 10.20, the values of η required to reduce q_W to the limiting levels shown there can be estimated as simply

$$1 - \eta_{req} = (q_W)_{limiting}/q_W.$$

Such estimates are plotted in figure 10.24 as a function of M_0 and x/D for two combustor pressures. We see that near the inlet of the combustor and at the higher Mach numbers the $\eta_{required}$ is fairly large, while further down the combustor it decreases substantially.

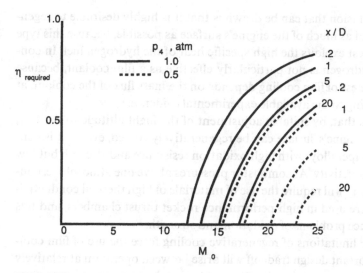

Figure 10.24
Heat-transfer effectiveness required to limit wall heat flux to the values shown in figure 10.20 as limits for regenerative cooling of superalloys.

Finally, from figure 10.24 and figures 10.22 and 10.23, we can estimate the equivalence ratios required in the film cooling to reduce the wall heat flux to the limits set in figure 10.20 for combustor channels of specified L/D.

To see how this is done, consider as an example a combustor with $L/D = 5$ and $p_4 = 1$ atm at $M_0 = 20$. From figure 10.24, $\eta_{required} = 0.47$. From figure 10.23, this requires $(x/D)/(x/D)_{\eta=1} = 2.5$, so that $(x/D)_{\eta=1} = 5/2.5 = 2.0$, and from figure 10.22 this requires $\phi_{cs} = 4.1$.

Such estimates are shown in figure 10.25 for two values of p_4 and several values of L/D from 20 down to 0.2. The dashed lines refer to $p_4 = 0.5$ atm and the solid lines to $p_4 = 1.0$ atm.

It is quite apparent from these results that for the limiting heat flux of 400 BTU/sec ft^2 the required film-cooling equivalence ratios would become excessive beyond about $M_0 = 18$ for $p_4 = 1$ atm and for reasonable values of L/D (say on the order of 10–15). On the other hand, they are quite reasonable even up to $M_0 = 25$ for $p_4 = 0.5$ atm.

10.7.5   Conclusions about Cooling of Scramjets

Although the approaches outlined here are certainly not accurate enough to be used as design tools, they do provide a qualitative description of the approaches to the cooling of scramjet engines.

One conclusion that can be drawn is that it is highly desirable to regeneratively cool as much of the engine's surface as possible, because this type of cooling best exploits the high specific heat of the hydrogen fuel. In contrast, the hydrogen is not particularly effective as a film coolant, because the effectiveness of film cooling depends on the mass flux of the coolant, at least according to the available experimental evidence.

It appears that, by suitable adjustment of the flight altitude to limit p_4, most of the engine's surface can be regeneratively cooled, even if it is constructed of superalloys with high oxidation resistance and strength but low thermal conductivity. At combustor pressures above one atmosphere, convective cooling will require the use of materials of high thermal conductivity, such as are used in high-performance rocket thrust chambers, and this may introduce problems of oxidation and strength.

When the limitations of regenerative cooling force the use of film cooling, an important design tradeoff will arise between operation at relatively high combustor pressures with film cooling and operation at lower pressures with regenerative cooling. In general, the use of film cooling will

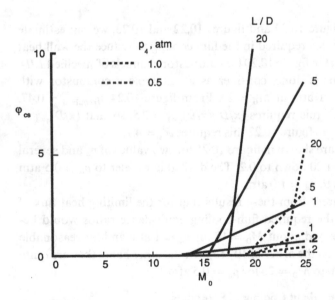

Figure 10.25
Film-cooling mass flows required to limit the wall heat flux to the values shown in figure 10.22 as limits for regenerative cooling of superalloys, expressed as an equivalence ratio.

degrade the specific impulse of the scramjet, but operation at higher pressures will allow higher vehicle accelerations. This is a design tradeoff that can be made only through a trajectory study.

10.8 The Air Turborocket

The turborocket engine has been proposed for use in very-high-speed vehicles, such as Earth-orbit launchers. It offers a high thrust-to-weight ratio at low speeds, and good performance at high Mach numbers. One version uses two liquid propellants, a fuel and an oxidizer, which are pumped to high pressure and burned in a rocket-like combustion chamber, as indicated in the upper half of figure 10.26. The products of combustion are expanded through a turbine, which drives the air compressor, and then combusted with the compressed air.

In another version, shown in the lower half of figure 10.26 , liquid hydrogen is pumped to high pressure, vaporized and heated by exchange with combustion products, and then expanded through a turbine, which drives the compressor.

In either version, this concept has several appealing characteristics. The compressor has fewer stages and can be lighter than that of a turbojet that produces a comparable nozzle pressure ratio and hence a comparable thrust per unit of airflow. Because the turbine can operate at higher pres-

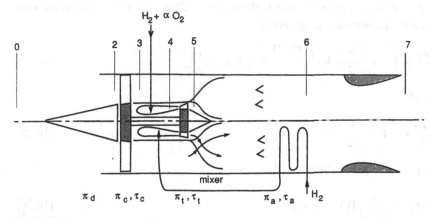

Figure 10.26
Air turborocket-bipropellant gas generator cycle (upper half of figure) and H_2 expander cycle (lower half of figure).

sure and produces less power than that of a turbojet, it can be more compact and lighter. Since the turbine inlet temperature is set by the fuel/oxidizer mixture or hydrogen heat exchanger, it does not set a limit on the flight Mach number, as is the case for the turbojet. In the air turborocket this limit is set by the compressor; because the combustion occurs downstream of it, the materials-limited Mach number is higher than for the turbojet.

There are, however, some negative aspects to the air turborocket, or at least some limitations. In the first version, it must be provided with oxidizer as well as fuel, and the low weight of the turbine and the compressor is offset by the weight of the oxidizer to a degree that depends on the operating time. The mixture ratio of fuel and oxidizer is constrained by the temperature limit of the turbine, so the energy of the combustion products is less per unit mass than for a rocket. One attractive propellant combination is liquid hydrogen and liquid oxygen, with which the combustor would be operated very hydrogen-rich to limit the turbine inlet temperature. The excess hydrogen is then available to burn with the compressed air, so that complete combustion of the fuel can be realized.

The second version eliminates the penalty of oxidizer weight, but the energy available in the hydrogen then places a limit on the compressor pressure ratio which is lower than that for the bipropellant system.

The analysis of either version of the air turborocket proceeds similarly to that of the turbojet, but some qualitative differences arise from the separation of the compressor and turbine flows. Thus, the nozzle pressure and temperature ratios are given by

$$\frac{p_{t7}}{p_7} = \left(1 + \frac{\gamma - 1}{2} M_7{}^2\right)^{\gamma/(\gamma-1)} = \delta_0 \pi_d \pi_c \pi_a \frac{p_0}{p_7} \qquad (10.34)$$

and

$$\frac{T_{t7}}{T_7} = \left(1 + \frac{\gamma - 1}{2} M_7{}^2\right) = \theta_a \frac{T_0}{T_7}, \qquad (10.35)$$

so that for an ideally expanded nozzle

$$M_7{}^2 = \frac{2}{\gamma - 1} [(\delta_0 \pi_d \pi_c \pi_a)^{(\gamma-1)/\gamma} - 1] \qquad (10.36)$$

and

$$\frac{T_7}{T_0} = \frac{\theta_0 \tau_c \tau_a}{1 + [(\gamma - 1)/2]M_7{}^2} = \frac{\theta_a}{(\delta_0 \pi_d \pi_c \pi_a)^{(\gamma-1)/\gamma}} = \frac{\theta_a}{\theta_0 \tau_c}.$$

The last equality applies only for the ideal cycle, without diffuser or combustor pressure losses.

The thrust per unit of airflow is given by

$$\frac{F}{\dot{m}_a a_0} = M_0 \left(\frac{M_7}{M_0} \sqrt{\frac{T_7}{T_0}} - 1 \right) = \sqrt{\frac{2}{\gamma - 1} \frac{(\theta_0 \tau_c - 1)\theta_a}{\theta_0 \tau_c}} - M_0. \qquad (10.37)$$

The specific impulse is

$$\frac{I}{a_0/g} = \frac{F/\dot{m}_a a_0}{\dot{m}_p/\dot{m}_a}. \qquad (10.38)$$

To compute the thrust and the specific impulse we must then determine the afterburner exit temperature ratio θ_a, the compressor temperature rise τ_c, and the ratio of propellant to air mass flows. The interrelationship of these is different for the bipropellant and H_2 expander versions, but they have some features in common.

In most cases, the performance of the turborocket will be limited by the pressure ratio available from the compressor, which is set by the available turbine power, so it will be desirable to operate the turbine at as high an inlet pressure as is practical. The limit will be set by stresses and temperatures in the turbine and in the combustion chamber. Thus, we regard p_{t4} and T_{t4} as design parameters, in the same sense that T_{t4} is a design parameter for the turbine engines.

The available compressor temperature rise is then determined by the turbine-compressor work balance:

$$\dot{m}_p \frac{p_{t4} - p_0}{\rho_p} + \dot{m}_a c_{pa}(T_{t3} - T_{t0}) = \dot{m}_p c_{pp}(T_{t4} - T_{t5}),$$

where the first term on the left is the power required to pump the propellants, ρ_p being a mean density for the propellant mixture. Normally this term is small in relation to the compressor work, but it will be retained for now. Solving for the available compressor temperature rise, we get

$$\tau_c = \frac{1 + \dfrac{\dot{m}_p}{\dot{m}_a} \left[\dfrac{c_{pp}}{c_{pa}} \dfrac{\theta_t}{\theta_0} - \dfrac{p_0}{\rho_p} \dfrac{\gamma - 1}{\gamma} \dfrac{1}{\theta_0} \left(\dfrac{p_{t4}}{p_0} - 1 \right) \right]}{1 + \dfrac{\dot{m}_p}{\dot{m}_a} \dfrac{c_{pp}}{c_{pa}} \theta_t \left(\dfrac{p_0}{p_{t4}} \right)^{(\gamma-1)/\gamma}},$$

which can be simplified considerably by noting that the second term in the denominator will normally be much less than unity, and that p_{t4}/p_0 is much greater than unity. Thus,

$$\tau_c = 1 + \frac{\dot{m}_p}{\dot{m}_a}\left(\frac{c_{pp}}{c_{pa}}\frac{\theta_t}{\theta_0} - \frac{\gamma-1}{\gamma}\frac{p_{t4}}{p_p R_a T_{t0}}\right). \tag{10.39}$$

Since we regard θ_t and p_{t4}/p_0 as design parameters, we are left with the determination of $\dot{m}_p/\dot{m}_a$. From this point on, it is convenient to treat the bipropellant and expander cycles separately.

10.8.1 Bipropellant ATR

The afterburner heat balance gives T_{t6}:

$$(\dot{m}_a + \dot{m}_p)c_p T_{t6} = \dot{m}_a c_{pa} T_{t3} + \dot{m}_p c_{pp} T_{t5} + h_{H_2}\mu\dot{m}_p,$$

where μ is the fraction of the propellant mass flow that is hydrogen available for combustion in the afterburner and h_{H_2} is the heating value of hydrogen in the usual sense. Letting $c_p \approx c_{pa}$ and assuming $\dot{m}_p \ll \dot{m}_a$, we find

$$\frac{\dot{m}_p}{\dot{m}_a} = \frac{c_{pa}}{c_{pp}}\frac{\theta_a - \theta_0\tau_c}{\theta_t\theta_0\tau_c\left(\dfrac{p_0}{p_{t4}}\right)^{(\gamma-1)/\gamma} + \mu\dfrac{h_{H_2}}{c_{pp}T_0} - \dfrac{c_{pa}}{c_{pp}}\theta_a}, \tag{10.40}$$

where the first term in the denominator is rather small if the pressure ratio is large.

We need a description of the process in the combustion chamber, where hydrogen and oxygen react to produce water plus excess hydrogen. Schematically,

$$H_{2l} + \alpha O_{2l} \rightarrow (1 - 2\alpha)H_{2g} + 2\alpha(H_2O)_g,$$

where α is the number of moles of O_2 supplied per mole of H_2. The subscripts indicate that the H_2 and the O_2 on the left are liquids, while the H_2 and H_2O on the right are gases. The heat balance gives

$$H_{H_2l} + \alpha H_{O_2l} = (1 - 2\alpha)H_{H_2,g}(T_{t4}) + 2\alpha[H_{H_2O,g}(T_{t4})],$$

where each H denotes the complete enthalpy of the species, as explained earlier in this chapter. This expression can be solved for the α required to give any T_{t4}:

$$\alpha = \frac{H_{H_{2}g}(T_{t4}) - H_{H_{2}l}}{2[H_{H_{2}g}(T_{t4}) - H_{H_{2}Og}(T_{t4})] + H_{O_{2}l}}.$$ (10.41)

Now that we have the value of α, we can find the amount of H_2 available for combustion with air. Thus,

$$\mu = \frac{\dot{m}_{H_{2}}}{\dot{m}_{H_{2}} + \dot{m}_{H_{2}O}} = \frac{2(1 - 2\alpha)}{18(2\alpha) + 2(1 - 2\alpha)} = \frac{1 - 2\alpha}{1 + 16\alpha}.$$ (10.42)

10.8.2 H_2-Expander ATR

Because there is a release of chemical energy only in the afterburner in this version, only an overall heat balance is required, in the form

$$\frac{\dot{m}_{p}}{\dot{m}_{a}} = (\theta_{a} - \theta_{0})\frac{c_{p}T_{0}}{h_{H_{2}}}.$$ (10.43)

This relation may be regarded as determining θ_a if $\dot{m}_p/\dot{m}_a$ is prescribed, or $\dot{m}_p/\dot{m}_a$ if θ_a is prescribed.

For either the H_2 expander or the H_2-O_2 gas generator cycle, the afterburner temperature sets the hydrogen fuel flow, and then the temperature and pressure in the compressor drive turbine determine the compressor pressure ratio. Thus, the thrust per unit of airflow is the same for the two cycles. It is plotted in figure 10.27 for certain values of the parameters. The specific impulse differs for the two cycles, being lower for the H_2-O_2 cycle because of the extra mass flow of oxidizer. Results are shown in figure

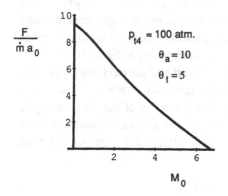

Figure 10.27
Thrust per unit of airflow for air-turborocket engines.

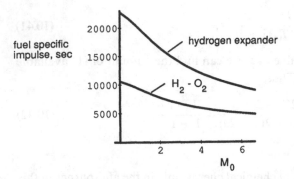

Figure 10.28
Fuel specific impulse for air-turborocket engines with hydrogen expander and H_2-O_2 gas
generator cycles.

10.28. The values of specific impulse are high, about the same as those for
a hydrogen-fueled turbojet at low speeds, but sustained to higher Mach
numbers.

10.9   The Liquid-Air Collection Engine

The liquid-air collection engine was conceived to eliminate or at least miti-
gate the difficulties with diffusion, supersonic mixing, and cooling associated with scramjets at high Mach numbers. It would take advantage of
the high heat capacity and low temperature of liquid hydrogen fuel to
liquefy the air captured by the inlet in a heat exchanger cooled by the liquid
hydrogen. The liquid air would be pumped as liquid to high pressure, used
to precool the airflow, then injected into a rocket-like combustion chamber
along with the hydrogen. This is shown schematically in figure 10.29.

Although such an engine is complex in detail and probably quite difficult
in execution, its advantages and disadvantages may be readily understood
at the conceptual level. We note first that pumping of the air as a liquid
requires a fuel-air ratio sufficiently high that the air can be liquefied. With
reference to the station numbers in figure 10.29, this implies the heat
balance

$$\dot{m}(H_0 - H_3) = \dot{m}(H_4 - H_3) + \dot{m}_{H_2}(H_6 - H_5),$$

where each H is the complete enthalpy of the fluid at the indicated station.

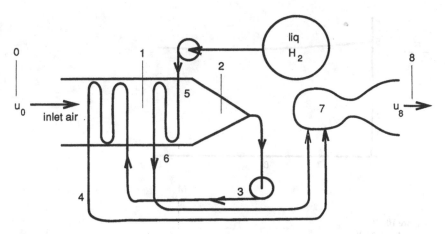

Figure 10.29
Schematic of liquid-air collection engine.

Approximating H by $c_p T + u^2/2$, we find the required equivalence ratio to be

$$\phi = \frac{1 + \frac{1}{2}(\gamma - 1)M_0^2 - T_4/T_0}{0.0293(c_{pH_2} T_6/c_p T_0)},\qquad(10.44)$$

where we may think of T_4 and T_6 as limited by the materials of the two heat exchangers. As the flight Mach number increases, the required equivalence ratio also increases, because of the increasing enthalpy of the captured air.

The reaction process in the rocket chamber may be represented schematically as

$$1N_2 + 0.266O_2 + 0.532\phi H_2 \rightarrow 1N_2 + 0.532(\phi - 1)H_2 + 0.532H_2O$$

so long as $\phi > 1$ so that the combustion products are primarily H_2 and H_2O. An overall energy balance for the engine indicates that the total energy flux in the exhaust must equal the sum of the kinetic energy and enthalpy of the inlet air and that of the liquid hydrogen fuel, augmented by the heat of formation of the water in the exhaust. If the nozzle pressure ratio is large, the nozzle energy flux will be mainly in the form of kinetic energy, because the temperature of the exhaust gases will be low. Thus, we find for the exhaust velocity the simple expression

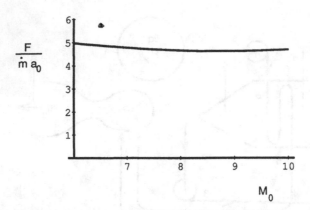

Figure 10.30
Thrust of LACE.

$$\frac{u_8{}^2}{2} = \frac{\dot{m}}{\dot{m}_8}\left(c_p T_0 + \frac{u_0{}^2}{2}\right) - \frac{\dot{m}_{H_2O}}{\dot{m}_8}\Delta H^\circ_{f,\,H_2O}, \tag{10.45}$$

where, from the above stoichiometric relationship,

$$\frac{\dot{m}}{\dot{m}_8} = \frac{28 + 0.266(32)}{28 + 0.266(32) + 0.532(2)\phi}$$

and

$$\frac{\dot{m}_{H_2O}}{\dot{m}_8} = \frac{0.532(18)}{28 + 0.266(32) + 0.532(2)\phi}.$$

The thrust then follows from

$$\frac{F}{\dot{m}a_0} = \frac{\dot{m}_8}{\dot{m}}\frac{u_8}{a_0} - M_0 \tag{10.46}$$

and the specific impulse from

$$I_{sp} = \frac{\dot{m}}{\dot{m}_{H_2}}\frac{a_0}{g}\frac{F}{\dot{m}a_0}. \tag{10.47}$$

Typical results for the thrust and the specific impulse are shown in figures 10.30 and 10.31 for the range of Mach numbers where the equivalence ratio given by equation 10.44 is greater than unity and the specific impulse is high enough to be interesting.

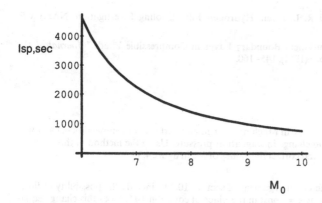

Figure 10.31
Specific impulse of LACE.

References

10.1 W. H. Avery and G. L. Dugger, "Hypersonic Airbreathing Propulsion." *Astronautics and Space Engineering*, June 1964: 42–47.

10.2 D. P. Hearth and A. E. Preyss, "Hypersonic Technology: Approach to an Expanded Program". *Astronautics and Aeronautics*, December 1976: 20–37.

10.3 J. L. Kerrebrock, *Aircraft Engines and Gas Turbines*, first edition. MIT Press, 1977.

10.4 R. H. Petersen, X-15 to Aero-Space Plane, A Perspective on Hypersonic Aircraft Technology in the United States. Sam Bloomfield Distinguished Engineer-in-Residence Lecture, College of Engineering, Wichita State University, 1987.

10.5 M. Martinez-Sanchez, Fundamentals of Hypersonic Airbreathing Propulsion. Notes for AIAA Professional Study, 1988.

10.6 Committee on Hypersonic Technology for Military Application, Air Force Studies Board, National Research Council, *Hypersonic Technology for Military Application*. National Academy Press, 1989.

10.7 G. Anderson, A. Kumar, and J. Erdos, Progress in Hypersonic Combustion and Experiment. AIAA-90-5254, AIAA Second International Aerospace Planes Conference, 1990.

10.8 S. Serdengecti, A Procedure for Predicting Fuel Specific Impulse of Supersonic Combustion Ramjet Engines. Daniel and Florence Guggenheim Jet Propulsion Center, California Institute of Technology, 1991.

10.9 D. Papamoschou and A. Roschko, Observations of Supersonic Free-Shear Layers. AIAA Paper 86-0162, 1986.

10.10 P. E. Dimotakis, Turbulent Free Shear Layer Mixing and Combustion. ISABE 89-7006, 1989. (See also AIAA Paper 89-0262, 1989.)

10.11 F. E. Marble, E. E. Zukoski, J. Jacobs, G. Hendricks, and I. Waitz, "Shock Enhancement and Control of Hypersonic Mixing and Combustion." AIAA, SAE, ASME, ASEE Joint Propulsion Conference, Orlando, 1990.

10.12 M.T. Yeung, personal communication, 1990.

10.13 D. C. Rousar and R. L. Ewen, Hydrogen Film Cooling Investigation. NASA CR 121235, 1973.

10.14 E. Van Driest, "Turbulent Boundary Layer in Compressible Fluids." *Journal of the Aerospace Sciences* 18, no. 3 (1951): 145–160.

Problems

10.1   In computing the change in Mach number in an afterburner in subsection 4.4.4, it was assumed that there was no change in stagnation pressure. Using the method of this chapter, evaluate that assumption and correct the values of M_6 in figure 4.35.

10.2   Generalize the differential argument of section 10.2 to include the possibility of flow-area variation by putting $\rho u A = $ const in the place of equation 10.5. Does this change equation 10.9? How is the "thermal choking" condition modified?

10.3   Using the approximate method of section 10.3 to analyze the performance of the combustor, find the specific impulse of a scramjet operating at $M_0 = 8$ with stoichiometric combustion beginning at $M_2 = 3$. Take $\gamma \approx 1.2$ for the nozzle flow. Compare with figure 10.1.

10.4   One proposed turborocket engine uses hydrazine (N_2H_4) as a fuel. The N_2H_4 is chemically unstable, having a heat of formation of $\Delta H_f^\circ = 12.05$ kcal/mole. It is decomposed in a catalyst bed to produce a hot mixture of N_2 and H_2, which is expanded through a turbine. The turbine drives a compressor; its discharge mixes with the turbine exhaust and burns before exiting through the nozzle. Develop expressions for the thrust per unit of airflow and the specific impulse of this engine. Plot its performance as a function of M_0, choosing the altitude so that the compressor inlet pressure is constant at 1 atm.

10.5   Following the approximate argument of section 10.5, plot as a function of M_0 the altitude at which a scramjet vehicle must cruise for radiative cooling plus fuel cooling to be sufficient to maintain it in thermal equilibrium. Also, plot the diffuser exit pressure p_2 for this altitude schedule.

11 Propulsion Systems Analysis

The objective of propulsion systems analysis is to determine the best propulsion system for some application. Because an engine generally takes longer to design and develop than an aircraft, the needs that will be associated with future aircraft systems must be anticipated by the engine manufacturer well in advance of the commitment to the design of a new aircraft. Thus, "preliminary design" of engines to meet the needs of new aircraft (and in some cases to make possible the design of new families of aircraft) is an essential part of engine design. Components must be designed and developed on the basis of preliminary engine designs before an engine manufacturer can commit to a final engine design, and long before an aircraft system is committed to design and development.

In the preliminary design phase the criteria for optimization of the engine will usually be stated in rather general terms; for example, it may be desired to determine the "best" engine for a next-generation commercial transport to serve medium-range hub-and-spoke routes, with a capacity of about 150 passengers. In this case, "best" would mean the engine that would lead to the most profitable aircraft while meeting environmental constraints. But the best engine design depends on the aircraft design, so the engine manufacturer must include in his engine studies all those aircraft possibilities most likely to be optimum. For a high-performance military fighter, "best" might at one time have been defined as minimum takeoff gross weight to carry a given load of weapons up to specified speeds and altitudes, with some minimum range or endurance. More recently the criteria have become quite complex, but in essence they attempt to select for minimum life-cycle cost on the basis of specified performance in the context of a mission profile. The mission profile is specified in terms of typical sorties, which involve takeoff, climb, subsonic cruise, supersonic flight, and combat maneuvering. The engine must be capable of specified numbers of throttle excursions simulating these sorties, with some predicted level of maintenance. Its performance against these criteria is carefully evaluated in the fiercely competitive process of selecting an engine and a manufacturer for a new aircraft program. Since the basic philosophy of the engine design will have been defined long before this competition occurs, it is extremely important that the preliminary design process take into account as many of the controlling factors as possible. Clearly an integrated treatment of the engine and the airframe is essential. It is not possible within the scope of this book to deal with this process in any detail, but it is hoped that the following brief discussion will convey some of the ideas at a conceptual level.

The first step in attacking such a preliminary design problem is to identify the *criteria* by which the system is to be judged and the *context* in which the system's performance is to be evaluated against these criteria. In the above examples, the criterion for the commercial engine would be minimum seat-mile cost and the context would be serving hub-and-spoke routes. In the case of the fighter the criterion would be the minimum life-cycle cost and the context would be the mission profile.

Next, one defines a set of *model propulsion systems* and *model vehicles*, whose characteristics can be determined in terms of sets of *engine parameters* and *airframe parameters*. A flight plan or mission is chosen, which may also involve parameters to be determined. In a mathematical sense the problem is then to determine the optimum set of these many parameters. In general, this includes all the parameters that influence the performance, initial cost, and durability of the engine.

Of necessity, the following discussion will be limited to the aspects of performance that are understandable on the basis of the earlier chapters of this book. For these purposes the important performance variables of the engine are thrust/engine weight, thrust/frontal area or thrust/mass flow, and specific impulse. At the level of thermodynamic modeling, the engine cycle parameters that might be varied are π_c, θ_t, α, M_T, and perhaps others. At the next level of specificity, the number of compressor stages, the type of turbine cooling, the number of spools, and other features controlling the engine design would be represented in the evaluation. At the lowest level, the airframe variables might be the lift/drag ratio and the ratio of structural weight to gross weight, both of which depend on many design parameters, such as wing loading, aspect ratio, materials, and maximum M_0.

Whatever the level of treatment, most aircraft missions can be constructed of the following elements: takeoff, climb and acceleration, cruise, maximum speed, maneuver, loiter, and land. The next few sections will deal with some of these at the level at which they might be treated in the simplest engine-aircraft systems analyses.

11.1 Takeoff

Assume that during the takeoff roll $F = F(0)$ and $D = \frac{1}{2}\gamma p_0 M_0^2 A_W C_{D0}$, where A_W is the wing area on which the drag coefficient at zero lift, C_{D0}, is based. Also assume a rolling friction coefficient, C_f. Then, with the change in mass of the aircraft during takeoff neglected,

$$a_0 m(0) \frac{dM_0}{dt} = F(0) - m(0)gC_f - \tfrac{1}{2}\gamma p_0 A_w C_{D0} M_0^2.$$

Writing this as

$$\frac{dM_0}{dt} = a - bM_0^2,$$

where $a = F(0)/a_0 m(0) - gC_f/a_0$ and $b = \tfrac{1}{2}\gamma p_0 A_f C_{D0} a_0 m(0)$, and integrating, gives

$$M_0 = \sqrt{a/b} \tanh \sqrt{ab}\, t. \tag{11.1}$$

Now, if C_{LT} is the lift coefficient at takeoff, the Mach number at takeoff, M_{0T}, will be determined by

$$\tfrac{1}{2}\gamma p_0 M_{0T}^2 A_w C_{LT} = m(0)g,$$

where A_w is the wing area; inserting this value of M_0 in equation 11.1 gives the time required for the takeoff roll. More important usually is the length of the takeoff roll, which is

$$x_T = a_0 \int_0^{t_T} M_0\, dt = a_0 \int_0^{t_T} \sqrt{a/b} \tanh \sqrt{ab}\, t\, dt$$

or

$$x_T = \frac{a_0}{b} \ln(\cosh \sqrt{ab}\, t_T) = \frac{-a_0}{2b} \ln\left(1 - \frac{M_{0T}^2 b}{a}\right). \tag{11.2}$$

The term

$$\frac{M_{0T}^2 b}{a} = \frac{\gamma p_0 A_w C_{D0} M_{0T}^2}{F(0) - gC_f m(0)}$$

is the ratio of aerodynamic drag at the end of takeoff roll to the net accelerating force at the beginning.

A civil transport aircraft must be able to continue to take off and to climb out if one engine fails at a point on the runway beyond that at which it is possible to brake to a stop before the end of the runway. This requires that the aircraft be designed with excess thrust in normal operation—the smaller the number of engines, the more excess thrust. A fighter aircraft might be required to be able to land in a distance equal to its takeoff distance. For a high-performance aircraft this might imply thrust reversal on landing. There are many such elaborations of the basic requirement, which is to accelerate the vehicle to takeoff velocity on the runway.

11.2   Climb and Acceleration

For a given airframe and engine, an infinite number of paths in altitude-velocity coordinates can be followed to any desired final altitude and speed. But if some criterion such as minimum time or minimum fuel consumption is specified, then there is an optimum path. It is most easily deduced using the calculus of variations and the *total energy* formulation of the aircraft dynamic problem (references 11.1 and 11.2).

The total energy E is defined as the sum of the potential and kinetic energies of the aircraft:

$$E = m\left(gh + \frac{u_0{}^2}{2}\right).$$ (11.3)

Its variation with time is due to the net work done on the aircraft by the thrust minus the drag, so

$$\frac{d}{dt}\left(\frac{E}{m}\right) = \frac{F - D}{m}u_0.$$ (11.4)

Contours of constant E/m appear as inverted parabolas on h and u_0 coordinates, as shown in figure 11.1, and the problem of optimizing climb and acceleration is then to determine the best path for going from a low initial E/m, say at point 1, to a higher value at point 2, according to some criterion.

11.2.1   Minimum Time to Climb

Suppose, for example, that minimum time to climb and accelerate is desired. From equation 11.4 the time to climb, t_c, is

$$t_c = \int_{(E/m)_1}^{(E/m)_2} \frac{m}{(F - D)u_0}\, d\left(\frac{E}{m}\right),$$ (11.5)

and the problem is to determine the path on h and u_0 (or h and M_0)

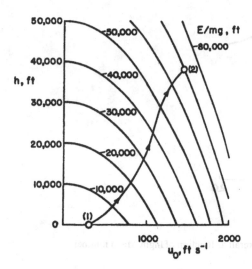

Figure 11.1
Total energy E/mg as a function of altitude and flight velocity with possible initial and final points of climb path.

coordinates that minimizes the integral. This requires knowledge of the integrand as a function of (h, M_0). Since minimum t_c is desired, F should have its maximum value at any given (h, M_0) (the engine should operate at full throttle), and we can consider $F = F(h, M_0)$. But D depends on the vehicle's mass m and on the lift or normal acceleration as well as on (h, M_0), so in complete generality the problem is quite complex. To illustrate the approach, we may assume the lift equals the weight and estimate the thrust of an afterburning turbojet from figure 2.5, assuming operation at constant $W_2 \sqrt{\theta_2}/\delta_2$. The drag coefficient is approximately

$$C_D = C_{D0} + \frac{C_L{}^2}{\pi AR},$$

where C_{D0} is the drag at zero lift, C_L is the lift coefficient, and AR is the aspect ratio of the wing. A typical variation of C_{D0} with M_0 is given in figure 11.2. If we then choose A_2/A_w (that is, the engine size) and AR, we can plot curves of $(F - D)(u_0/m(0))$ as shown in figure 11.3.

The vehicle's mass m depends on its entire flight history, so in general it cannot be represented as a function of (h, M_0). It is necessary to integrate along the flight path to find m at any point. Again for the sake of simplicity

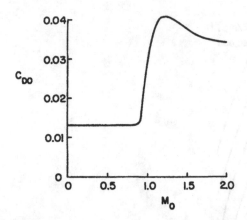

Figure 11.2
Typical drag at zero lift of interceptor aircraft as function of flight Mach number.

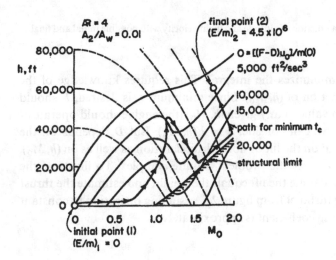

Figure 11.3
Acceleration margin as function of altitude and Mach number for interceptor aircraft,
showing path for minimum climb time.

it will be assumed here that m is approximately equal to $m(0)$, the initial mass. With this additional assumption,

$$\frac{m(0)}{(F - D)u_0} = f(h, E/m) = g(u_0, E/m);$$

that is, the integrand of equation 11.5 is a function of h and E/m, and a *different* but equally well defined function of u_0 and E/m. To determine the optimum path, let $\eta(E/m)$ be any function of E/m such that $\eta(E/m)_1 = \eta(E/m)_2 = 0$, and let ε be an arbitrary small constant. Then put

$$t_c(\varepsilon) = \int_{(E/m)_1}^{(E/m)_2} f(h + \varepsilon\eta, E/m)d(E/m),$$

where $h = h(E/m)$ is the altitude. Now, if h is the desired optimizing function, which minimizes t_c, then $t_c(\varepsilon)$ has its minimum for $\varepsilon = 0$, and

$$\frac{dt_c}{d\varepsilon} = 0 = \int_{(E/m)_1}^{(E/m)_2} \left(\frac{\partial f}{\partial h}\right)_{E/m} \eta d(E/m)$$

(where the subscript emphasizes that the derivative is taken at constant E/m). This is the condition on $h(E/m)$ that minimizes t_c. But since η is arbitrary, this condition can be satisfied only if

$$\left(\frac{\partial f}{\partial h}\right)_{E/m} = 0. \tag{11.6}$$

Similarly, if $f(h, E/m)$ is replaced by $g(u_0, E/m)$, one finds the condition

$$\left(\frac{\partial g}{\partial u_0}\right)_{E/m} = 0. \tag{11.7}$$

These conditions state that, along the trajectory $h(M_0)$, f should not change with h along curves of constant E/m, so the trajectory must be such that the curves of f (or g) are tangent to those of E/m. The desired trajectory is therefore as sketched in figure 11.3. For the engine-aircraft combination depicted, minimum time to climb is achieved by *diving* through the transonic speed range, where the drag coefficient is high, accelerating at low altitude at a speed just below the structural limit of the vehicle (maximum permissible p_{t0}), then "zooming" to the desired final point at constant E/m.

11.2.2 Minimum Fuel to Climb

The fuel consumed in climbing from $(E/m)_1$ to $(E/m)_2$ is

$$m_f = \int_{(E/m)_1}^{(E/m)_2} \frac{dm}{d(E/m)} d(E/m) = \int_{(E/m)_1}^{(E/m)_2} \frac{Fm}{I(F-D)u_0} d(E/m).$$

Since the minimum fuel consumption probably will not result from full-throttle operation, it is necessary in this case to recognize that the integrand is a function of $\delta = F/F_{max}$ as well as of h and u_0. Thus we put

$$\frac{Fm(0)}{I(F-D)u_0} = f(h, E/m, \delta) = g(u_0, E/m, \delta).$$

We must now find the optimum path $h(E/m)$ and the optimum throttle schedule $\delta(E/m)$, so we put

$$m_f(\varepsilon_1, \varepsilon_2) = \int_{(E/m)_1}^{(E/m)_2} f(h + \varepsilon_1 \eta_1, E/m, \delta + \varepsilon_2 \eta_2) d(E/m),$$

where ε_1 and ε_2 are two arbitrary constants and η_1 and η_2 are two arbitrary functions such that $\eta(E/m)_1 = \eta(E/m)_2 = 0$. Since m_f must be a minimum with respect to both ε_1 and ε_2, the conditions for minimum fuel consumption in climb become

$$\left(\frac{\partial f}{\partial h}\right)_{E/m} = 0, \quad \left(\frac{\partial f}{\partial \delta}\right)_{E/m} = 0;$$

$$\left(\frac{\partial g}{\partial u_0}\right)_{E/m} = 0, \quad \left(\frac{\partial g}{\partial \delta}\right)_{m/E} = 0. \tag{11.8}$$

These do not admit of quite such a simple graphical interpretation as was possible for the previous example, but the procedure can be extended to allow determination of schedules for any number of variables.

11.2.3 An Example: Boost Performance of Scramjets

This is a good example of use of the energy method, because the principle requirement of a booster is to increase the energy of the vehicle to that corresponding to the altitude and velocity of low Earth orbit, the exact trajectory that is flown being rather inconsequential so long as it allows efficient operation of the propulsion system and does not place unacceptable requirements on the vehicle. The estimates of scramjet performance

developed in chapter 10 will be used to carry this case through as an
illustration of the energy approach.

Following the above argument, the rate of change of the total energy per
unit mass, E/m, is equated to the power per unit mass delivered to the
airframe, i.e.,

$$\frac{d(E/m)}{dt} = \frac{(F - D)u_0}{m},$$

where F is the thrust, D is the drag, u_0 is the flight velocity, and m is the
vehicle's mass. The rate of change of m is given by

$$\frac{dm}{dt} = -\frac{F}{gI_{sp}};$$

thus, the rate of change of the specific energy with respect to mass is

$$\frac{d(E/m)}{dm} = -\left(1 - \frac{D}{F}\right)\frac{gu_0 I_{sp}}{m}. \tag{11.9}$$

The propulsion system's performance is contained in the group $u_0 I_{sp}$, and
this energy method has the advantage of collecting all the vehicle's characteristics into the single ratio D/F. In evaluating this ratio it is important to
note first that in this formulation D is the drag associated with lift other
than that produced by the propulsive streamtube, since the viscous and
shock losses in the engine are included in the specific impulse calculation.
In computing this drag by an approximate L/D, we will equate L to the
weight of the vehicle, corrected for the centrifugal force due to the vehicle's
velocity. Thus,

$$D = m\left(\frac{D}{L}\right)\left(g - \frac{u_0^2}{R}\right),$$

where R is the radius of the earth and g is the acceleration of gravity. The
propulsive lift (i.e., the force normal to the flight direction due to the propulsive streamtube) has been set to zero. This may seem questionable in
view of the apparently large vertical force on the inlet ramp of an engine
such as that in figure 10.2, but if the inlet flow returns to the axial direction
in the engine the net vertical force on the inlet must be zero, since the air
has no momentum flux perpendicular to the flight direction at the entrance
to the combustion chamber. The vertical force on the ramp must be just

offset by a pressure force on the inside of the cowl that acts on a smaller area than that of the ramp, but at a much higher pressure level. In the same spirit, if the nozzle is assumed to be fully expanded, as it was in the calculations of chapter 10, the net vertical force due to it may be assumed to be zero. This will not be true in practice unless the expansion is confined; certainly the nozzle sketched in figure 10.2 would generate a substantial lift (and nose-down pitching moment), but since the details of the nozzle flow were not treated there this effect cannot be included consistently.

The thrust will be expressed in terms of the capture streamtube area and the free-stream mass flux as

$$F = \rho_0 u_0 A_0 I_{sp\,air} = \rho_0 u_0 A_0 (0.0292)\phi I_{sp},$$

where $I_{sp\,air}$ is the thrust per unit of air mass flow, ϕ is the equivalence ratio, and 0.0292 is the stoichiometric fuel/air ratio for hydrogen.

With these expressions for F and D, equation 11.9 becomes

$$\frac{d(E/m)}{d(m/m_i)} = -\frac{gu_0 I_{sp}}{m/m_i} + \frac{m_i(D/L)(g - u_0^2/R)}{\rho_0 A_0 (0.0292)\phi}, \qquad (11.10)$$

where m_i is the initial mass of the vehicle.

Here u_0, r_0, M_0 and ϕ are treated as variables along the flight path (although ϕ will be taken as constant at 1.2, as it was for the calculations of chapter 10). The variation of these quantities is set by the choice of the flight path, and the specific impulse then follows as well. The remaining quantities in the second term on the right-hand side of equation 11.10 can be lumped into a group, which may be thought of as an effective drag:

$$D_{eff} = \frac{m_i(D/L)}{A_0(0.0292)} = \frac{\bar{\rho}(\text{length})A_0(D/L)}{2A_0(0.0292)},$$

where $\bar{\rho}$ is the average initial density of the vehicle and where the 2 in the denominator represents the idea that the volume of the vehicle (per unit width) is approximately $A_0(\text{length})/2$. Clearly this is a crude estimate, but this group will be treated parametrically anyway. For the illustrative calculations to be described below, the length of the vehicle will be set at 150 feet, the density at 1.15 times the density of liquid hydrogen, and D/L at 1/3, resulting in $D_{eff} = 3600\,\text{lb/ft}^2$.

The equation describing the variation of total energy per unit mass along the trajectory is then, finally,

$$\frac{d(E/m)}{d(m/m_i)} = -\frac{gu_0 I_{sp}}{m/m_i} + \frac{D_{eff}(g - u_0^2/R)}{\rho_0 \phi}. \tag{11.11}$$

Since I_{sp} can be given only numerically, the integration along the trajectory must be performed numerically. This is usually done using a technique such as the Runge-Kutta method. In this case a standard fourth-order Runge-Kutta procedure available in *Mathematica* has been used.

Suppose that the quantity of primary interest is the ratio of final mass to initial mass required to attain orbital energy. To obtain an estimate of this quantity, we can integrate equation 11.11 along some trajectory. As a first guess, we might take the trajectory for which the I_{sp} is maximized. Alternatively, we might take the trajectory defined by the maximum dynamic pressure the vehicle can stand for structural or thermal reasons. These two cases have been carried out using the corresponding specific-impulse estimates given in figure 10.16.

The flight velocity is plotted in figure 11.4 as a function of the fraction of mass consumed in accelerating from 6000 ft/sec to the indicated velocity for the case of M_4 chosen for maximum specific impulse, and for two nozzle flow assumptions treated in figure 10.16 (that where the flow freezes in the nozzle at 1.0 atm and that where it freezes at 0.1 atm). The dynamic pressure corresponding to this trajectory is also shown. It becomes quite large at the high flight velocities, making such a trajectory quite problematic.

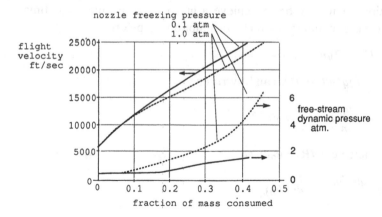

Figure 11.4
Flight velocity as a function of mass ratio for the trajectory that maximizes I_{sp}, and for the dynamic pressure implied by this trajectory. For discussion of specific impulse see subsection 10.6.5.

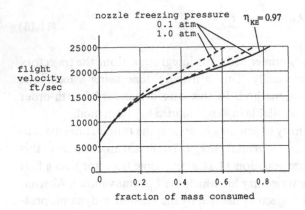

Figure 11.5
Flight velocity as a function of mass ratio for trajectories with $q_0 = 0.5$. For discussion of
specific impulse see subsection 10.6.5.

If the trajectory is limited to a dynamic pressure of 0.5 atm, the results
shown in figure 11.5 are found. We see that the propellant consumption is
significantly greater for this case.

11.3 Cruise

Here the objective may be to attain the maximum range for a given fuel
consumption, or it may be to achieve a given range in minimum time.
Consider the case of minimum fuel consumption. As before,

$$d(E/m) = [(F - D)/m]u_0 dt,$$

but $dm = -(F/gI)dt$; so this can be written

$$d(E/m) = -gI\frac{u_0}{m}dm - \frac{D}{m}u_0 dt.$$

The range increment $dR = u_0 dt$, so

$$dR = -Iu_0 \frac{gm}{D}\frac{dm}{m} - \frac{m}{D}d(E/m).$$

If we neglect the change in E/m, and put $gm = W = L$, this integrates to the
classical Breguet range equation:

$$R = Iu_0 \frac{L}{D} \ln \frac{m(0)}{m}; \quad d(E/m) = 0,$$

which was discussed in section 1.3. However, for high-speed flight the change in E/m may be quite important. Furthermore, as the flight speed becomes large, the required lift is reduced by centrifugal force, so that $L = mg - mu_0^2/r$ where r is the radius to the earth's center. In this more general situation,

$$R = -\int_{m_1}^{m_2} Iu_0 \frac{L/D}{1 - (u_0^2/rg)} \frac{dm}{m} - \int_{(E/m)_1}^{(E/m)_2} \frac{L}{gD} \frac{1}{1 - (u_0^2/rg)} d(E/m). \quad (11.12)$$

In the classical Breguet case, the first integral is maximized by maximizing $Iu_0 L/D$; at hypersonic speeds the optimum u_0 will be larger because the centrifugal lift term increases R. An idea of the magnitude of this effect and of the E/m correction can be had by assuming u_0, L/D, and Iu_0 constant; then

$$R = \frac{Iu_0(L/D)}{1 - (u_0^2/rg)} \ln\left(\frac{m_1}{m_2}\right) - \frac{L/D}{1 - (u_0^2/rg)} \frac{(E/m)_2 - (E/m)_1}{g}. \quad (11.13)$$

In order to maintain constant L/D, h must increase from point 1 to point 2 so that $(p_0)_2/(p_0)_1 = m_2/m_1$. Taking $p_0 \propto e^{-ah}$ gives

$$\frac{(E/m)_2 - (E/m)_1}{g} = h_2 - h_1 = \frac{1}{\alpha} \ln\left(\frac{m_1}{m_2}\right).$$

Finally,

$$R = \frac{Iu_0(L/D) \ln(m_1/m_2) - [(L/D) \ln(m_1/m_2)]/\alpha}{1 - u_0^2/rg}. \quad (11.14)$$

From this result we can compute R versus u_0, for any m_1/m_2, using the I dependence of figure 10.1 and estimates for L/D as a function of M_0. Some typical results for $m_1/m_2 = 2$ are given in table 11.1.

Such results have been interpreted by some to mean that hypersonic transports could be more efficient than subsonic aircraft for long-distance travel. In fact, when the structural requirements are met for the high temperatures associated with hypersonic flight, the apparent advantage probably more than disappears, owing to the large structural weight fraction of the hypersonic aircraft, so that unless the greatly reduced travel time or the

Table 11.1

M_0	$I(s)$	L/D	Fuel	R (miles)
0.8	7000	20	JP-4	14,200
2.7	3000	8	JP-4	8,300
12.0	2000	4	H2	15,300
20.0	1200	3	H2	16,500

Table 11.2

	M_0			
	1	2	3	20
r (miles)	3.2	13	29	742

increased utilization is sufficient to outweigh this effect the hypersonic aircraft is unlikely to prove efficient as a transport. The situation is somewhat less clear for vehicles that accelerate to orbital velocity, as was discussed in subsection 11.2.3.

11.4 Maneuvering

The turning radius of a military aircraft in sustained air combat is limited by the thrust, because additional drag results from the lift required for turning. If r is the turning radius, the centrifugal force that must be overcome is mu_0^2/r; thus, if the turn is in a horizontal plane, the total lift is

$$L = \sqrt{(mg)^2 + (mu_0^2/r)^2}.$$

The ratio of total lift to weight, which is termed the *number of g's*, is then

$$\text{g's} = \sqrt{1 + (u_0^2/rg)^2}.$$

Thus, the turning radius is

$$r = \frac{u_0^2/g}{(\text{g's})^2 - 1}.$$

Some typical values for 2 g's are given in table 11.2, from which it is clear that the pilot of an hypersonic transport would have to plan ahead.

References

11.1 E. S. Rutowski, "Energy Approach to the General Aircraft Performance Problem." *Journal of the Aeronautical Sciences* 21, no. 3 (1954): 187–195.

11.2 A. E. Bryson, Jr., M. N. Desai, and W. C. Hoffman, "Energy-State Approximation in Performance Optimization of Supersonic Aircraft." *Journal of Aircraft* 6, no. 6 (1969): 481–488.

Problems

11.1   Show that if the aerodynamic drag is small relative to the net accelerating force during the takeoff roll, equation 11.2 reduces to the simple statement that accelerating force times X_T equals vehicle kinetic energy at end of takeoff roll.

11.2   A transport aircraft powered by turbofan engines with $\theta_t = 7.5$ has $L/D = 15$. Estimate the amount of fuel it will use in ratio to its takeoff mass in taking off, climbing to an altitude of 10 km and a speed of $M_0 = 0.8$, and cruising on a Breguet path a distance of 4000 km, as a function of bypass ratio α. Use the data of figure 3.6 for the engine performance.

11.3   A fighter aircraft powered by an afterburning turbofan engine with $\alpha = 1$, $\theta_t = 7.5$, and $\pi_c = 24$ engages in air combat at about $M_0 = 1$, in which it maneuvers through ten full turns at 4 g's. Estimate the amount of fuel in ratio to the initial total mass consumed during these maneuvers. Assume the aircraft's L/D is 10. How far could the aircraft cruise without after-burning at $M_0 = 0.8$ with this same fuel expenditure?

11.4   A transport aircraft is to be powered by either a high-bypass turbofan ($\alpha = 5$) or a turbojet. In either case, the engine is sized by the takeoff requirement that the takeoff roll be 1500 m. Using the simple cycle analysis of chapter 2, estimate the ratios of air mass flow/takeoff mass for the two engine types. Assume $\theta_t = 6$ and $\pi_c = 24$ for both. Then estimate the ratio of thrust to maximum available thrust for each engine when cruising at $M_0 = 0.8$ and $h = 10$ km. For the takeoff roll, take $C_{D0} = 0.01$, $C_f = 0.02$, and $m(0)/A_f \approx 5000 \text{ kg m}^{-2}$.

Index

Printed in the United States
by Baker & Taylor Publisher Services